3D Printing Technology

3D Printing Technology

Umakanta Swain
Rudranarayan Kandi
Ajit Kumar Pattanaik
Dulu Patnaik
Dilip Kumar Bagal

CWP

This edition has been published by Central West Publishing PTY LTD, (ABN 13 683 898 722) Australia

For more information about the books published by Central West Publishing PTY LTD, please visit https://centralwestpublishing.com

Disclaimer
Every effort has been made by the publisher, editors and authors while preparing this book, however, no warranties are made regarding the accuracy and completeness of the content. The publisher, editors and authors disclaim without any limitation all warranties as well as any implied warranties about sales, along with fitness of the content for a particular purpose. Citation of any website and other information sources does not mean any endorsement from the publisher, editors and authors. For ascertaining the suitability of the contents contained herein for a particular lab or commercial use, consultation with the subject expert is needed. In addition, while using the information and methods contained herein, the practitioners and researchers need to be mindful for their own safety, along with the safety of others, including the professional parties and premises for whom they have professional responsibility. To the fullest extent of law, the publisher, editors and authors are not liable in all circumstances (special, incidental, and consequential) for any injury and/or damage to persons and property, along with any potential loss of profit and other commercial damages due to the use of any methods, products, guidelines, procedures contained in the material herein.

NATIONAL LIBRARY OF AUSTRALIA

A catalogue record for this book is available from the National Library of Australia

ISBN (print): 978-1-922617-70-5

About Authors

Umakanta Swain is a dedicated professional in the field of additive manufacturing, currently serving as Manager at the Center of Excellence-AM, Govt. College of Engineering, Kalahandi. His research focuses on advancing metal additive manufacturing techniques and developing custom metal alloys for 3D printing applications. Umakanta completed his Master's degree in Additive Manufacturing from the National Institute of Technology Warangal in 2023, where he specialized in the innovative applications and development of additive technologies. Following his postgraduate studies, he took on the role of Manager - Special Projects and Additive Manufacturing at Saveer MatrixNano in Hyderabad. In this position, he led numerous projects involving additive manufacturing and the R&D of custom metal alloys. His academic journey began with a Bachelor's degree in Manufacturing Engineering and Technology from the Central Institute of Plastics Engineering and Technology, Bhubaneswar, Odisha, which he completed in 2017. Prior to this, Umakanta obtained a Diploma in Mechanical Engineering from the Institute of Textile Technology, Choudwar, in 2014. Umakanta Swain's career is marked by a strong commitment to advancing manufacturing technologies and contributing to the academic and industrial development of additive manufacturing and custom alloy materials.

Dr. Rudranarayan Kandi is presently working as an Assistant Professor in the Department of Mechanical Engineering at the National Institute of Technology Rourkela. His present area of research work is on the novel fabrication of scaffolds for soft and hard tissue regeneration using 3D bio fabrication, tubular scaffolds, biomaterials, shape-memory materials, and 4D Printing. He received his Ph. D degree from the Indian Institute of Technology Delhi in 2022. He earned his master's degree from the National Institute of Technology Rourkela in the Department of Mechanical Engineering in 2017. After his master's, he worked as a full-time lecturer at the Govt. Polytechnic Phulbani, Odisha. He received his bachelor's degree from the National Institute of Technology Rourkela from the Department of Mechanical Engineering in 2012. Further, he worked as an executive engineer for two years in Larsen and Toubro in the Construction Equipment Business (CMB), Bangalore.

Ajit Kumar Pattanaik is an Assistant Professor in the Department of Mechanical Engineering at the Government College of Engineering (GCE), Kalahandi. He holds a B. Tech. and an M. Tech. from Veer Surendra Sai University of Technology (VSSUT), Burla. His research interests include manufacturing and production engineering, and he has contributed to various studies and publications in these fields. Some of his notable works involve parametric optimization in CNC-drilling, mechanical characterization of materials, and optimization techniques in manufacturing processes.

Dr. Dulu Patnaik is the Principal of the Government College of Engineering (GCE), Kalahandi. He has a strong academic and research background, with significant contributions in the fields of instrumentation, management, circular economy, optimization, and data science. Dr. Patnaik has published numerous research papers and has a notable presence on platforms like Google Scholar, where his work is frequently cited. His research includes studies on the parameters of the SIR model for COVID-19, optimization techniques in manufacturing processes, and the mechanical behavior of hybrid natural fiber composites.

Dilip Kumar Bagal is an Assistant Professor-II in the Department of Mechanical Engineering at the Government College of Engineering (GCE), Kalahandi. He earned his M. Tech. (Research) degree from the National Institute of Technology (NIT) Rourkela, specializing in mechanical engineering. Dilip's research interests include metal machining, non-traditional machining, and data analytics. He has made significant contributions to advanced optimization and soft computing in mechanical engineering. His work has been recognized with the Outstanding Research Faculty Award 2021 from Biju Patnaik University of Technology, Rourkela. He is a member of the

International Association of Engineers and the Institution of Engineers (India) and has served as a peer reviewer for numerous academic publications. Dilip has published more than 130 academic papers, which can be accessed on platforms like Google Scholar and ResearchGate.

Preface

The field of additive manufacturing, more commonly known as 3D printing, has experienced exponential growth and evolution over the past few decades. What once began as a niche technology, limited to prototyping and hobbyist applications, has now permeated numerous industries, revolutionizing the way we approach design, production, and material science. This book, 3D Printing Technology, aims to provide a comprehensive overview of the principles, processes, and applications of this transformative technology.

As educators and researchers in the field of mechanical engineering, we have witnessed firsthand the significant impact that 3D printing has had on both academia and industry. The technology's ability to fabricate complex geometries with high precision, its versatility in using various materials, and its potential to reduce waste and production time have made it a pivotal tool in modern manufacturing. Recognizing the need for a detailed and structured resource on 3D printing, we have collaborated to compile this extensive text.

This book is divided into five parts, each meticulously designed to cover the essential aspects of 3D printing technology:

1. Introduction to 3D Printing Technology: This section provides a foundational understanding of additive manufacturing. It traces the evolution of 3D printing, discusses its fundamental principles, and examines the current state and future prospects of the technology.
2. 3D Printing Technologies: Here, we delve into the various technologies that fall under the umbrella of 3D printing. From Vat Photopolymerization to Laminated Object Manufacturing, each chapter explores specific methods, their processes, applications, and limitations.
3. Materials and Design: The focus of this part is on the materials used in 3D printing and the design considerations necessary for additive manufacturing. We discuss polymers, composites, metals, and ceramics, along with the design principles that optimize the use of these materials in 3D printing processes.
4. Applications and Industries: This section highlights the diverse applications of 3D printing across various industries.

From aerospace and healthcare to consumer products and construction, we examine how additive manufacturing is being leveraged to innovate and solve complex challenges.

5. Business and Economics: The final part addresses the business and economic implications of 3D printing. It discusses business models, economic benefits and challenges, supply chain implications, and future trends in the industry.

Each chapter is equipped with questions to encourage critical thinking and deeper engagement with the material. Our goal is to provide readers with a thorough understanding of 3D printing technology, preparing them to leverage its potential in their respective fields.

We would like to extend our gratitude to our co-authors and colleagues, whose expertise and insights have been invaluable in the creation of this book. We also thank our students and industry partners, whose curiosity and innovative spirit continue to drive advancements in this exciting field. We hope that 3D Printing Technology serves as a valuable resource for students, researchers, and professionals alike, fostering further exploration and innovation in additive manufacturing.

Umakanta Swain
Rudranarayan Kandi
Ajit Kumar Pattanaik
Dulu Patnaik
Dilip Kumar Bagal

Table of Contents

Part 1

Introduction to 3D Printing Technology

Chapter 1

Introduction to Additive Manufacturing

1.1 The Evolution of 3D Printing

Additive manufacturing (AM), colloquially known as 3D printing, has undergone a profound evolution since its inception in the 1980s. Originating as a specialized technology for rapid prototyping, AM has matured into a multifaceted manufacturing method capable of producing intricate and functional components. This transformation is underpinned by advancements in materials science, computer-aided design (CAD), and computational algorithms, alongside escalating demands for tailored, environmentally sustainable, and efficient production techniques. The amalgamation of these factors has catalyzed the expansion of AM's applications beyond prototyping to encompass diverse industries, ranging from aerospace and automotive sectors to medical and consumer goods. Figure 1 visually illustrates a contemporary 3D printer mid-process, showcasing the technology's ability to fabricate complex geometries and intricate structures with unprecedented accuracy and detail, a capability that was once unimaginable mere decades ago.

Figure 1: Evolution of 3D printing

The genesis of 3D printing can be traced back to the pioneering work of Chuck Hull in 1984, who developed stereolithography (SLA), utilizing a laser to solidify liquid resin layer by layer to create three-dimensional objects. This marked the advent of additive manufacturing (AM), initially confined to rapid prototyping and model making. The late 1980s saw the commercial introduction of SLA systems, followed by a significant breakthrough in the 1990s with fused deposition modeling (FDM) by Scott Crump, also known as fused filament fabrication (FFF), which revolutionized the industry by employing melted plastic to build objects layer by layer at a lower cost compared to SLA. The democratization of 3D printing gained momentum in the 2000s through the RepRap project, which aimed to develop self-replicating 3D printers, thereby enhancing accessibility and fostering a community-driven approach to innovation.

The 2010s witnessed a surge in 3D printing adoption across diverse sectors driven by advances in materials science. Innovations such as carbon fiber-reinforced polymers and metal alloys expanded the scope of applications, spanning aerospace, automotive, healthcare, and consumer goods. The technology's ability to fabricate intricate geometries and custom designs catered to the growing demand for personalized products and sustainable manufacturing practices. Integral to these advancements was the evolution of computer-aided design (CAD) software, empowering designers to model complex structures and simulate material behaviors, thus accelerating the prototyping and testing phases. Concurrently, machine learning and artificial intelligence (AI) algorithms optimized print parameters, predicted material properties, and enhanced defect detection, bolstering the reliability and precision of 3D printing processes.

The transformative impact of 3D printing extends beyond technical innovation to reshaping business models and supply chain strategies. Companies leverage additive manufacturing to reduce inventory costs, minimize waste through on-demand production, and achieve agile manufacturing capabilities. This shift towards decentralized production models holds potential to profoundly influence global economic dynamics. Looking ahead, the continued evolution of 3D printing promises further advancements in materials, CAD capabilities, and AI integration, facilitating the fabrication of increasingly complex and functional components across sectors such as bioprinting, nanotechnology, and space exploration.

In conclusion, the evolution of 3D printing represents a transformative journey characterized by technological breakthroughs in materials, design, and manufacturing processes. Initially confined to prototyping, 3D printing has evolved into a pivotal force across industries, enabling customized production and sustainable practices. As the field continues to innovate, the future of additive manufacturing appears promising, poised to redefine product development and manufacturing paradigms on a global scale.

1.2 Definition and Principles of AM

Additive manufacturing (AM), often referred to as 3D printing, represents a transformative industrial process wherein 3D objects are fabricated by depositing successive layers of material onto a substrate. This method stands in contrast to traditional subtractive manufacturing techniques, which involve the removal of material through machining or drilling. The key advantage of AM lies in its ability to produce intricate geometries, internal structures, and customized components with unparalleled precision. This capability is facilitated by the fundamental principle of layer-by-layer material deposition, enabled by advancements in computer-aided design (CAD), computer-controlled machinery, and materials science.

The additive manufacturing process typically follows several distinct steps. First, a digital model of the desired object is created using CAD software, which is then sliced into thin layers to guide the printing process. Subsequently, the 3D printer interprets this design file and begins depositing material—whether in powder, liquid, or solid form—onto a build platform or substrate. The deposited layers are then fused together using various techniques such as heat, pressure, or light, thereby solidifying the part. This layer-by-layer buildup continues until the entire object is formed to specification.

Post-processing steps, such as machining or surface treatments, may be applied to achieve the final desired properties and finish of the printed part.

Figures 2 and 3 illustrate key aspects of the additive manufacturing process. Figure 2 highlights the extruder head depositing material layer by layer, while Figure 3 delineates the sequential steps involved in additive processing methodology. These visual aids exemplify the intricate and methodical nature of AM, showcasing its capability to produce detailed and complex structures efficiently.

The advantages of additive manufacturing over traditional methods are manifold. Firstly, it enables the fabrication of geometries and internal features that are impractical with subtractive techniques. Secondly, AM facilitates rapid customization and prototyping with minimal setup costs, making it particularly advantageous for iterative design processes. Additionally, AM promotes material efficiency by depositing only necessary amounts of material, thereby reducing waste. Furthermore, the speed of AM processes allows for swift production cycles, with some printers capable of generating functional parts within hours.

In conclusion, additive manufacturing represents a paradigm shift in manufacturing technology, offering unprecedented capabilities in complexity, customization, material efficiency, and production speed. As advancements in AM technology continue to unfold, its applications across industries—from aerospace and automotive to healthcare and consumer goods—are poised to expand, driving further innovation and reshaping manufacturing practices globally.

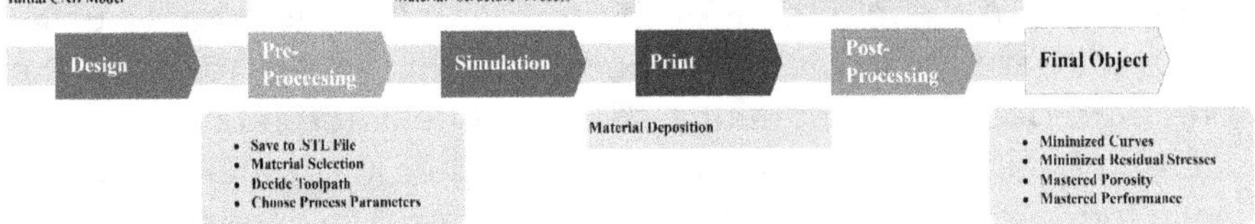

Figure 2: Stages of additive manufacturing process

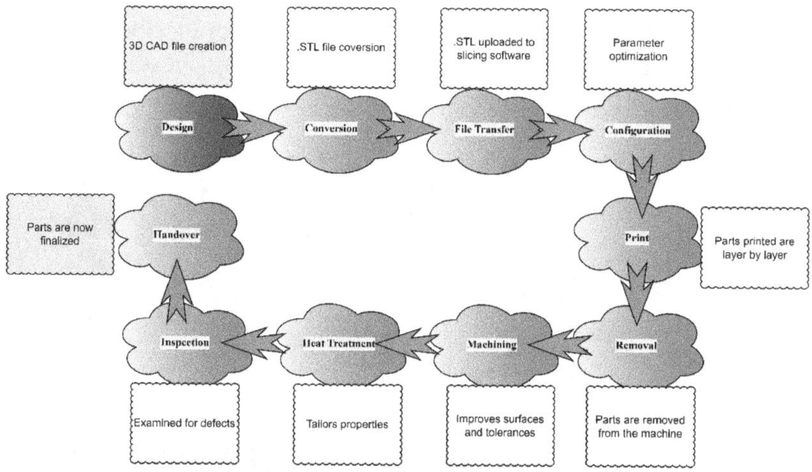

Figure 3: Steps of additive manufacturing process

Despite its transformative potential, additive manufacturing (AM) also presents several limitations that influence its widespread adoption. Chief among these is the current constraints on available materials, although ongoing research and development are expanding this repertoire. Another significant challenge is scalability; AM processes tend to be slower compared to traditional manufacturing methods, which poses difficulties in achieving large-scale production volumes efficiently. Moreover, the initial costs associated with AM equipment and materials can be prohibitive, particularly for small businesses and individual users. However, these limitations are balanced by AM's ability to revolutionize product design, manufacturing workflows, and consumer interaction. As AM technology advances, it promises to unlock new opportunities and applications across diverse industries.

Figure 4 provides a visual representation of additive manufacturing in operation. The image captures a close-up view of a 3D printer's extruder head meticulously depositing material layer by layer onto a build platform, constructing a structured object. The precision of the process is evident in the clearly defined layers being laid down, showcasing the fundamental additive manufacturing technique. The presence of heated components and intricate wiring in the printer suggests its capability to handle high-temperature materials, enabling the production of durable and intricately detailed parts. This illustration not only highlights the technological advancements in 3D printing but also underscores its potential applications in rapid prototyping and tailored manufacturing scenarios.

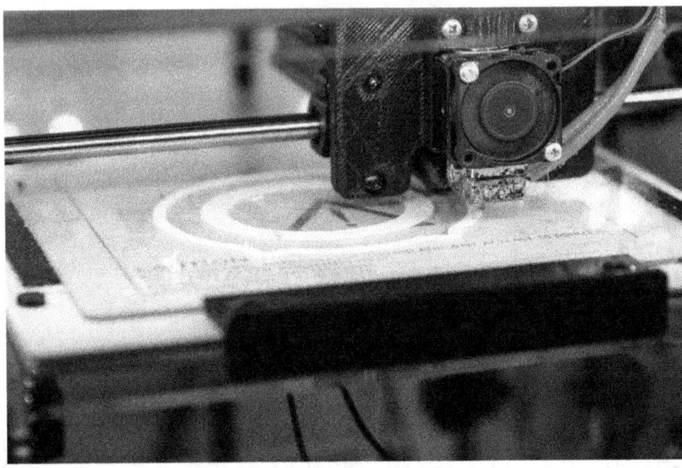

Figure 4: Additive manufacturing process
[Source: https://pixabay.com/photos/printer-3d-print-3d-printing-white-2416269/]

1.3 Processes of Additive Manufacturing

Additive manufacturing (AM), commonly known as 3D printing, is revolutionizing the field of manufacturing with its unique principles and capabilities. The core principles of AM can be summarized as follows:

1.3.1 Layer-by-Layer Fabrication

Material is deposited sequentially, building up the part layer by layer. This fundamental principle of AM enables the creation of complex geometries and internal structures that would be difficult or impossible to produce using traditional manufacturing methods. The layer-by-layer approach allows for precise control of material deposition, enabling the creation of parts with high accuracy and precision. The layer thickness can vary depending on the specific AM technology being used, ranging from a few microns to several millimeters. This approach also facilitates the creation of parts with varying material properties, such as density, texture, and color, by adjusting the material deposition parameters. Consequently, designers can achieve a high degree of customization and functionality in their products.

1.3.2 Additive Process

Unlike subtractive manufacturing methods, which involve removing material to create a part, AM is an additive process. Material is added to the part, rather than being removed. This approach enables the creation of complex shapes and structures without the need for molds, dies, or tooling. The additive process reduces material waste, as only the required amount of material is deposited layer by layer. Additionally, it allows for the creation of parts with complex internal structures, such as hollow cavities, lattice structures, and internal channels. These features are particularly beneficial for applications requiring lightweight components with high strength-to-weight ratios, such as in the aerospace and automotive industries.

1.3.3 Digital Design

AM relies on digital CAD models and computer-aided engineering (CAE) to design and optimize parts. The digital design process enables the creation of complex geometries and shapes that would be challenging or impossible to produce using traditional manufacturing methods. Digital design also allows for the optimization of part performance, such as structural integrity, thermal conductivity, and fluid dynamics. CAE software can simulate the behavior of materials under various conditions, enabling the prediction of part performance and the identification of potential design flaws. This integration of digital tools ensures that parts are not only manufacturable but also meet the required performance standards.

1.3.4 Material Versatility

AM can process a wide range of materials, including metals, polymers, ceramics, and composites. This material versatility enables the creation of parts with unique properties, such as high strength-to-weight ratios, corrosion resistance, and biocompatibility. The range of materials available for AM is rapidly expanding, with new materials being developed and commercialized regularly. This includes advanced materials, such as nanomaterials, metamaterials, and smart materials, which can be used to create parts with unique properties and functionalities. The ongoing development of new materials is crucial for broadening the application scope of AM and enhancing the performance characteristics of AM-produced parts.

1.4 Key Factors Influencing AM

In addition to these core principles, several other key factors are critical to the success and advancement of AM. Computer-controlled machinery is integral to AM, enabling the precise deposition of material and the creation of complex shapes and structures. The machinery's accuracy and repeatability are essential for producing high-quality parts consistently. Advanced materials science plays a pivotal role in AM, as it drives the development of new materials with unique properties and functionalities. Understanding the properties of these materials and how they interact during the AM process is essential for achieving desired outcomes.

Post-processing techniques are also vital in AM. Despite the precision of the AM process, parts often require additional machining, grinding, and surface treatment to achieve the desired level of precision and finish. These post-processing steps are crucial for meeting specific tolerance and surface quality requirements, particularly in high-performance applications.

The combination of these core principles and key factors enables the creation of complex parts and products with unprecedented precision, accuracy, and functionality. As AM technology continues to evolve, it is expected to unlock new and innovative applications across a wide range of industries. From customized medical implants to lightweight aerospace components, AM is poised to transform manufacturing processes and product capabilities, driving forward the next industrial revolution.

1.5 Brief History of AM

The history of additive manufacturing (AM) dates back to the 1960s, when the first 3D printing technologies were conceptualized. However, it wasn't until the 1980s that the first commercial AM systems were introduced. Figure 5 displays a timeline of the development of 3D printing, starting with the 1960s and continuing to the present day. This timeline highlights key moments in the advancement of 3D printing technologies.

1.5.1 The 1960s: The Concept of 3D Printing Emerges

The concept of 3D printing emerged in the 1960s when researchers began exploring the idea of creating three-dimensional objects from digital models. This was a time of significant innovation in the field of computer-aided design (CAD), which saw the development of the first CAD software. This software enabled the creation of complex digital models, laying the groundwork for future advancements in 3D printing. One of the pioneers of 3D printing, Ivan Sutherland, developed the first virtual reality (VR) headset in 1965. Sutherland's work on VR was instrumental in advancing 3D printing technologies as it facilitated the creation of immersive, interactive environments that could be used to design and visualize complex objects.

1.5.2 The 1980s: The First SLA Printer is Invented

The 1980s marked a significant milestone in the development of 3D printing with the invention of the first stereolithography (SLA) printer by Chuck Hull. The SLA printer used a laser to solidify liquid resin, layer by layer, to create a three-dimensional object. This innovation marked the beginning of commercial 3D printing, enabling rapid prototyping of complex parts and products. The initial applications of the SLA printer were in the aerospace and automotive industries, where it was used to create prototype parts and models. This era also saw the filing of the first patent for 3D printing technology, solidifying its importance in the manufacturing landscape.

1.5.3 The 1990s-2000s: Advancements and Commercialization Amid High Costs

During the 1990s and 2000s, 3D printing technology continued to evolve with the development of new printing methods such as selective laser sintering (SLS) and fused deposition modeling (FDM). Despite these advancements, 3D printers remained prohibitively expensive, often costing around $50,000, which limited their accessibility to large corporations and niche industries. However, the high cost was justified in fields such as aerospace and healthcare, where the complexity and customization of parts warranted the investment. This period also saw significant advancements in materials and printing techniques, which expanded the capabilities and applications of 3D printing technology.

1.5.4 The 2010s: Increased Accessibility and Industrial Adoption

The 2010s marked a pivotal turning point for 3D printing technology, as the cost of 3D printers significantly decreased, making them more affordable and accessible to a broader range of individuals and businesses. The introduction of low-cost 3D printing technologies, particularly FDM, enabled hobbyists and small businesses to explore the potential of 3D printing. As a result, the technology found new applications across various industries, including healthcare, aerospace, and consumer products. In healthcare, 3D printing was used to create customized prosthetics, implants, and medical devices, while in aerospace, it facilitated the production of complex components. The consumer market also saw innovations in product customization, allowing for the creation of personalized items such as jewelry and toys.

1.5.5 The 2020s and Present Day: Advanced Capabilities and Expanding Horizons

In the 2020s, 3D printing technology continues to advance, enabling the creation of fully functional robots and expanding into new areas such as bioprinting and nanotechnology. New printing technologies, such as digital light processing (DLP) and multi-material printing, have emerged, allowing for the creation of objects with unprecedented precision and complexity. These advancements have also spurred the development of new business models, including 3D printing services and product customization platforms. These platforms empower individuals and businesses to design and print customized products, ranging from consumer electronics to intricate art pieces. The evolution of 3D printing technology, as depicted in Figure 5, underscores its transition from a conceptual idea in the 1960s to a transformative force in modern manufacturing. The journey includes the pioneering efforts in the 1960s, the invention and commercialization of the SLA printer in the 1980s, technological advancements in the 1990s and 2000s despite high costs, the democratization of 3D printing in the 2010s, and the cutting-edge developments of the present day. Today, 3D printing is not only highly convenient and accessible but also continually evolving, driving innovation across various industries and shaping the future of manufacturing.

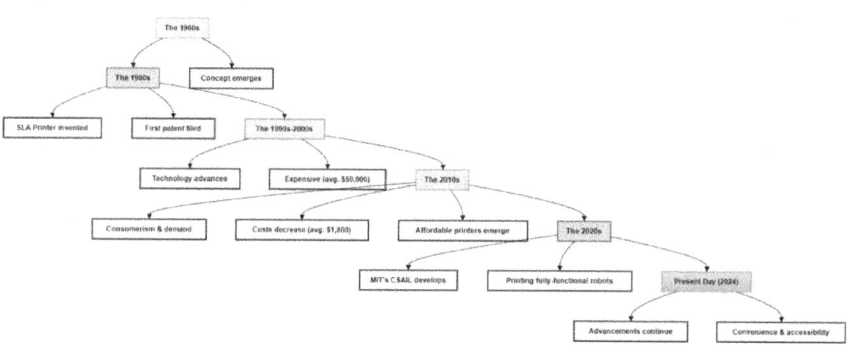

Figure 5: A timeline illustrating the evolution of 3D printing from the 1960s to the present day, highlighting key milestones and technological advancements

1.6 Current State and Future Prospects

Today, additive manufacturing (AM) is a rapidly growing industry, with a projected market size exceeding $20 billion by 2025. The current state of AM is characterized by several key trends and developments that are driving its adoption and growth.

1.6.1 Increasing Adoption in Industries

One of the most significant trends in AM is its increasing adoption across various industries, including aerospace, automotive, and healthcare. These sectors are leveraging AM to produce complex parts and products with unprecedented precision and accuracy. In the aerospace industry, AM is used to fabricate lightweight, intricate components such as engine parts and satellite structures. The automotive industry employs AM to manufacture customized car parts, including engine components and interior trims. Meanwhile, the healthcare sector utilizes AM to create bespoke implants, prosthetics, and surgical models, enhancing patient-specific treatments and outcomes.

1.6.2 Advancements in Materials Science

Advancements in materials science are enabling the development of high-performance materials tailored for AM. These materials include advanced metals, polymers, ceramics, and composites, which offer enhanced strength, durability, and thermal resistance. Researchers are also innovating new materials with unique properties, such as self-healing materials, shape-memory alloys, and metamaterials. These novel materials hold the potential to revolutionize various industries by providing unprecedented functionality and performance. For instance, self-healing materials can extend the lifespan of aerospace components, while shape-memory alloys are valuable in medical devices due to their ability to return to a predefined shape after deformation.

1.6.3 Improvements in Machine Learning and Artificial Intelligence

Improvements in machine learning (ML) and artificial intelligence (AI) are significantly enhancing the design and optimization of AM parts. AI algorithms can analyze complex datasets and optimize part designs for performance, weight, and cost efficiency. ML algorithms can predict the behavior of materials under various conditions, facilitating the development of new materials with superior properties. Additionally, AI-powered design tools enable the creation of complex geometries and structures that are difficult or impossible to produce using traditional manufacturing methods. This integration of AI and AM is fostering innovation in design and manufacturing processes, driving efficiency and performance.

1.6.4 Growing Demand for Customized, Sustainable, and Efficient Production Methods

There is a growing demand for customized, sustainable, and efficient production methods, driven by consumer preferences and environmental concerns. AM is well-positioned to meet this demand, as it enables the production of tailored parts and products with minimal waste and energy consumption. Compared to traditional manufacturing methods, which often involve significant material waste and energy use, AM offers a more sustainable alternative. Furthermore, AM allows for the production of products with complex geometries and structures, reducing material usage and weight, which can lead to further sustainability and efficiency gains.

1.6.5 Future Prospects

The future prospects for AM are both exciting and promising. As the technology continues to evolve, new and innovative applications of AM across various industries are anticipated. One of the most promising areas of research is the development of 4D printing, which enables the creation of shape-shifting structures and materials. This technology has the potential to revolutionize industries such as aerospace, energy, and healthcare by introducing dynamic, responsive components. Another groundbreaking area of research is bioprinting, which allows for the creation of living tissues and organs. This technology could transform the field of medicine by enabling the production of customized organs and tissues for transplantation, addressing critical shortages and improving patient outcomes.

Figure 6 presents the projected future of additive manufacturing, highlighting significant market growth. As technology continues to advance, the AM industry is expected to expand rapidly. Increased adoption across sectors such as aerospace, healthcare, automotive, and consumer goods is a key driver of this growth. Innovations in materials, precision, and speed of 3D printing processes are likely to enhance the capabilities and applications of AM, solidifying its role as a transformative force in modern manufacturing. The continual evolution of AM technology will likely lead to more efficient production methods, new business models, and expanded applications, further integrating AM into the fabric of contemporary industry.

Figure 6: Future of additive manufacturing with market growth [Source: https://www.youtube.com/watch?v=mfh56xQ7H84]

1.6.6 Looking Ahead: Future Prospects of Additive Manufacturing

The future of additive manufacturing (AM) holds tremendous promise, with potential applications poised to revolutionize various fields. Among the most exciting prospects are those in space exploration and satellite manufacturing. AM's ability to produce lightweight, complex components is invaluable for space missions, where reducing weight can significantly lower costs. The technology also enables the in-situ manufacturing of parts in space, potentially supporting long-duration missions and the establishment of extraterrestrial colonies.

In the realm of personalized medicine and biomedical devices, AM is set to make profound impacts. The capability to create patient-specific implants and prosthetics tailored to individual anatomical needs can drastically improve patient outcomes and reduce recovery times. Moreover, advancements in bioprinting may one day enable the fabrication of living tissues and organs, offering solutions to organ shortages and transforming transplantation medicine.

Sustainable energy and environmental solutions represent another critical area where AM is making strides. The technology's precision and ability to minimize waste align well with the goals of sustainability. For instance, AM can be used to produce optimized components for renewable energy systems, such as wind turbines and solar panels, enhancing their efficiency and durability. Additionally, AM offers the potential for innovative environmental solutions, such as the creation of custom filtration systems or sustainable packaging alternatives.

Advanced robotics and artificial intelligence also stand to benefit significantly from AM. The production of highly intricate and customized robotic components can lead to the development of more versatile and efficient robots. This, coupled with AI, can drive advancements in robotics, enabling the creation of machines capable of performing complex tasks with high precision. The synergy between AM and AI could lead to unprecedented innovations in automation and machine learning.

In conclusion, the current state of AM is marked by increasing adoption across various industries, significant advancements in materials science, notable improvements in machine learning and AI, and a growing demand for customized, sustainable, and efficient production methods. As the technology continues to evolve, we can expect to see a proliferation of new and innovative applications of AM, driving growth and innovation across numerous sectors. The future of AM is not only bright but also integral to the ongoing technological revolution, promising to reshape industries and enhance the quality of life in myriad ways.

1.7 Questions

Long Response Questions

1. What are the key differences between additive and subtractive manufacturing?
2. How has the evolution of 3D printing led to the development of new industries and applications?
3. What are the core principles of additive manufacturing, and how do they enable the creation of complex parts and products?
4. What are some of the current challenges and limitations of additive manufacturing, and how are they being addressed?
5. What are some potential future applications of additive manufacturing, and how might they impact various industries and society as a whole?

1.7.2 Multiple Choice Questions

1) What does AM stand for?
a) Advanced Manufacturing
b) Additive Manufacturing
c) Automated Manufacturing
d) Applied Manufacturing
Solution: - b) Additive Manufacturing

2) Who developed the first stereolithography (SLA) printer?
a) Scott Crump
b) Chuck Hull

c) Ivan Sutherland
d) Charles Darwin
Solution: - b) Chuck Hull

3) **In which decade did 3D printing technologies first emerge?**
a) 1960s
b) 1970s
c) 1980s
d) 1990s
Solution: - a) 1960s

4) **What does FDM stand for in 3D printing?**
a) Fused Design Modeling
b) Fused Deposition Modeling
c) Fabricated Deposition Manufacturing
d) Functional Design Manufacturing
Solution: - b) Fused Deposition Modeling

5) **What is the primary advantage of additive manufacturing?**
a) High material waste
b) Rapid prototyping
c) Requires molds and dies
d) High cost
Solution: - b) Rapid prototyping

6) **Which project aimed to create self-replicating 3D printers?**
a) SLA Project
b) RepRap Project
c) CAD Project
d) MakerBot Project
Solution: - b) RepRap Project

7) **Which material is commonly used in FDM technology?**
a) Metal alloys
b) Liquid resin
c) Melted plastic
d) Ceramics
Solution: - c) Melted plastic

8) **What technology uses a laser to solidify liquid resin?**
a) FDM
b) SLS

c) SLA
d) DLP
Solution: - c) SLA

9) **What does CAD stand for?**
a) Computer-Aided Design
b) Computer-Assisted Design
c) Computer-Aided Drawing
d) Computer-Assisted Drawing
Solution: - a) Computer-Aided Design

10) **Which decade saw the introduction of open-source 3D printing?**
a) 1980s
b) 1990s
c) 2000s
d) 2010s
Solution: - c) 2000s

11) **Which 3D printing technology was introduced by Scott Crump?**
a) SLA
b) SLS
c) FDM
d) DLP
Solution: - c) FDM

12) **What is the primary principle of additive manufacturing?**
a) Material removal
b) Layer-by-layer deposition
c) Subtractive machining
d) Mold casting
Solution: - b) Layer-by-layer deposition

13) **What role does CAD software play in 3D printing?**
a) It creates physical molds.
b) It designs digital models.
c) It provides laser cutting.
d) It solidifies materials.
Solution: - b) It designs digital models.

14) **What advantage does AM offer in terms of material usage?**
a) High material waste
b) Precise material removal

c) Reduced material waste
d) Increased production costs
Solution: - c) Reduced material waste

15) What is one of the limitations of additive manufacturing?
a) Low material costs
b) Limited scalability
c) High-speed production
d) Simple geometries only
Solution: - b) Limited scalability

16) What was a significant impact of the RepRap project?
a) Reduced the cost of SLA printers
b) Enabled 3D printers to self-replicate
c) Introduced metal 3D printing
d) Developed CAD software
Solution: - b) Enabled 3D printers to self-replicate

17) Which industry was one of the first to adopt 3D printing technology?
a) Textile
b) Aerospace
c) Agriculture
d) Construction
Solution: - b) Aerospace

18) Which decade saw 3D printers becoming more affordable and accessible?
a) 1980s
b) 1990s
c) 2000s
d) 2010s
Solution: - d) 2010s

19) What is one of the key benefits of using AI in 3D printing?
a) Increased material waste
b) Reduced print quality
c) Real-time defect detection
d) Slower print speeds
Solution: - c) Real-time defect detection

20) What is a common post-processing technique in AM?
a) Layer deposition
b) Material fusion
c) Machining

d) Slicing

Solution: - c) Machining

21) What is one of the unique properties of materials developed in the 2010s for 3D printing?

a) High electrical conductivity
b) Carbon fiber reinforcement
c) High thermal expansion
d) Low melting point

Solution: - b) Carbon fiber reinforcement

22) What challenge does the scalability of additive manufacturing face?

a) High-speed production
b) Production of large quantities
c) High material costs
d) Simple geometries only

Solution: - b) Production of large quantities

23) Which printing technology involves the use of powder as the material form?

a) SLA
b) FDM
c) SLS
d) DLP

Solution: - c) SLS

24) What does the layer thickness in AM affect?

a) Color of the part
b) Strength of the part
c) Print speed and resolution
d) Cost of the material

Solution: - c) Print speed and resolution

25) What innovation did the RepRap project introduce in 3D printing?

a) Printing with liquid resin
b) Self-replicating printers
c) Laser sintering
d) Metal alloy printing

Solution: - b) Self-replicating printers

26) What is the significance of digital light processing (DLP) in 3D printing?

a) It uses powder-based materials.

b) It creates parts with high precision and accuracy.
c) It is the oldest 3D printing technology.
d) It is exclusively used for metal printing.
Solution: - b) It creates parts with high precision and accuracy.

27) What role does machine learning play in 3D printing?
a) Simplifies the design process
b) Manages inventory costs
c) Optimizes print parameters
d) Replaces CAD software
Solution: - c) Optimizes print parameters

28) How has AM impacted the aerospace industry?
a) By increasing production costs
b) By reducing the complexity of parts
c) By enabling the production of complex components
d) By eliminating the need for prototyping
Solution: - c) By enabling the production of complex components

29) Which aspect of AM allows for on-demand production capabilities?
a) High production speeds
b) Digital design and CAD models
c) Material waste reduction
d) Limited material options
Solution: - b) Digital design and CAD models

30) What future application of AM holds immense promise?
a) Traditional manufacturing
b) Subtractive machining
c) Bioprinting and space exploration
d) High-speed drilling
Solution: - c) Bioprinting and space exploration

31) What is one of the main drivers for the adoption of 3D printing in various industries?
a) Limited customization
b) Increased production times
c) Demand for sustainable manufacturing practices
d) High setup costs
Solution: - c) Demand for sustainable manufacturing practices

32) What was one of the initial uses of 3D printing in the 1980s?

a) Mass production of consumer goods
b) Rapid prototyping and model making
c) Printing complex metal parts
d) Creating textiles

Solution: - b) Rapid prototyping and model making

33) What is one of the critical challenges faced by the adoption of AM in mass production?

a) Limited design flexibility
b) High material waste
c) Slow production speeds
d) High energy consumption

Solution: - c) Slow production speeds

34) How has the accessibility of 3D printing changed in recent years?

a) It has become more restricted.
b) It has remained the same.
c) It has increased significantly.
d) It has decreased.

Solution: - c) It has increased significantly.

35) What is one of the environmental benefits of additive manufacturing?

a) Increased energy consumption
b) High material waste
c) Reduced material usage and waste
d) Use of non-recyclable materials

Solution: - c) Reduced material usage and waste

36) What does the integration of AI in 3D printing help achieve?

a) Slower printing speeds
b) Real-time defect detection and correction
c) Increased manual intervention
d) Higher material costs

Solution: - b) Real-time defect detection and correction

37) What is the role of simulation software in AM?

a) To reduce material costs
b) To create physical prototypes

c) To predict and optimize print outcomes
d) To design simple parts
Solution: - c) To predict and optimize print outcomes

38) Which material development has expanded the applications of AM?
a) Liquid resin
b) Metal powders
c) Wood fibers
d) Paper sheets
Solution: - b) Metal powders

39) What is a common challenge in the AM process related to part quality?
a) Simple geometries
b) Material removal
c) Warping and layer adhesion
d) Slow print speeds
Solution: - c) Warping and layer adhesion

40) What future trend is anticipated for additive manufacturing?
a) Decrease in industrial use
b) Increase in traditional manufacturing techniques
c) Growth in bioprinting and custom manufacturing
d) Reduction in material options
Solution: - c) Growth in bioprinting and custom manufacturing

Chapter 2

Fundamentals of 3D Printing

2.1 Layer-by-Layer Fabrication

The core principle of 3D printing, or additive manufacturing (AM), is the layer-by-layer fabrication process, wherein a part or product is meticulously constructed by sequentially depositing material. This method facilitates the creation of intricate geometries and internal structures that would be challenging or unfeasible to produce using traditional manufacturing techniques.

2.1.1 Detailed Steps of the Layer-by-Layer Process

The layer-by-layer process encompasses several critical steps:

- **Design**: The initial phase involves designing the part or product using computer-aided design (CAD) software. This digital model is then sliced into thin layers, forming a series of cross-sectional profiles that guide the subsequent manufacturing stages.

- **Layer Formation**: Each layer of the part or product is then formed by depositing material—such as plastic, metal, or ceramic—onto a build platform or substrate. The deposition pattern is meticulously controlled and dictated by the design specifications of the digital model.

- **Fusion**: After each layer is deposited, it is fused together using a suitable method—heat, pressure, or light—to bond the material and form a solid layer. This step is critical for ensuring the structural integrity of the final product.

- **Build Platform Adjustment**: Depending on the specific type of 3D printing technology employed, the build platform is either lowered or raised to create space for the next layer.

This adjustment is crucial for maintaining the precise layering required for accurate part formation.

- **Repetition**: The process of layer formation and fusion is repeated iteratively until the entire part or product is completed. This step-by-step approach ensures the accurate realization of complex designs.

2.1.2 Advantages of Layer-by-Layer Fabrication

The layer-by-layer fabrication process offers numerous advantages over traditional manufacturing methods:

- **Complex Geometries**: This technique enables the production of complex geometries and intricate internal structures that are often unattainable with conventional manufacturing processes.

- **Customization**: The ability to easily modify designs at any stage of the process allows for high levels of customization, catering to specific requirements and preferences.

- **Reduced Material Waste**: By using only the necessary amount of material for each layer, this method significantly reduces material waste, promoting sustainability.

- **Increased Accuracy**: The precision control over each layer's deposition enhances the overall accuracy of the final product, ensuring high-quality outcomes.

2.1.3 Types of Layer-by-Layer Fabrication

Several types of layer-by-layer fabrication methods are prevalent in AM, each with unique characteristics and applications:

- **Fused Deposition Modeling (FDM)**: Involves depositing melted plastic onto a build platform, where it cools and solidifies to form each layer.

- **Stereolithography (SLA)**: Utilizes a laser or light source to cure liquid resin on a build platform, forming solid layers through photopolymerization.
- **Selective Laser Sintering (SLS)**: Employs a laser to fuse powdered material on a build platform, creating solid layers through sintering.

- **Binder Jetting**: Involves depositing a liquid binder onto a bed of powdered material, bonding the powder to form solid layers.

2.1.4 Challenges and Limitations

Despite its advantages, the layer-by-layer fabrication process is not without challenges and limitations:

- **Speed**: The process can be time-consuming, particularly for large parts or products, due to the sequential nature of layer deposition and fusion.

- **Material Properties**: The material properties of the final product may differ from those of traditionally manufactured parts, potentially impacting performance and durability.

- **Post-Processing Requirements**: Often, the fabricated parts require additional post-processing, such as machining or grinding, to achieve the desired surface finish and dimensional accuracy.

Figure 7 provides a visual illustration of the layer-by-layer fabrication process, underscoring its centrality to additive manufacturing. In this approach, a digital 3D model is dissected into thin cross-sectional layers, which the 3D printer sequentially builds by precisely depositing material. This method enables the realization of complex geometries that would be challenging to achieve with traditional manufacturing techniques. The layer-by-layer process is fundamental to 3D printing, offering high precision, customization, and efficient material usage, thus minimizing waste and enhancing overall manufacturing efficiency.

Figure 1: Layer-by-Layer fabrication

2.2 Material Properties and Selection

The properties of the materials used in 3D printing are crucial in determining the final product's characteristics. Material selection is influenced by the specific application, desired properties, and the printing technology employed. In this section, we will explore key material properties and selection criteria essential for 3D printing.

2.2.1 Material Properties

Material properties in 3D printing can be broadly classified into three categories: mechanical, thermal, and physical.

Mechanical Properties

Mechanical properties pertain to a material's ability to withstand external forces, such as tension, compression, and shear. Key mechanical properties include:

- **Strength**: The maximum stress a material can withstand without failing. This property is critical for applications requiring load-bearing capabilities.
- **Stiffness**: The material's resistance to deformation under stress. High stiffness is essential for parts that must maintain their shape under load.
- **Toughness**: The material's ability to absorb energy before failing. Tough materials are desirable for applications subject to impact or dynamic loads.
- **Ductility**: The material's ability to deform without breaking. Ductile materials can undergo significant shape changes before fracturing, which is useful for components that experience bending or stretching.

Thermal Properties

Thermal properties describe how a material responds to temperature changes. Key thermal properties include:

- **Melting Point**: The temperature at which a material transitions from a solid to a liquid state. Materials with high melting points are suitable for high-temperature applications.
- **Glass Transition Temperature (T_g)**: The temperature at which a material's molecular structure shifts from rigid to flexible. This property is important for thermoplastics used in 3D printing.
- **Thermal Conductivity**: The material's ability to conduct heat. Materials with high thermal conductivity are essential for applications requiring efficient heat dissipation.

- **Specific Heat Capacity**: The amount of heat energy required to change the material's temperature. This property affects the material's thermal stability during the printing process.

Physical Properties

Physical properties refer to the inherent characteristics of a material, such as its density, color, and texture. Key physical properties include:

- **Density**: The material's mass per unit volume. Density influences the weight and structural integrity of the printed part.

- **Color**: The material's visible color, which can be important for aesthetic applications.

- **Texture**: The material's surface roughness or smoothness. Texture affects the final finish and feel of the printed product.

- **Optical Properties**: The material's ability to transmit, absorb, or reflect light. These properties are crucial for applications involving lenses, light guides, or transparent components.

2.2.2 Material Selection Criteria

Selecting the appropriate material for 3D printing depends on several factors, including:

- **Application**: The intended use of the final product dictates the required material properties. For instance, medical implants require biocompatibility, while aerospace components need high strength and low weight.

- **Desired Properties**: The specific mechanical, thermal, and physical properties required for the application guide material selection.

- **Printing Technology**: Different 3D printing technologies, such as FDM, SLA, or SLS, are compatible with specific materials. The chosen technology influences material compatibility and performance.
- **Cost**: The cost and availability of materials can impact the overall feasibility of a 3D printing project. Budget constraints may necessitate the use of more affordable materials.

- **Sustainability**: The environmental impact and recyclability of the material are increasingly important considerations. Sustainable materials help reduce waste and environmental footprint.

The selection of materials for 3D printing is a complex process that requires a thorough understanding of the material properties and the specific requirements of the intended application. Figure 8 provides a visual representation of various types of additive manufacturing practices, highlighting the diverse range of materials and technologies employed in the field. This understanding enables the optimization of material performance and cost-effectiveness, paving the way for innovative applications across various industries.

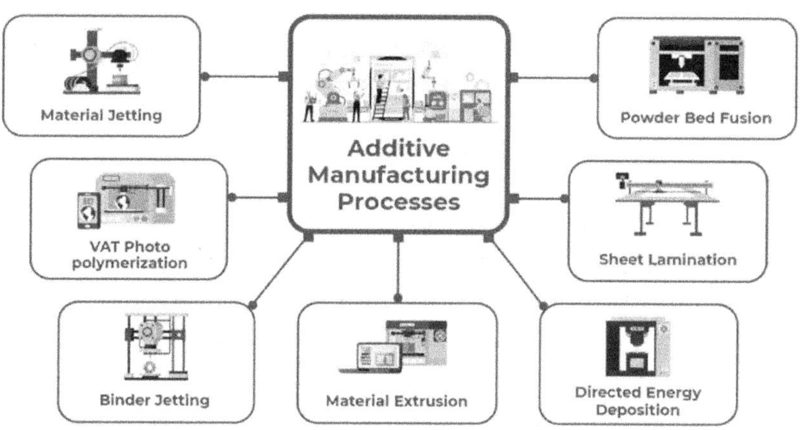

Figure 2: Type of additive manufacturing

2.2.3 Different Material Selection

In the realm of 3D printing, material selection is critical to achieving the desired characteristics and performance of the final product. Some commonly used materials in 3D printing include PLA (Polylactic Acid), ABS (Acrylonitrile Butadiene Styrene), resin, metal powders, and ceramics. Each of these materials offers unique properties and is suited to specific applications and printing technologies.

PLA (Polylactic Acid)

PLA is a biodegradable and renewable thermoplastic that is widely used in Fused Deposition Modeling (FDM) printing. It is favored for its ease of use, low warpage, and ability to produce detailed prints. PLA exhibits moderate strength, stiffness, and toughness, along with high ductility, making it suitable for a variety of applications, including prototyping and educational projects. Its relatively low melting point of around 180°C and glass transition temperature of approximately 60°C make it easy to print, though it may not be suitable for high-temperature applications. Additionally, PLA's low thermal conductivity and specific heat capacity contribute to its stability during the printing process. The material is available in various colors and can produce smooth textures and transparent finishes.

ABS (Acrylonitrile Butadiene Styrene)

ABS is a strong, impact-resistant thermoplastic also used in FDM printing. It is known for its high strength, stiffness, and toughness, coupled with moderate ductility. ABS has a higher melting point (around 220°C) and glass transition temperature (approximately 100°C) compared to PLA, which allows it to withstand higher temperatures. However, these properties also make it more challenging to print due to higher warpage and shrinkage tendencies. ABS exhibits low thermal conductivity and specific heat capacity, moderate density, typically opaque colors, and a slightly rough texture. It is not transparent, making it ideal for durable and functional parts such as automotive components and electronic housings.

Resin

Resin is a photosensitive material used in Stereolithography (SLA) printing, where it is cured using a laser or light source. The properties of resin can vary significantly depending on the type, but generally, it shows varying strengths, stiffness, and toughness, with low ductility. Resin, being a liquid before curing, does not have a melting point or glass transition temperature. After curing, resins exhibit low thermal conductivity and specific heat capacity, varying densities, various colors, and smooth textures. Resins can produce highly detailed and smooth prints, making them suitable for applications requiring fine details, such as jewelry, dental models, and prototypes.

Metal Powders

Metal powders are used in Selective Laser Sintering (SLS) and Selective Laser Melting (SLM) to create strong, dense metal parts. These powders have high strength, stiffness, and toughness, but generally low ductility. The melting point of metal powders is specific to each metal, but they generally possess high thermal conductivity and moderate specific heat capacity. Metal powders are dense, metallic in color, and have smooth textures with reflective optical properties. They are ideal for high-performance applications requiring durability and strength, such as aerospace and automotive components.

Ceramic

Ceramic materials are used in SLA and SLS printing to create parts with high thermal resistance. Ceramics are known for their high strength and stiffness but exhibit low toughness and very low ductility. Certain ceramics used in 3D printing do not have a melting point as they are photosensitive liquids before curing. They possess moderate to high thermal conductivity and low specific heat capacity. Ceramics typically have moderate to high density, varying colors (often white or beige), smooth textures, and varying optical properties, ranging from opaque to translucent. These properties make ceramics suitable for applications requiring high thermal resistance and electrical insulation, such as in electronics and biomedical devices.

2.2.4 Material Limitations and Challenges

Despite the numerous benefits of 3D printing, there are several material limitations and challenges that must be considered. These include material inconsistencies, warpage and shrinkage, layer adhesion issues, and the need for post-processing.

Material Inconsistencies

Variations in material properties can significantly affect the quality of the final product. Inconsistencies in material composition, particle size (for powders), and thermal properties can lead to defects and variations in mechanical performance.

Warpage and Shrinkage

During the printing process, materials can warp or shrink, especially when cooling from the molten state to solid. This can result in dimensional inaccuracies and distortions in the final product. Warpage and shrinkage are particularly challenging for high-temperature materials like ABS and metal powders.

Layer Adhesion

The bonding between layers in 3D printing can sometimes be weak, leading to delamination and failure of the printed part. Ensuring strong interlayer adhesion is critical for the structural integrity and mechanical performance of the printed object.

Post-Processing

Many 3D printed parts require additional processing, such as machining, grinding, or heat treatment, to achieve the desired surface finish and mechanical properties. Post-processing can add significant time and cost to the production process.

Table 1 provides a detailed summary of the key properties and selection criteria for various 3D printing materials, including PLA, ABS, resin, metal powders, and ceramics. This comprehensive information assists in selecting appropriate materials based on their mechanical, thermal, and physical properties for specific 3D printing applications. Understanding these properties and challenges is essential for optimizing the performance and cost-effectiveness of 3D printed products in diverse industries.

Table 1: Material properties and selection

Material	Mechanical Properties	Thermal Properties	Physical Properties
PLA (Poly-lactic Acid)	- Strength: Moderate - Stiffness: Moderate - Toughness: Moderate - Ductility: High	- Melting Point: ~180°C - Tg: ~60°C - Thermal Conductivity: Low - Specific Heat Capacity: Low	- Density: Low - Color: Various - Texture: Smooth - Optical Properties: Transparent (usually)
ABS (Acrylonitrile Butadiene Styrene)	- Strength: High - Stiffness: High - Toughness: High - Ductility: Moderate	- Melting Point: ~220°C - T_g: ~100°C - Thermal Conductivity: Low - Specific Heat Capacity: Low	- Density: Moderate - Color: Typically opaque - Texture: Slightly rough - Optical Properties: Not transparent
Resin	- Strength: Varies (depends on resin type) - Stiffness: Varies - Toughness: Varies - Ductility: Low	- Melting Point: Not applicable (photosensitive liquid) - Tg: Not applicable - Thermal Conductivity: Low - Specific Heat Capacity: Low	- Density: Varies - Color: Varies - Texture: Smooth - Optical Properties: Transparent (after curing)
Metal Powders	- Strength: High -	- Melting Point: Material-specific (varies by metal)	- Density: High - Color: Metallic - Texture:

	Stiffness: High - Toughness: High - Ductility: Low	- Tg: Not applicable - Thermal Conductivity: High - Specific Heat Capacity: Moderate	Smooth - Optical Properties: Reflective
Ceramic	- Strength: High (for certain ceramics) - Stiffness: High - Toughness: Low - Ductility: Very low	- Melting Point: Not applicable (photosensitive liquid) - T_g: Not applicable - Thermal Conductivity: Moderate to high - Specific Heat Capacity: Low	- Density: Moderate to high - Color: Varies (often white or beige) - Texture: Smooth - Optical Properties: Varies (opaque or translucent)

2.2.5 Materials Demands of Different Applications

In the field of 3D printing, a diverse range of materials are utilized to meet the varying demands of different applications. These materials include metals, polymers, ceramics, and composites, each offering distinct properties that make them suitable for specific uses. This section delves into the characteristics and applications of these common 3D printing materials.

Metals

Metals are extensively used in 3D printing, particularly in high-performance applications that require excellent mechanical properties. Commonly used metals include aluminum, steel, titanium, and copper alloys. Aluminum is favored for its lightweight nature and high strength-to-weight ratio, making it ideal for aerospace and automotive applications. Steel, known for its durability and strength, is used in manufacturing and construction industries. Titanium, renowned for its exceptional strength, corrosion resistance, and biocompatibility, finds applications in aerospace, medical implants, and high-end sports equipment. Copper alloys are valued for their excellent thermal and electrical conductivity, making them suitable for electrical components and heat exchangers.

Polymers

Polymers are among the most versatile and widely used materials in 3D printing. Common polymers include ABS (Acrylonitrile Butadiene Styrene), PLA (Polylactic Acid), nylon, and polycarbonate. ABS is known for its strength, toughness, and impact resistance, making it suitable for durable parts and functional prototypes. PLA, a biodegradable thermoplastic derived from renewable resources, is popular for its ease of use and low warpage, making it ideal for educational purposes and prototyping. Nylon offers excellent mechanical properties, including high strength, flexibility, and abrasion resistance, making it suitable for functional parts and industrial applications. Polycarbonate is valued for its high impact resistance, transparency, and thermal stability, making it ideal for applications requiring durability and clarity, such as eyewear lenses and protective gear.

Ceramics

Ceramics are used in 3D printing for their high thermal resistance, hardness, and stability. Common ceramic materials include silica, alumina, and zirconia. Silica is used in applications requiring thermal insulation and chemical resistance, such as refractory components and aerospace parts. Alumina is known for its high hardness, wear resistance, and electrical insulation properties, making it suitable for electronic components and wear-resistant parts. Zirconia offers exceptional thermal and chemical stability, as well as high fracture toughness, making it ideal for dental implants, cutting tools, and high-temperature applications.

Composites

Composite materials combine the properties of different materials to achieve enhanced performance characteristics. In 3D printing, common composites include carbon fiber, glass fiber, and metal matrix composites. Carbon fiber composites are known for their high strength-to-weight ratio, stiffness, and fatigue resistance, making them suitable for aerospace, automotive, and sports equipment applications.

Glass fiber composites offer good mechanical properties, thermal resistance, and cost-effectiveness, making them ideal for construction and industrial applications. Metal matrix composites combine metal with reinforcing fibers or particles, resulting in materials with improved mechanical properties, wear resistance, and thermal conductivity, suitable for aerospace, automotive, and industrial applications.

2.2.6 Material Limitations and Challenges

Despite the advantages of 3D printing materials, there are several limitations and challenges that need to be addressed. These include material inconsistencies, warpage and shrinkage, layer adhesion issues, and the need for post-processing. Variations in material properties can affect the quality and performance of the final product, while warpage and shrinkage during the printing process can lead to dimensional inaccuracies. Ensuring strong interlayer adhesion is crucial for the structural integrity of the printed parts, and many materials require additional post-processing to achieve the desired surface finish and mechanical properties.

Understanding the properties and applications of different 3D printing materials is essential for selecting the appropriate material for a specific application. Metals, polymers, ceramics, and composites each offer unique advantages and are used in a wide range of industries, from aerospace and automotive to medical and industrial manufacturing. By considering the mechanical, thermal, and physical properties of these materials, as well as their limitations and challenges, it is possible to optimize the performance and cost-effectiveness of 3D printed products.

2.3 Energy Sources and Binding Mechanisms

The energy source used in 3D printing plays a pivotal role in determining the binding mechanism, which in turn affects the final properties of the printed product. The binding mechanism refers to the process by which the material is fused or bonded together to form a solid part. This section delves into the common energy sources and binding mechanisms employed in various 3D printing technologies, providing a comprehensive understanding of their applications and limitations.

2.3.1 Thermal Energy

Thermal energy is a widely utilized energy source in several 3D printing technologies, including Fused Deposition Modeling (FDM), Selective Laser Sintering (SLS), and Selective Laser Melting (SLM).

Fused Deposition Modeling (FDM)

FDM technology uses thermal energy to melt and extrude thermoplastic materials such as PLA (Polylactic Acid) and ABS (Acrylonitrile Butadiene Styrene). The melted material is deposited layer by layer onto a build platform, where it cools and solidifies, bonding with the previously deposited layers. This method is favored for its cost-effectiveness and ease of use, making it popular for prototyping and low-volume production.

Selective Laser Sintering (SLS)

SLS employs thermal energy to fuse powdered materials, such as nylon or aluminum, into solid objects. A laser is directed at the powder bed, selectively sintering the particles together to form each layer of the object. This process allows for the creation of complex geometries and is often used in the production of functional parts and prototypes.

Selective Laser Melting (SLM)

Similar to SLS, SLM uses thermal energy to melt and fuse metal powders like titanium or steel. The laser melts the metal particles, forming a fully dense and solid layer. SLM is particularly valued in industries requiring high-performance metal parts, such as aerospace and medical implants, due to its ability to produce components with superior mechanical properties.

2.3.2 Photonic Energy

Photonic energy is harnessed in several 3D printing technologies, including Stereolithography (SLA), Digital Light Processing (DLP), and Laser-Induced Forward Transfer (LIFT).

Stereolithography (SLA)

SLA technology uses a laser to cure liquid resin, layer by layer. The laser selectively polymerizes the resin, bonding it to the previous layer to build up the final part. This method is renowned for its high precision and ability to produce parts with fine details, making it suitable for applications such as jewelry, dental implants, and intricate prototypes.

Digital Light Processing (DLP)

DLP also uses photonic energy to cure liquid resin but employs a projector to direct light onto the resin. This approach enables faster curing times compared to SLA, while still maintaining high resolution and accuracy. DLP is often used for producing detailed and complex parts with high surface quality.

Laser-Induced Forward Transfer (LIFT)

LIFT utilizes photonic energy to transfer material from a donor film to a substrate. A laser vaporizes the material on the donor film, depositing it onto the substrate in a controlled manner. This technique is particularly useful for depositing materials in precise patterns, often used in electronics and biomedical applications.

2.3.3 Mechanical Energy

Mechanical energy is utilized in several 3D printing technologies, including Binder Jetting and Laminated Object Manufacturing (LOM).

Binder Jetting

In binder jetting, mechanical energy is used to deposit a liquid binder onto a bed of powdered material. The binder selectively bonds the powder particles together to form each layer of the object. This process is capable of producing parts with a variety of materials, including metals, ceramics, and polymers, and is valued for its ability to create complex geometries and large parts.

Laminated Object Manufacturing (LOM)

LOM involves bonding together layers of sheet material, such as paper or metal, using mechanical energy. A roller or press applies pressure to bond the layers, forming a solid part. This method is appreciated for its ability to produce large and durable parts quickly, making it suitable for applications in aerospace and automotive industries.

2.4 Comparison of Energy Sources and Binding Mechanisms

Each energy source and binding mechanism offers distinct advantages and limitations. Thermal energy-based methods, such as FDM and SLS, are often preferred for their high speed and cost-effectiveness in prototyping and production. Photonic energy-based techniques, like SLA and DLP, are chosen for applications requiring high accuracy and intricate details, such as in the production of jewelry and dental implants. Mechanical energy-based methods, such as binder jetting and LOM, are utilized for their capability to produce strong and durable parts, particularly in aerospace and automotive applications.

2.5 Challenges and Limitations

While the choice of energy source and binding mechanism is critical to the success of 3D printing, several challenges and limitations must be addressed. These include material limitations, process control, and scalability. The choice of energy source and binding mechanism is often constrained by the properties of the materials used. Additionally, achieving consistent results requires precise control over the energy source and binding process. Scalability is another significant challenge, as the energy source and binding mechanism must be adaptable to produce large parts and accommodate high-volume production.

Understanding the energy sources and binding mechanisms in 3D printing is essential for optimizing the performance and quality of printed parts. Thermal, photonic, and mechanical energy sources each offer unique advantages and are suitable for different applications based on their specific properties. By carefully selecting the appropriate energy source and binding mechanism, it is possible to enhance the efficiency, accuracy, and durability of 3D printed products. Figure 9 illustrates the different energy sources and binding mechanisms used in 3D printing, providing a visual summary of the various processes and their categorization based on energy sources.

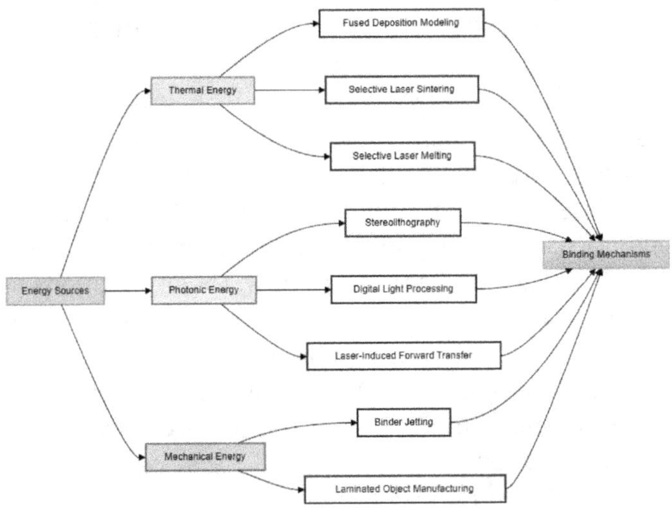

Figure 3: Energy sources and binding mechanisms

2.6 Process Parameters and Control

Process parameters such as temperature, pressure, and speed are fundamental in determining the properties and quality of 3D printed products. Effective control and optimization of these parameters are crucial to achieving consistent and high-quality results. This section elaborates on the significance of these process parameters and their influence on the 3D printing process.

2.6.1 Temperature

Temperature is a critical process parameter in 3D printing as it directly affects the material's properties and its behavior during the printing process. Key temperature-related parameters include:

- **Melt Temperature**: This is the temperature at which the material transitions from a solid to a molten state, allowing it to be extruded through the nozzle. The melt temperature is vital as it influences the material's viscosity and flow rate, which in turn impacts the print quality. If the melt temperature is not adequately controlled, it can lead to inconsistencies in layer deposition and bonding.

- **Bed Temperature**: The temperature of the build platform, or print bed, is crucial for material adhesion and the prevention of warpage. A properly regulated bed temperature ensures that the material adheres correctly to the build surface and minimizes the risk of deformation or shifting during the printing process.

- **Chamber Temperature**: In some advanced 3D printing systems, the temperature of the entire printing chamber is controlled. This parameter affects the curing and drying of the material, influencing the overall print quality and dimensional accuracy. Consistent chamber temperature helps in reducing thermal gradients that can cause warpage and layer separation.

Temperature control is essential to mitigate issues such as:

- **Warpage and Shrinkage**: Fluctuations in temperature can cause the material to warp or shrink, affecting the dimensional accuracy of the printed part. Uniform temperature control helps in maintaining the shape and size of the final product.

- **Layer Bonding**: Temperature impacts the bonding between successive layers. Insufficient bonding can lead to delamination, which compromises the structural integrity of the part.

- **Material Properties**: The thermal properties of the material, such as its mechanical strength and thermal stability, are influenced by temperature. Proper temperature management ensures that these properties meet the requirements of the final application.

2.6.2 Pressure

Pressure is another vital process parameter that influences the material flow, layer bonding, and overall print quality. Key pressure-related parameters include:

- **Build Pressure**: This refers to the pressure applied to the build platform or the material bed. It affects the adhesion of the material to the platform and contributes to the formation of a solid layer. Proper build pressure is necessary to prevent issues such as layer separation or uneven build quality.

- **Layer Pressure**: This is the pressure applied to each layer as it is deposited. It influences how well each layer bonds to the previous one, affecting the material's flow and layer bonding. Inconsistent layer pressure can lead to weak spots or structural failures in the final product.

- **Extrusion Pressure:** The pressure applied during the extrusion of the material through the nozzle affects the flow rate and consistency of the deposited material. Control of extrusion pressure is critical to ensure that the material is extruded uniformly and adheres properly to the build platform.

Pressure control is crucial for:

- **Layer Bonding**: Adequate pressure ensures proper bonding between layers, minimizing the risk of delamination and enhancing the overall strength of the printed part.

- **Material Flow**: Proper pressure management affects the flow rate of the material, which is vital for achieving accurate and consistent print results.

- **Part Strength**: The strength and durability of the final part are influenced by the pressure parameters. Consistent pressure contributes to the structural integrity of the product.

2.6.3 Speed

Speed is a significant process parameter that impacts material flow, layer bonding, and the overall quality of the print. Key speed-related parameters include:

- **Print Speed**: This is the rate at which the print head moves during the printing process. Print speed affects the material flow rate and can influence the surface finish and accuracy of the printed part. Higher speeds may reduce print time but can lead to reduced print quality if not managed properly.

- **Layer Speed**: The speed at which each layer is printed affects the bonding between layers and the overall finish of the part. Consistent layer speed is important for achieving uniform layer deposition and maintaining part quality.

- **Extrusion Speed**: The speed at which the material is extruded through the nozzle impacts the material flow rate and the overall print quality. Proper extrusion speed ensures that the material is deposited evenly and consistently.

Speed control is essential to prevent:

- **Layer Bonding Issues**: Inconsistent speeds can lead to poor layer bonding, resulting in weak or delaminated parts.

- **Material Flow Problems**: Variations in speed can affect the flow rate of the material, leading to inconsistencies in the printed part.

- **Part Quality**: Speed influences the overall quality and accuracy of the printed part. Optimal speed settings help achieve a balance between print time and part quality.

2.6.4 Control and Monitoring

Effective control and monitoring of process parameters are vital and essential for ensuring consistent and high-quality results. This can be achieved through:

- **Sensors and Feedback Loops**: Sensors are employed to monitor critical process parameters such as temperature, pressure, and speed. Feedback loops enable real-time adjustments to these parameters, maintaining consistency and quality throughout the printing process.

- **Algorithmic Control**: Advanced algorithms analyze process data to optimize parameters and enhance print quality. These algorithms can adapt to variations in material properties and environmental conditions to achieve the best possible results.

- **Operator Intervention**: Human operators play a crucial role in monitoring and adjusting process parameters as needed. Their expertise and intervention are vital for addressing issues that automated systems may not detect.

2.7 Further Issues

Despite the importance of process parameters and control, several other challenges and limitations exist:

- **Material Variability**: Variations in material properties can affect the control and optimization of process parameters, leading to inconsistencies in print quality.

- **Machine Variability**: Differences in machine performance and calibration can impact the effectiveness of process parameter control, requiring regular maintenance and calibration to ensure optimal performance.

- **Scalability**: As the size of the printed parts and the volume of production increase, scaling the process parameters and control systems becomes more complex. Ensuring that the parameters can be adapted for larger or higher-volume production is a significant challenge.

Figure 10 provides a systematic breakdown of process parameters and control elements related to manufacturing and production processes. The flowchart categorizes the key parameters—temperature, pressure, and speed—and their interactions, offering a visual representation of how these factors influence the 3D printing process.

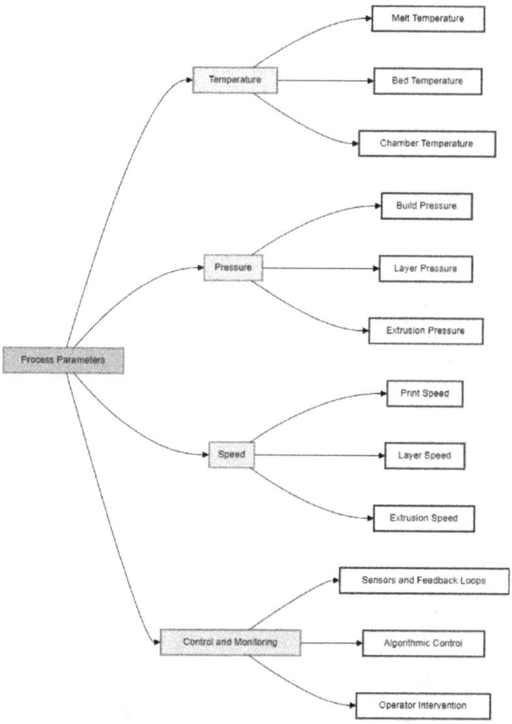

Figure 4: Process parameters and control

2.8 Questions

2.8.1 Long Response Questions

1. What are the key steps involved in the layer-by-layer fabrication process?
2. How do material properties affect the final product's characteristics in 3D printing?
3. What are the different energy sources and binding mechanisms used in 3D printing?
4. What are the key process parameters that need to be controlled in 3D printing?
5. How do process parameters affect the final product's properties in 3D printing?

2.8.2 Multiple Choice Questions

1) **What is the first step in the layer-by-layer fabrication process?**
   a) Layer Formation
   b) Design
   c) Fusion
   d) Build Platform
   Solution: - b) Design
2) **Which of the following is a common 3D printing material used in FDM technology?**
   a) Resin
   b) Ceramic
   c) PLA (Polylactic Acid)
   d) Metal Powder
   Solution: - c) PLA (Polylactic Acid)
3) **Which energy source is used in Stereolithography (SLA)?**
   a) Thermal Energy
   b) Photonic Energy
   c) Mechanical Energy
   d) Chemical Energy
   Solution: - b) Photonic Energy
4) **Which process is used to fuse materials together in SLS?**
   a) Heating

b) Cooling
c) Laser Sintering
d) Pressurizing

Solution: - c) Laser Sintering

5) **Which type of 3D printing is known for its high precision and smooth surface finish?**
a) FDM
b) SLS
c) SLA
d) Binder Jetting

Solution: - c) SLA

6) **What does the acronym "FDM" stand for?**
a) Fused Deposition Modeling
b) Fast Digital Manufacturing
c) Fused Design Machine
d) Functional Design Manufacturing

Solution: - a) Fused Deposition Modeling

7) **Which 3D printing technology uses a liquid photopolymer that is cured layer by layer?**
a) FDM
b) SLA
c) SLS
d) Binder Jetting

Solution: - b) SLA

8) **Which material is not typically used in 3D printing?**
a) PLA
b) ABS
c) Nylon
d) Wood

Solution: - d) Wood

9) **Which of the following is an advantage of using 3D printing for manufacturing?**
a) High cost
b) Long production time
c) Ability to create complex geometries
d) Limited material choices

Solution: - c) Ability to create complex geometries

10) **What is a key benefit of using PLA in 3D printing?**
a) High strength

b) Low melting point
c) Biodegradability
d) High flexibility
Solution: - c) Biodegradability

11) What does the term "sintering" refer to in the context of SLS technology?
a) Melting material completely
b) Heating material without melting to bond particles
c) Cooling material to solidify it
d) Applying pressure to shape material
Solution: - b) Heating material without melting to bond particles

12) In which 3D printing process are materials deposited layer by layer using a nozzle?
a) SLS
b) FDM
c) SLA
d) Binder Jetting
Solution: - b) FDM

13) Which material property is crucial for ensuring the durability of 3D printed parts?
a) Color
b) Texture
c) Strength
d) Flexibility
Solution: - c) Strength

14) Which technology uses a powder bed and a binding agent to create objects?
a) FDM
b) SLS
c) SLA
d) Binder Jetting
Solution: - d) Binder Jetting

15) What is the primary function of the build platform in a 3D printer?
a) To supply material
b) To fuse layers together
c) To provide a base for the object being printed
d) To cool the printed object

Solution: - c) To provide a base for the object being printed

16) Which factor can affect the resolution of a 3D printed object?

a) Material color
b) Layer height
c) Build platform temperature
d) Object weight

Solution: - b) Layer height

17) What is one of the main challenges of 3D printing with metal?

a) High cost of metal powders
b) Limited design flexibility
c) Low strength of metal parts
d) Slow printing speed

Solution: - a) High cost of metal powders

18) Which of the following is a post-processing technique used in 3D printing?

a) Printing
b) Slicing
c) Sanding
d) Layering

Solution: - c) Sanding

19) Which parameter is important for determining the detail of 3D printed objects?

a) Print speed
b) Nozzle diameter
c) Material color
d) Build platform size

Solution: - b) Nozzle diameter

20) What is a typical application of SLA technology?

a) Printing food
b) Creating jewelry prototypes
c) Building large structures
d) Manufacturing car parts

Solution: - b) Creating jewelry prototypes

21) How does Binder Jetting differ from other 3D printing technologies?

a) It uses a liquid binder to bond powder particles

b) It melts material to form layers
c) It uses photopolymerization to cure resin
d) It deposits material through a heated nozzle

Solution: - a) It uses a liquid binder to bond powder particles

22) Which factor can influence the mechanical properties of 3D printed parts?

a) Print bed temperature
b) Print speed
c) Infill density
d) Filament color

Solution: - c) Infill density

23) What is a common issue when 3D printing with ABS material?

a) Low melting point
b) Warping during cooling
c) High flexibility
d) Limited availability

Solution: - b) Warping during cooling

24) Which method is used to improve the surface finish of SLS printed parts?

a) Sanding
b) Polishing
c) Painting
d) Heat treatment

Solution: - b) Polishing

25) What is the purpose of a support structure in 3D printing?

a) To enhance material properties
b) To provide a base for overhanging parts
c) To increase printing speed
d) To reduce material usage

Solution: - b) To provide a base for overhanging parts

26) Which 3D printing technology is most suitable for creating functional metal parts?

a) FDM
b) SLA
c) SLS
d) Binder Jetting

Solution: - c) SLS

27) What does the term "infill pattern" refer to in 3D printing?

a) The color of the printed object
b) The pattern used to fill the interior of the object
c) The outer shell of the printed object
d) The layer height of the printed object

Solution: - b) The pattern used to fill the interior of the object

28) How can the adhesion of the first layer be improved in 3D printing?

a) Increasing print speed
b) Lowering the nozzle temperature
c) Using a heated bed
d) Reducing infill density

Solution: - c) Using a heated bed

29) Which factor is crucial for ensuring dimensional accuracy in 3D printed parts?

a) Material color
b) Print bed leveling
c) Layer height
d) Nozzle diameter

Solution: - b) Print bed leveling

30) What is the main advantage of using SLA over FDM for prototyping?

a) Lower material cost
b) Faster print speed
c) Higher resolution and surface finish
d) Easier material handling

Solution: - c) Higher resolution and surface finish

31) Which process is used to harden the liquid resin in SLA printing?

a) Heating
b) Cooling
c) UV light exposure
d) Pressurizing

Solution: - c) UV light exposure

32) What is a potential drawback of using photopolymers in 3D printing?

a) High melting point
b) Limited mechanical strength

c) High flexibility
d) Long curing time

Solution: - b) Limited mechanical strength

33) Which 3D printing technology is known for its high build volume capabilities?

a) FDM
b) SLA
c) SLS
d) Binder Jetting

Solution: - a) FDM

34) What is the main purpose of post-processing in 3D printing?

a) To change the material properties
b) To enhance the appearance and finish of the printed object
c) To speed up the printing process
d) To reduce the cost of printing

Solution: - b) To enhance the appearance and finish of the printed object

35) How does the layer height affect the strength of a 3D printed object?

a) Higher layer height generally increases strength
b) Lower layer height generally increases strength
c) Layer height does not affect strength
d) Higher layer height generally decreases strength

Solution: - b) Lower layer height generally increases strength

36) In the context of 3D printing, what does "slicing" refer to?

a) Cutting the printed object into parts
b) Creating a digital 3D model from a scan
c) Dividing the 3D model into thin horizontal layers for printing
d) Removing excess material from the object

Solution: - c) Dividing the 3D model into thin horizontal layers for printing

37) Which technology would be most appropriate for creating detailed dental implants?

a) FDM
b) SLA

c) SLS
d) Binder Jetting
Solution: - b) SLA

38) What factor is typically used to determine the optimal printing temperature for a material?
a) Material density
b) Material viscosity
c) Melting point
d) Material color
Solution: - c) Melting point

39) Which method is commonly used to improve the mechanical properties of 3D printed composites?
a) Using higher layer heights
b) Incorporating reinforcement fibers
c) Lowering the printing speed
d) Increasing the build volume
Solution: - b) Incorporating reinforcement fibers

40) What is the primary challenge of using FDM technology for creating complex geometries?
a) Limited material selection
b) Difficulties with support structures and overhangs
c) High cost of materials
d) Low printing speed
Solution: - b) Difficulties with support structures and overhangs

Part 2

3D Printing Technologies

Chapter 3

3D Printing Technologies: An Overview

3.1 Vat Photopolymerization

Vat photopolymerization is a sophisticated 3D printing technology that utilizes light to cure liquid resin layer by layer. This technique, which encompasses methods such as stereolithography (SLA) and digital light processing (DLP), is renowned for its exceptional accuracy, resolution, and surface finish. The vat photopolymerization process is integral to various industrial applications, including rapid prototyping and medical device manufacturing.

3.1.1 The Vat Photopolymerization Process

The vat photopolymerization process is a multi-step procedure that involves the precise management of resin, light, and build platform movement. This process can be described in detail as follows:

- **Resin Preparation**: The initial step involves the preparation of a liquid resin, which is poured into a vat. This resin is a photopolymer, a polymer that solidifies upon exposure to light. The formulation of the resin is critical and must be optimized for specific properties such as viscosity, density, and curing time. The resin's formulation directly impacts the quality and characteristics of the final printed part.

- **Light Exposure**: The resin is selectively solidified using a light source, which can be a laser, an LED, or a projector, depending on the specific vat photopolymerization technique employed. The light source is directed onto precise areas of the resin, initiating the polymerization process that solidifies the material layer by layer. The light intensity and exposure duration must be meticulously controlled to ensure uniform curing and high-resolution detail.

- **Platform Movement**: Following the solidification of each layer, the build platform is adjusted either upwards or downwards to prepare for the deposition of the subsequent layer. The movement of the platform is executed in incremental steps, allowing for precise control over layer thickness and part geometry. This step is crucial for maintaining the accuracy and consistency of the printed part.

- **Repetition**: The process of light exposure and platform movement is repeated until the entire part is constructed. The duration of this process can vary significantly based on the size and complexity of the part being printed. Larger or more intricate parts generally require extended printing times.

3.1.2 Types of Vat Photopolymerization

Vat photopolymerization encompasses several techniques, each with distinct characteristics and advantages:

- **Stereolithography (SLA)**: SLA is a well-established technique that utilizes a laser to cure the resin. The laser's focused beam allows for high precision and resolution, making SLA suitable for applications requiring intricate details and complex geometries.

- **Digital Light Processing (DLP)**: DLP employs a digital projector to expose the resin to light, enabling rapid curing of entire layers simultaneously. This method is known for its fast print speeds and high resolution, which are beneficial for applications demanding efficient production and fine detail.

- **Liquid Crystal Display (LCD)**: LCD-based vat photopolymerization uses a liquid crystal display to selectively expose the resin. This technique offers high resolution and quick print speeds, combining aspects of both SLA and DLP technologies.

3.1.3 Advantages of Vat Photopolymerization

Vat photopolymerization presents several advantages that contribute to its widespread use:

- **High Accuracy**: The precision of vat photopolymerization enables the production of parts with intricate details and high dimensional accuracy. This capability is particularly valuable in fields requiring fine features and precise tolerances.

- **Smooth Surface Finish**: The technology produces parts with a smooth surface finish, which is advantageous for applications where aesthetic quality and surface smoothness are critical.

- **Fast Print Speeds**: The efficiency of vat photopolymerization facilitates rapid prototyping and production, reducing the time required to create functional prototypes and final products.

3.1.4 Limitations of Vat Photopolymerization

Despite its benefits, vat photopolymerization also has several limitations:

- **Material Limitations**: The technology is restricted to photopolymer resins, which can be costly and may offer limited material properties compared to other 3D printing technologies. This constraint can impact the range of applications and the performance of the printed parts.

- **Support Material**: Vat photopolymerization often requires the use of support structures to stabilize the part during printing. The removal of these supports can be time-consuming and may require additional post-processing steps.

- **Post-Curing:** The process necessitates post-curing to fully solidify the resin and achieve optimal material properties. This additional step can be labor-intensive and may require specialized equipment.

3.1.5 Applications of Vat Photopolymerization

The versatility of vat photopolymerization makes it suitable for various applications:

- **Rapid Prototyping:** The technology is widely used for rapid prototyping, allowing designers and engineers to quickly create and test prototypes. This accelerates the development cycle and facilitates iterative design improvements.

- **Production:** Vat photopolymerization is employed in production settings to manufacture complex parts with high accuracy and resolution. Its capabilities are particularly useful for creating detailed components in industries such as aerospace and automotive.

- **Medical Applications:** The technology is utilized in medical fields for producing custom implants, prosthetics, and surgical models. Its precision and ability to create complex geometries are advantageous for medical device applications.

3.1.6 Future of Vat Photopolymerization

The future of vat photopolymerization holds significant promise, with ongoing research focused on advancing the technology:

- **New Materials:** Researchers are developing advanced photopolymer resins with enhanced properties such as increased strength, durability, and biocompatibility. These innovations aim to expand the range of applications and improve the performance of printed parts.

- **Improved Print Speeds**: Efforts are underway to enhance the speed of vat photopolymerization processes, which would enable faster production and reduce costs. Innovations in light sources and curing techniques are key areas of focus.

- **Increased Resolution:** - Advancements in resolution are being pursued to enable the production of even more detailed and accurate parts. Improvements in optical systems and control algorithms are contributing to this goal.

Figure 11 illustrates the vat photopolymerization process by breaking it down into three primary components: Resin Preparation, Light Exposure, and Build Platform. This visual representation elucidates the interplay between these elements, highlighting the critical role of each in the overall manufacturing process. The diagram emphasizes the connection between material preparation, light application, and platform movement, providing a clear overview of the complex technology involved.

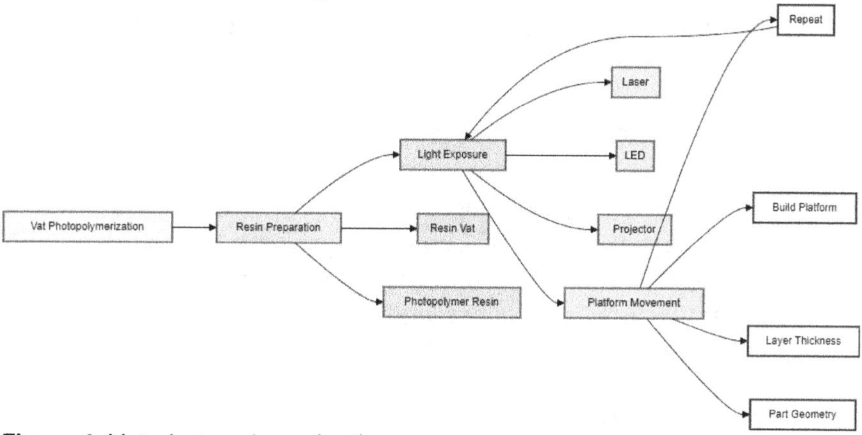

Figure 1: Vat photopolymerization

3.2 Powder Bed Fusion

Powder bed fusion (PBF) represents a sophisticated 3D printing technology that employs a thermal energy source, such as a laser or an electron beam, to selectively melt and fuse particles of a powdered material. This technique is commonly referred to as selective laser sintering (SLS) or selective electron beam melting (SEBM), depending on the type of energy source used. Powder bed fusion is highly valued for its capacity to produce robust, durable parts with intricate geometries and fine details, making it a preferred choice in various advanced manufacturing applications.

3.2.1 The Powder Bed Fusion Process

The PBF process involves several critical steps, each essential for achieving high-quality and accurate parts:

- **Powder Preparation**: The process begins with the preparation of a fine powder, which can be made from various materials, including polymers, metals, and ceramics. The powder is distributed evenly across a build platform, forming a thin layer. The characteristics of the powder, such as particle size, shape, and distribution, are crucial for the quality of the final part.

- **Layer-by-Layer Fusion:** A heat source, typically a laser or electron beam, is directed onto the powder bed to selectively melt and fuse the particles according to the digital model. The thermal energy source is precisely controlled to ensure accurate melting and bonding of the powder particles. This selective fusion process solidifies each layer of the part, building it up layer by layer.

- **Platform Movement**: After each layer is fused, the build platform is lowered slightly to accommodate the next layer of powder. This movement is performed with high precision to maintain the desired layer thickness and overall part geometry. The process continues, with the heat source selectively melting and fusing successive layers until the part is fully built.

- **Post-Processing**: Once the build is complete, the part is typically allowed to cool within the powder bed. The surrounding unfused powder acts as a support material, reducing the risk of warping and distortion during cooling. After cooling, the part is removed from the powder bed, and any residual powder is cleaned off through various post-processing techniques.

3.2.2 Types of Powder Bed Fusion

Powder bed fusion encompasses several specific technologies, each with distinct operational characteristics:

- **Selective Laser Sintering (SLS)**: SLS utilizes a laser to selectively sinter polymer or metal powders. The laser's focused beam heats the powder particles to just below their melting point, causing them to fuse together. This technique is renowned for its versatility in material selection and its ability to produce complex geometries with high mechanical strength.

- **Selective Electron Beam Melting (SEBM)**: SEBM employs an electron beam as the heat source to fully melt metal powders. The electron beam provides a high-energy density, allowing for the creation of fully dense metal parts with superior mechanical properties. SEBM is particularly useful for producing components in high-performance industries, such as aerospace and medical implants.

- **Direct Metal Laser Sintering (DMLS)**: DMLS is a variant of SLS that specifically focuses on metal powders. Like SLS, DMLS uses a laser to selectively melt the metal powder, but it is optimized for producing high-density metal parts with complex geometries and excellent mechanical properties.

3.2.3 Advantages of Powder Bed Fusion

Powder bed fusion offers several significant advantages that contribute to its widespread adoption:

- **Complex Geometries**: PBF enables the creation of highly complex and intricate geometries that would be challenging or impossible to achieve using traditional manufacturing methods. This capability allows for greater design freedom and innovation.

- **Material Versatility**: The technology supports a wide range of materials, including various polymers, metals, and ceramics. This versatility makes PBF suitable for diverse applications across different industries.

- **High Mechanical Strength**: Parts produced through powder bed fusion often exhibit excellent mechanical properties, including high strength and durability. This makes the technology suitable for producing functional components that can withstand demanding conditions.

3.2.4 Limitations of Powder Bed Fusion

Despite its advantages, powder bed fusion also has several limitations:

- **Powder Handling**: The handling and processing of powder materials can be challenging. Issues such as powder contamination, inconsistent particle size, and powder recycling need to be managed to ensure consistent part quality.

- **Surface Finish**: The surface finish of parts produced by PBF may require additional post-processing to achieve the desired level of smoothness and aesthetic quality. This can add time and cost to the overall manufacturing process.

- **Build Volume**: The build volume of PBF systems can be limited, which may restrict the size of parts that can be produced in a single build. Larger parts may require multiple build cycles or additional assembly steps.

3.2.5 Applications of Powder Bed Fusion

Powder bed fusion technology is utilized in a variety of applications, each benefiting from its unique capabilities:

- **Aerospace**: PBF is used to manufacture high-performance aerospace components, including turbine blades and structural parts, due to its ability to produce complex geometries and high-strength materials.

- **Medical**: The technology is employed in the production of customized medical implants and prosthetics, where precise geometries and biocompatibility are crucial.

- **Automotive**: In the automotive industry, PBF is used for producing functional prototypes and end-use parts with complex shapes and high mechanical performance.

3.2.6 Future of Powder Bed Fusion

The future of powder bed fusion is promising, with ongoing advancements aimed at enhancing the technology:

- **Material Innovation**: Research is focused on developing new powder materials with improved properties, such as enhanced strength, toughness, and functionality. Innovations in material science will expand the applications and performance of PBF technology.

- **Process Optimization**: Advances in process control and monitoring are expected to improve the efficiency and reliability of PBF systems. Enhanced control over parameters such as laser power, scanning speed, and layer thickness will contribute to higher quality and consistency.

- **Scalability**: Efforts are being made to increase the scalability of PBF systems, allowing for larger build volumes and higher production rates. This will enable the technology to meet the demands of high-volume manufacturing and large-scale applications.

Figure 12 provides a visual representation of the powder bed fusion process, illustrating the key components and stages involved. The diagram highlights the interaction between the powder layer, heat source, and build platform, offering a clear overview of the technology's operational principles. This visual aid aids in understanding the complex processes underlying powder bed fusion and underscores the technology's role in modern additive manufacturing.

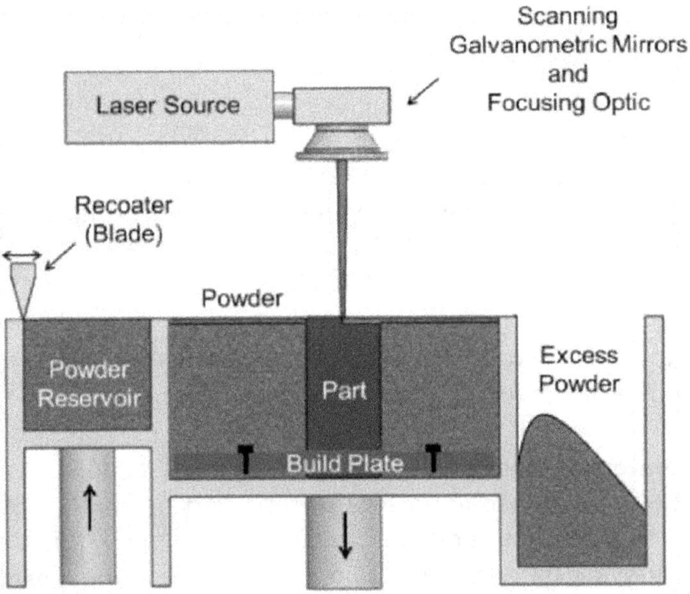

Figure 2: Working principle of powder bed fusion

Powder bed fusion (PBF) is a prominent 3D printing technology characterized by its use of thermal energy sources to selectively fuse powdered material into solid three-dimensional objects. This process is integral to several advanced manufacturing techniques, including Selective Laser Sintering (SLS), Selective Electron Beam

Melting (SEBM), and Electron Beam Additive Manufacturing (EBAM). Each variant of PBF operates on the fundamental principles of powder deposition, selective melting, and layer-by-layer construction.

3.2.7 Detailed Process Description

- **Powder Layering**: The initial stage of the PBF process involves the precise deposition of a thin layer of powdered material onto a build platform. This powder, which can be composed of various materials including polymers, metals, or ceramics, is uniformly spread using mechanical means such as a roller or blade. The even distribution of powder is crucial for ensuring consistent layer thickness and achieving uniform part quality throughout the build process.

- **Fusion**: In the fusion stage, a focused thermal energy source, such as a laser or electron beam, is directed onto the powder bed. This heat source selectively melts or sinters the powder particles at specified locations according to the digital model of the part. The laser or electron beam's precise control over energy delivery allows for high-resolution detailing and complex geometric features, crucial for producing intricate and accurate parts. The fusion process effectively transforms the powder into a solid layer, bonding the particles together and establishing the structural integrity of the part.

- **Platform Movement**: Following the fusion of each layer, the build platform is incrementally lowered to prepare for the deposition and fusion of the subsequent layer of powder. This movement is carefully controlled to maintain the desired layer thickness and part geometry. The iterative nature of this step, combined with the precision of the platform adjustments, ensures that the part's dimensions and features are accurately replicated layer by layer.

- **Repeating the Process**: The steps of powder layering, fusion, and platform movement are sequentially repeated until the part is fully constructed. Depending on the complexity and size of the part, this iterative process can extend over several hours. Each cycle of these steps builds upon the previous layers, culminating in the creation of a completed three-dimensional object with the desired specifications.

3.2.8 Types of Powder Bed Fusion Technologies

- **Selective Laser Sintering (SLS)**: SLS employs a laser to selectively sinter the powdered material. This technique is renowned for its ability to produce parts with high accuracy and resolution, leveraging the laser's focused energy to precisely control the fusion of powder particles. SLS is versatile in material selection, accommodating a range of polymers and metals, and is widely used in both prototyping and production applications.

- **Selective Electron Beam Melting (SEBM)**: SEBM utilizes an electron beam as the heat source to fully melt metal powders. The high-energy electron beam enables the creation of dense metal parts with superior mechanical properties. SEBM is particularly effective for producing components that require high strength and durability, making it a valuable technology in sectors such as aerospace and biomedical engineering.

- **Electron Beam Additive Manufacturing (EBAM)**: EBAM, similar to SEBM, employs an electron beam for melting powdered metal. This technology is optimized for producing large-scale metal parts with complex geometries and high-performance characteristics. EBAM's capability to handle large build volumes and its efficiency in material usage make it suitable for high-value applications.

3.2.9 Advantages of Powder Bed Fusion

- **Strong and Durable Parts**: One of the primary benefits of powder bed fusion is its ability to produce parts with high mechanical strength and durability. The fusion of powder particles creates a dense and cohesive material structure, suitable for functional components that must withstand rigorous conditions.

- **High Accuracy**: PBF technologies are capable of achieving remarkable levels of precision and resolution. The fine control over the thermal energy source enables the production of detailed and intricate parts, which is essential for applications requiring exacting standards.

- **Fast Print Speeds**: The efficiency of powder bed fusion allows for relatively rapid production times, particularly for complex parts. The technology's ability to consolidate layers quickly and its capacity for high-speed operations contribute to accelerated prototyping and manufacturing processes.

3.2.10 Limitations of Powder Bed Fusion

- **Material Limitations**: The range of materials that can be used in powder bed fusion is constrained by the availability of suitable powders. While the technology supports various polymers and metals, the cost and properties of these powders can limit the technology's applicability and affordability.

- **Support Material**: In some PBF processes, support structures are necessary to stabilize overhanging or complex geometries during printing. The removal of these supports can be labor-intensive and may require additional post-processing steps to achieve the final part's desired shape and finish.

- **Post-Processing Requirements**: Parts produced through powder bed fusion often require additional post-processing to enhance surface finish and accuracy. Techniques such as machining, grinding, or polishing may be necessary to achieve the final aesthetic and functional qualities of the part.

3.2.11 Applications of Powder Bed Fusion

- **Aerospace**: In the aerospace industry, powder bed fusion is utilized to create lightweight, high-strength components that meet stringent performance requirements. The technology's ability to produce complex geometries is advantageous for manufacturing parts like turbine blades and structural elements.

- **Automotive**: The automotive sector benefits from powder bed fusion through the production of components with intricate designs and high mechanical properties. Applications include engine parts, transmission components, and custom dashboard trim, where precision and performance are critical.

- **Medical**: Powder bed fusion plays a significant role in the medical field by enabling the production of customized implants and surgical models. The technology's capability to produce patient-specific solutions with high accuracy is particularly valuable for medical applications.

3.2.12 Future Prospects

The future of powder bed fusion is marked by ongoing advancements aimed at enhancing the technology's capabilities. Research efforts are focused on developing new powder materials with improved properties, optimizing process parameters for faster and more efficient production, and expanding the scalability of PBF systems to accommodate larger build volumes. These advancements promise to further broaden the applications and increase the impact of powder bed fusion in modern manufacturing.

Figure 13 provides a comprehensive overview of the powder bed fusion process, illustrating the key stages involved and highlighting its application in various industries. This visual representation aids in understanding the complex interactions between powder layering, thermal fusion, and platform movement, and underscores the technology's role in advanced additive manufacturing.

Figure 3: Application of powder bed fusion
[Source: https://makeagif.com/i/ehBvzO]

3.3 Trajectory of Powder Bed Fusion (PBF) Technology

The trajectory of powder bed fusion (PBF) technology is poised for significant advancements, driven by continuous research and development efforts. The evolution of this technology is focusing on several key areas aimed at enhancing its capabilities and expanding its applications.

New Materials

One of the most promising avenues for future development in the powder bed fusion is the innovation of novel powdered materials. Researchers are actively exploring new compositions that offer enhanced properties, including increased strength, toughness, and biocompatibility. The development of advanced powder materials could lead to the creation of parts with superior mechanical performance and suitability for a wider range of applications, including biomedical implants and high-stress industrial components. Enhanced material properties will also address current limitations related to the durability and functionality of parts produced via PBF.

Improved Print Speeds

Another critical area of focus is the improvement of print speeds. Accelerating the printing process can significantly reduce production times and costs, making powder bed fusion more economically viable for high-volume manufacturing. Researchers are investigating various methods to optimize the thermal energy sources, improve powder bed handling, and streamline the overall process to achieve faster build rates. Enhanced print speeds will enable more efficient production cycles, facilitating rapid prototyping and reducing lead times for complex parts.

Increased Resolution

Increasing the resolution of powder bed fusion processes is essential for producing highly detailed and accurate parts. Advances in beam control technologies, powder handling techniques, and scanning methodologies are being explored to achieve finer resolution and higher precision. Improved resolution capabilities will enhance the technology's ability to create intricate geometries and fine features, which is particularly valuable for applications requiring high levels of detail, such as aerospace components and intricate medical devices.

3.3.1 Challenges and Limitations

Despite its potential, powder bed fusion faces several challenges and limitations that must be addressed to fully realize its capabilities.

Material Properties

The properties of powdered materials used in PBF can significantly impact the performance of the final parts. Limitations in the strength, toughness, and other mechanical properties of available powders can affect the functionality and durability of the printed components. Ongoing research is needed to develop powders with improved performance characteristics and to better understand how these materials interact with the fusion process to achieve desired outcomes.

Scalability

Scaling up powder bed fusion technology to produce larger parts or to accommodate higher production volumes presents technical and logistical challenges. The current systems are often optimized for smaller-scale production, and adapting these systems for larger builds or higher throughput requires significant advancements in equipment design and process optimization. Addressing scalability issues will be crucial for expanding the use of PBF in industrial applications and for meeting the demands of large-scale manufacturing.

Cost

The cost of powder bed fusion technology remains a significant barrier to widespread adoption, particularly for small businesses and individual users. The expenses associated with high-quality powder materials, specialized equipment, and post-processing can be prohibitive. Researchers and manufacturers are working to reduce these costs through advancements in technology, material efficiency, and process improvements. Achieving cost reductions will make PBF more accessible and competitive, fostering broader utilization and innovation.

Figure 14 provides an illustrative representation of the powder bed fusion process, highlighting the essential components such as the powder bed, laser or electron beam, and build platform. This visual aids in understanding the intricate interactions within the PBF process and underscores the technological complexity and precision required for successful implementation. The illustration also serves as a reference for examining the advancements and addressing the challenges associated with the future development of powder bed fusion technology.

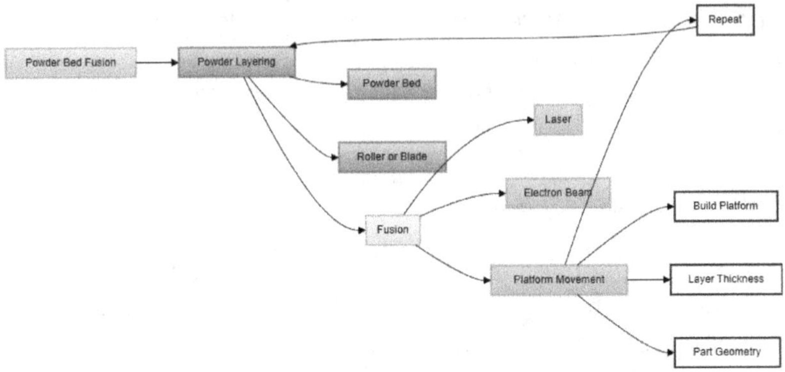

Figure 4: Powder bed fusion

3.4 Material Extrusion

Material extrusion, also known as fused deposition modeling (FDM) or fused filament fabrication (FFF), represents a foundational technique in additive manufacturing. This technology has gained widespread adoption due to its cost-effectiveness, simplicity of operation, and broad applicability across various industries. The process is characterized by its use of a heated nozzle to deposit melted thermoplastic material in a precise and controlled manner, enabling the construction of complex geometries layer by layer.

3.4.1 Process Overview

Material extrusion involves several critical steps:

- **Material Preparation**: The process begins with the preparation of thermoplastic filament or resin. This material is typically available in spools or cartridges and is selected based on the desired properties of the final product, such as strength, flexibility, or thermal resistance. Common materials used in material extrusion include acrylonitrile butadiene styrene (ABS), polylactic acid (PLA), and nylon.

- **Heating and Extrusion**: The thermoplastic material is fed into a heated nozzle where it is melted to a semi-liquid state.

The temperature of the nozzle is carefully controlled to ensure proper melting without degradation of the material. Once melted, the material is extruded through the nozzle in a controlled flow, allowing it to be deposited onto the build platform.

- **Layer-by-Layer Deposition**: The extruded material is deposited in thin, precise layers according to the digital model of the object. Each layer is laid down on top of the previous one, gradually building up the object from the bottom up. The build platform may move vertically to accommodate the increasing height of the printed object, while the nozzle typically moves horizontally and vertically to create the desired shape.

- **Cooling and Solidification**: As the material is deposited, it quickly cools and solidifies, adhering to the previous layers. This cooling process is crucial for maintaining the dimensional accuracy and structural integrity of the printed object.

3.4.2 Types of Material Extrusion

Material extrusion encompasses several variations, each with its own advantages and specific applications:

- **Fused Deposition Modeling (FDM)**: FDM is a widely recognized variant of material extrusion that employs a heated nozzle to extrude thermoplastic filaments. It is renowned for its versatility and is commonly used in both industrial and consumer-grade 3D printers. FDM is favored for its ability to produce robust and functional parts with relatively simple equipment.

- **Fused Filament Fabrication (FFF)**: FFF is essentially synonymous with FDM, with the primary difference being in the terminology used by different manufacturers and communities. FFF printers also utilize thermoplastic filaments but may vary in their specific designs and operational parameters.

3.4.3 Advantages of Material Extrusion

Material extrusion offers several notable benefits:

- **Affordability**: One of the primary reasons for the widespread use of material extrusion is its cost-effectiveness. The technology requires relatively inexpensive materials and equipment, making it accessible to a broad range of users, from hobbyists to industrial manufacturers.

- **Ease of Use**: Material extrusion systems are typically user-friendly, with straightforward setup and operation. The intuitive nature of these systems allows users to quickly learn and implement the technology for various applications.

- **Versatility**: Material extrusion is compatible with a wide array of thermoplastic materials, each offering different properties such as flexibility, strength, or resistance to environmental factors. This versatility makes it suitable for a diverse range of applications, including prototyping, product development, and functional part production.

3.4.4 Limitations of Material Extrusion

Despite its advantages, material extrusion has some limitations:

- **Surface Finish**: The layer-by-layer deposition method can result in visible layer lines and surface imperfections. While post-processing techniques can improve the surface finish, achieving a smooth and polished appearance may require additional effort

.

- **Build Volume Constraints**: Material extrusion systems are often limited by the size of the build platform. Large-scale objects may need to be printed in sections and assembled post-production, which can introduce challenges related to alignment and bonding.

- **Material Limitations**: Although material extrusion supports a variety of thermoplastics, the range of compatible materials is narrower compared to some other 3D printing technologies. This limitation may affect the choice of materials based on specific application requirements.

Figure 15 illustrates the material extrusion process, showcasing the heated nozzle, the build platform, and the layer-by-layer construction of the object. This visual representation highlights the fundamental components of the technology and provides insight into how the material is deposited and solidified to create complex three-dimensional structures.

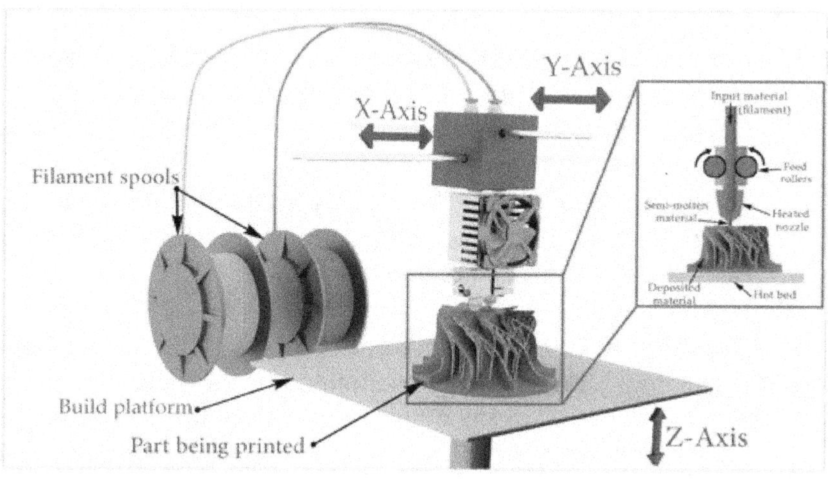

Figure 5: Material extrusion
[Source: https://3dprint.com/267642/fdm-3d-printing-effects-of-typical-parameters-on-functional-parts/]

Material extrusion is a prominent 3D printing technology characterized by its use of a heated nozzle to deposit thermoplastic material onto a build platform. This technique, widely recognized under the terms Fused Deposition Modeling (FDM) and Fused

Filament Fabrication (FFF), operates through a series of methodical steps to construct three-dimensional objects layer by layer.

3.4.5 Process Overview

The material extrusion process encompasses several critical phases:

- **Material Feeding**: The process initiates with the feeding of a thermoplastic filament into the extruder mechanism. These filaments, commonly composed of materials such as Polylactic Acid (PLA) or Acrylonitrile Butadiene Styrene (ABS), are designed to undergo melting and subsequent extrusion. The choice of filament material significantly impacts the properties of the final printed object, including its mechanical strength, flexibility, and thermal resistance.

- **Melting**: Once fed into the extruder, the filament is subjected to heating until it reaches a molten state. The nozzle of the extruder is precisely controlled to maintain a temperature above the material's melting point, ensuring smooth flow and consistent deposition. The molten material is then directed through the nozzle, which is engineered to achieve a controlled extrusion rate and path.

- **Deposition**: The molten material is deposited onto the build platform according to the predefined pattern determined by the digital model of the object. The deposition occurs in thin layers, which are sequentially built up to form the complete object. The nozzle's movement is coordinated to follow the contours of the design, enabling the creation of complex geometries and detailed features.

- **Repetition**: This layer-by-layer deposition continues iteratively, with the build platform typically moving downward to accommodate each new layer. The process persists until the entire part is fabricated. The duration of the printing process can vary significantly based on the size and complexity of the object being produced.

3.4.6 Types of Material Extrusion

Several variations of material extrusion exist, each with specific characteristics and applications:

- **Fused Deposition Modeling (FDM)**: FDM is a widely utilized material extrusion technique where a heated nozzle deposits melted thermoplastic material. This method is known for its versatility and is commonly employed in both consumer and industrial 3D printers. FDM systems are particularly valued for their ability to produce functional parts with acceptable precision and material properties.

- **Fused Filament Fabrication (FFF)**: FFF is essentially synonymous with FDM, differing primarily in terminology. Like FDM, FFF involves the extrusion of thermoplastic filaments to build objects layer by layer. This process is characterized by its ease of use and adaptability to various materials.

- **Material Jetting**: In contrast to FDM and FFF, material jetting involves the use of a heated nozzle combined with a binder material. The binder assists in holding the deposited material together, offering potential advantages in achieving higher resolution and detail. This method, however, involves additional complexity related to the management of binder and material interactions.

3.4.7 Advantages of Material Extrusion

Material extrusion offers several advantages that contribute to its widespread adoption:

- **Affordability**: Material extrusion is recognized for its cost-effectiveness. The technology typically requires less expensive equipment and materials compared to other 3D printing methods, making it accessible to hobbyists, educators, and small businesses.

- **Ease of Use**: The straightforward nature of material extrusion systems ensures that they are relatively easy to operate and maintain. Users benefit from minimal setup and calibration requirements, facilitating rapid adoption and integration into various applications.

- **Versatility**: Material extrusion supports a broad range of thermoplastic materials, enabling the production of objects with diverse properties. This versatility allows for applications spanning from simple prototypes to more complex, functional components.

3.4.8 Limitations of Material Extrusion

Despite its benefits, material extrusion presents several limitations:

- **Layer Adhesion**: One of the common challenges associated with material extrusion is the potential for inadequate layer adhesion. This issue can result in weak bonds between layers, potentially compromising the structural integrity and performance of the printed object.

- **Warpage**: Material extrusion can be prone to warping, where the object distorts or bends during printing due to uneven cooling or residual stresses. This phenomenon can impact the dimensional accuracy and overall quality of the part.

- **Resolution:** - The resolution achievable with material extrusion is often limited compared to other 3D printing technologies. Fine details and small features may be difficult to reproduce with high fidelity, which can be a constraint in applications requiring precise geometrical specifications.

3.4.9 Applications of Material Extrusion

Material extrusion is employed across various fields due to its flexibility and practical advantages:

- **Rapid Prototyping**: The technology is well-suited for rapid prototyping, allowing designers and engineers to quickly fabricate and test iterative designs. This capability accelerates the development process and supports iterative design improvements.

- **Production**: Material extrusion is also utilized in production environments to create end-use parts and products. Its ability to produce functional components makes it valuable for both low-volume manufacturing and custom production runs.

- **Education**: In educational settings, material extrusion serves as a hands-on tool for teaching 3D printing principles and design concepts. Its accessibility and ease of use make it an ideal choice for educational institutions and training programs.

3.4.10 Future of Material Extrusion

The future of material extrusion is marked by ongoing advancements aimed at enhancing the technology:

- New Materials: Research is focused on developing new thermoplastic materials with enhanced properties such as greater strength, toughness, and biocompatibility. These advancements are expected to expand the range of applications and improve performance outcomes.

- Improved Print Speeds: Efforts are underway to increase the speed of material extrusion processes, reducing production times and associated costs. Faster print speeds will enable more efficient manufacturing and rapid prototyping.

- Increased Resolution: - Enhancements in resolution are being pursued to achieve finer details and more accurate reproductions of digital models. Advances in nozzle design and material formulations are contributing to improvements in print resolution.

3.4.11 Challenges and Limitations

Despite its potential, material extrusion faces several challenges:

- **Material Properties**: Variations in material properties can influence print quality and consistency. Ensuring uniform material characteristics and achieving reliable results can be challenging, particularly with new or experimental materials.

- **Scalability:** Scaling material extrusion for large parts or high-volume production presents difficulties. The technology's limitations in build volume and production speed can constrain its applicability in certain industrial contexts.

- **Cost:** While generally affordable, the cost of material extrusion systems and materials can still be a barrier for some users. Ensuring cost-efficiency and accessibility remains a priority for continued adoption.

Figure 16 provides a visual representation of the material extrusion process, detailing the extruder mechanism, heated nozzle, and build platform. This illustration encapsulates the fundamental elements of the technology and aids in understanding the sequential steps involved in producing three-dimensional objects.

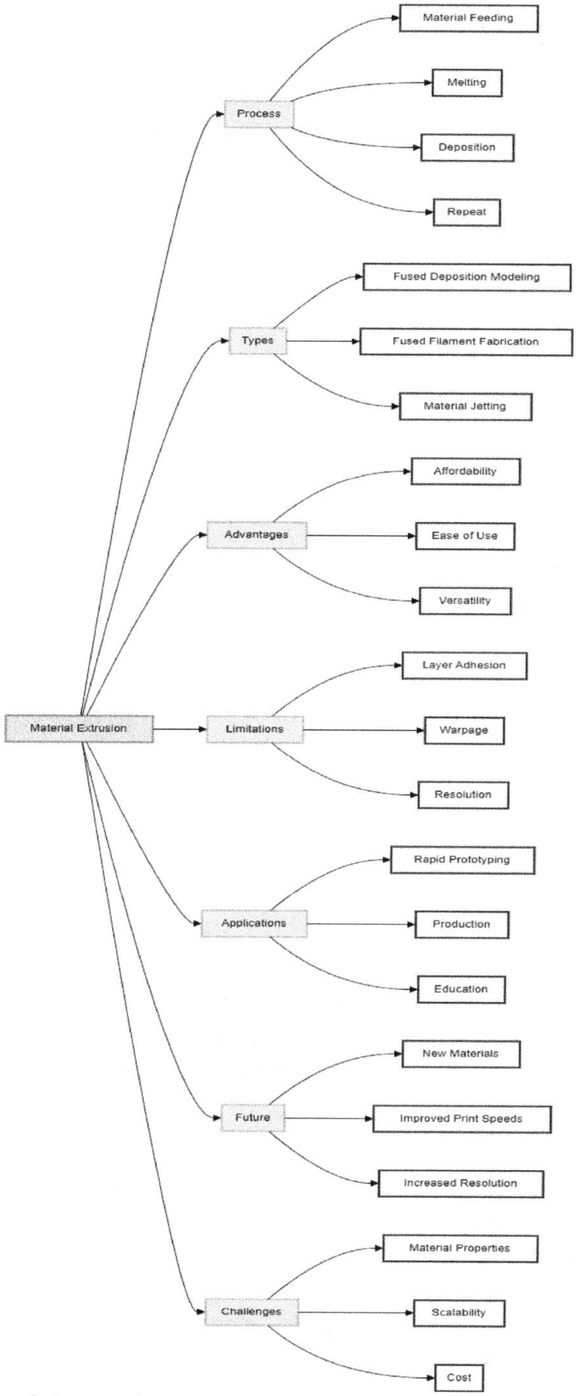

Figure 6: Material extrusion

3.5 Material Jetting

Material jetting is an advanced 3D printing technology that leverages a print head to deposit droplets of material onto a build platform with high precision. Also referred to as multi-material jetting or polyjetting, this technique is distinguished by its capability to create intricate geometries and multi-material components. Its versatility and precision make it a valuable technology in various industrial and research applications.

3.5.1 The Material Jetting Process

The material jetting process consists of several well-defined stages:

- **Material Loading**: Initially, the print head is loaded with a material that is typically a liquid photopolymer. This photopolymer is formulated to cure upon exposure to ultraviolet (UV) light. The selection of photopolymer material plays a crucial role in determining the mechanical properties and surface finish of the final printed object. The materials are often chosen for their specific attributes such as rigidity, flexibility, or transparency.

- **Droplet Deposition**: During the deposition phase, the print head precisely ejects droplets of the photopolymer onto the build platform. This stage is characterized by the accuracy of droplet placement, which is vital for constructing complex geometries. The droplets are strategically deposited according to the digital model, enabling the formation of detailed features and intricate structures.

- **Fusion**: Once deposited, the droplets undergo a curing process facilitated by UV light. This curing process solidifies the droplets, fusing them together to form a cohesive solid layer. The UV light exposure must be meticulously controlled to ensure proper curing and adhesion of the layers. This step is critical for achieving the desired mechanical properties and dimensional accuracy.

- **Repetition**: The process of material loading, droplet deposition, and curing is repeated iteratively until the complete part is fabricated. Depending on the complexity and size of the part, this iterative process can extend over several hours. Each layer builds upon the previous one, progressively constructing the final object.

3.5.2 Types of Material Jetting

Material jetting encompasses several variants, each offering unique capabilities:

- **Multi-Material Jetting:** This variant utilizes multiple print heads to simultaneously deposit different materials. The ability to incorporate multiple materials within a single print allows for the creation of parts with varied properties and functions. Multi-material jetting is particularly advantageous for applications requiring gradient properties or embedded functionalities.

- **Polyjetting:** Polyjetting employs a single print head to deposit droplets of material, often alternating between different materials within the same print job. This technique enables the production of high-resolution parts with fine details and smooth surfaces. Polyjetting is renowned for its ability to achieve high-quality finishes and complex geometrical features.

3.5.3 Advantages of Material Jetting

Material jetting offers several distinct advantages:

- **Complex Geometries:** The technology excels in producing complex geometries, including overhangs, undercuts, and intricate details that are challenging to achieve with other methods. This capability expands the possibilities for design innovation and functional part creation.

- **Multi-Material Parts**: Material jetting's ability to handle multiple materials in a single build allows for the production of multi-material components. This is beneficial for applications requiring parts with varying mechanical properties or aesthetic qualities.

- **High Resolution: -** The precision of material jetting enables the production of high-resolution parts with smooth surfaces and fine details. This high level of detail is particularly valuable in applications where surface finish and accuracy are paramount.

3.5.4 Limitations of Material Jetting

Despite its advantages, material jetting faces several limitations:

- **Material Limitations**: The technology predominantly relies on liquid photopolymers, which can have limitations in terms of material properties. These photopolymers may not always meet the requirements for mechanical strength or durability needed for certain applications.

- **Support Material**: To ensure stability during the printing process, support materials are often required. These supports must be removed post-printing, which can be time-consuming and may affect the surface finish of the final part.

- **Cost**: Material jetting systems and materials can be relatively expensive. The cost factor can be a barrier for small businesses and individual users, limiting the technology's accessibility.

3.5.5 Applications of Material Jetting

Material jetting is utilized in various fields, each benefiting from its unique capabilities:

- **Medical:** In the medical industry, material jetting is employed to produce customized implants and detailed surgical models. The ability to create patient-specific models and prosthetics enhances the precision of medical procedures and improves patient outcomes.

- **Jewelry**: The technology is also used in the jewelry industry for creating intricate designs and prototypes. Its high resolution and ability to handle complex geometries make it ideal for producing detailed jewelry pieces and casting molds.

- **Product Design**: Material jetting is extensively used in product design to develop high-resolution prototypes and models. This allows designers to evaluate form, fit, and function before committing to final production.

3.5.6 Future of Material Jetting

The future of material jetting holds significant promise, driven by ongoing research and development:

- **New Materials**: Research efforts are focused on developing new liquid photopolymers with enhanced properties, such as improved strength, toughness, and biocompatibility. These advancements are expected to broaden the range of applications and enhance performance outcomes.

- **Improved Print Speeds**: Enhancements in print speed are being explored to reduce production times and costs. Faster printing processes will contribute to more efficient manufacturing and rapid prototyping.

- **Increased Resolution: -** Ongoing research aims to further increase the resolution of material jetting, allowing for even finer details and higher precision in printed parts. Improvements in nozzle technology and curing methods are key areas of focus.

3.5.7 Challenges and Limitations

Material jetting also encounters several challenges:

- **Material Properties**: The properties of liquid photopolymers, such as strength and toughness, can impact the performance and suitability of printed parts. Addressing these limitations is essential for expanding the technology's applicability.

- **Scalability**: Scaling material jetting for larger parts or high-volume production presents challenges. The technology's limitations in build volume and production efficiency must be addressed to enhance its scalability.

- **Cost**: The expense associated with material jetting systems and materials remains a challenge. Making the technology more cost-effective is crucial for wider adoption.

In summary, material jetting represents a sophisticated 3D printing technology that excels in creating complex geometries, multi-material parts, and high-resolution components. While it has limitations related to material properties, support requirements, and cost, ongoing advancements are addressing these challenges. The technology finds valuable applications in industries such as medical, jewelry, and product design, where precision and complexity are critical. Figure 17 provides a visual representation of the material jetting process, illustrating the print head, material droplets, and build platform.

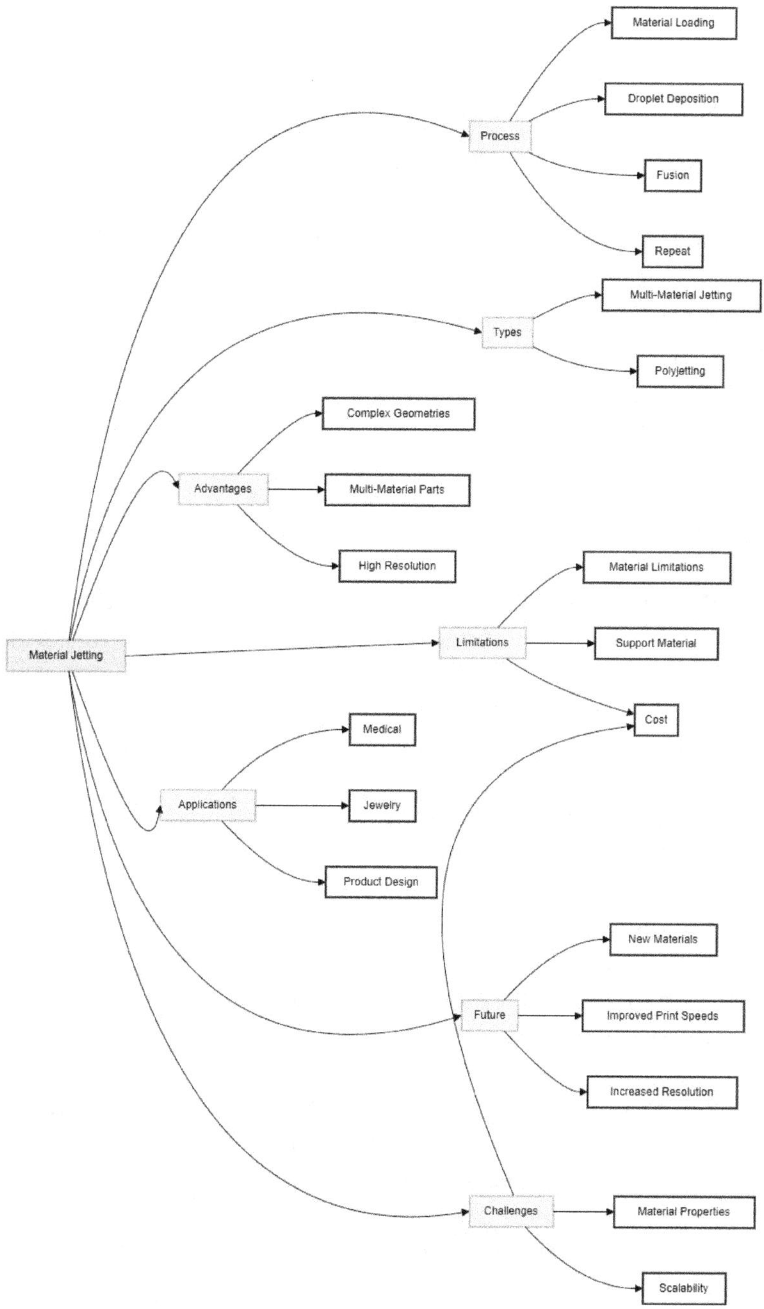

Figure 7: Material jetting

3.6 Binder Jetting

Binder jetting is an innovative 3D printing technology distinguished by its use of a liquid binder to consolidate powdered materials into a solid part. This process, also known as powder bed and inkjet head 3D printing, is renowned for its capacity to create complex geometries and large-scale components with impressive accuracy. Binder jetting is widely adopted in various industries due to its flexibility in material choices and its effectiveness in producing high-resolution parts.

3.6.1 The Binder Jetting Process

The binder jetting process comprises several sequential stages, each contributing to the successful fabrication of the final part

- **Powder Spreading**: Initially, a layer of powdered material is evenly spread across the build platform. This powder typically consists of materials such as metal, ceramic, or sand. The even distribution of powder is crucial for achieving uniform layer thickness and ensuring consistent part quality.

- **Binder Application**: Following powder spreading, a liquid binder is precisely deposited onto the powder bed using an inkjet print head. The binder acts as an adhesive, selectively bonding the powder particles together according to the digital model. The accuracy of the binder deposition is critical for defining the part's geometry and ensuring proper cohesion of the powder.

- **Layer Bonding**: Once the binder is applied, it is allowed to cure, solidifying the bonded powder particles into a solid layer. This step is essential for creating a stable and cohesive part. The curing process may involve heating or chemical reactions, depending on the binder and powder materials used.

- **Powder Removal**: After the binder has cured, the build platform is lowered, and a new layer of powder is spread over the previous layer. This process is repeated iteratively, building the part layer by layer until the entire object is completed. Excess unbound powder surrounding the part is then removed, often through manual or automated methods, leaving the final part intact.

- **Post-Processing**: Depending on the material and binder used, additional post-processing steps may be required. These can include sintering, infiltration, or coating to enhance the mechanical properties and finish of the part. Sintering, for instance, involves heating the part to fuse the powder particles more thoroughly, thereby increasing strength and durability.

3.6.2 Advantages of Binder Jetting

Binder jetting offers several notable advantages:

- **Complex Geometries**: The technology excels at producing intricate and complex geometries that are often challenging to achieve with other 3D printing methods. This capability is particularly valuable for creating detailed prototypes and customized parts with elaborate features.

- **Large-Scale Production**: Binder jetting is well-suited for manufacturing large parts or multiple parts simultaneously. The process's scalability allows for efficient production of sizable components, making it an attractive option for industrial applications.

- **Material Flexibility**: The technology supports a wide range of powdered materials, including metals, ceramics, and sands. This versatility enables the production of parts with diverse properties and functionalities, catering to various industry requirements.

- **Cost-Effectiveness**: Binder jetting is generally more cost-effective than other 3D printing technologies, particularly for large parts or high-volume production runs. The use of powdered materials and the ability to print multiple parts in a single build contribute to cost savings.

3.6.3 Limitations of Binder Jetting

Despite its advantages, binder jetting has certain limitations:

- **Material Properties**: The mechanical properties of binder jetting parts can be limited by the characteristics of the binder and powder materials used. Parts may require additional post-processing to achieve the desired strength and durability.

- **Surface Finish**: The surface finish of binder jetting parts may not always be as smooth as those produced by other methods. The need for post-processing to improve surface quality can add time and cost to the overall production process.

- **Dimensional Accuracy**: Although binder jetting provides good accuracy, achieving tight tolerances can be challenging, especially for complex geometries. Variations in powder distribution and binder application can affect dimensional precision.

3.6.4 Applications of Binder Jetting

Binder jetting is employed in a variety of applications across different industries:

- **Manufacturing**: In manufacturing, binder jetting is used for producing metal parts, prototypes, and casting molds. Its ability to handle metal powders makes it suitable for producing functional components with high precision.

- **Architecture**: The technology is utilized in architectural modeling to create detailed and large-scale models. Binder jetting's capacity to produce intricate designs makes it ideal for architectural visualization and planning.

- **Jewelry**: Binder jetting is also applied in the jewelry industry to produce intricate designs and prototypes. The technology's ability to handle fine details and create complex geometries is advantageous for designing and casting jewelry pieces.

3.6.5 Future of Binder Jetting

The future of binder jetting is marked by ongoing advancements and research aimed at enhancing the technology:

- **Material Development**: Research is focused on developing new binder formulations and powder materials that offer improved properties, such as higher strength, better thermal performance, and enhanced biocompatibility. These advancements will broaden the range of applications and improve part performance.

- **Process Optimization**: Efforts to optimize the binder jetting process include increasing print speeds, improving resolution, and reducing costs. Innovations in binder and powder handling, as well as improvements in curing and post-processing techniques, are key areas of development.

- **Integration with Other Technologies**: Combining binder jetting with other manufacturing technologies, such as additive and subtractive methods, may lead to enhanced capabilities and new applications. Integrated approaches can leverage the strengths of multiple technologies to achieve superior results.

3.6.6 Other Challenges

Binder jetting faces several other challenges:

- **Material Properties**: Ensuring consistent material properties and part performance remains a challenge. The mechanical characteristics of binder jetting parts can vary based on the materials used and the post-processing steps employed.

- **Scale-Up**: Scaling binder jetting technology for large-scale production or high-volume manufacturing can be complex. Addressing issues related to build size, powder handling, and binder application is crucial for achieving efficient and scalable production.

- **Cost**: While binder jetting is cost-effective for certain applications, the overall cost of materials, equipment, and post-processing can be significant. Reducing costs and improving cost-efficiency are ongoing challenges for broader adoption of the technology.

In conclusion, binder jetting is a powerful 3D printing technology capable of producing complex geometries and large parts with high accuracy. Despite its limitations related to material properties, surface finish, and dimensional accuracy, the technology offers significant advantages in terms of complexity, scalability, and material flexibility. Ongoing research and development are addressing these challenges and expanding the capabilities of binder jetting. Figure 18 illustrates the stages of the binder jetting process, highlighting the various steps involved in creating high-quality 3D printed parts.

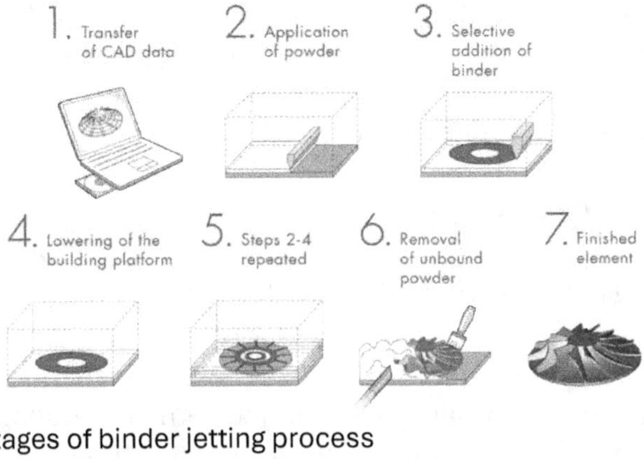

Figure 8: Stages of binder jetting process
[Source: https://engineeringproductdesign.com/knowledge-base/binder-jetting/]

Binder jetting is a sophisticated 3D printing technique that utilizes a methodical approach to transform powdered materials into solid parts. This process, characterized by its ability to handle diverse materials and create high-precision components, involves several key steps:

- **Powder Layering**: The initial stage of binder jetting involves the deposition of a layer of powdered material onto the build platform. This powder can be composed of various materials, including metals, ceramics, or glass, depending on the intended application and desired properties of the final part. The powder is typically spread using a roller or other dispensing mechanism to achieve a uniform and consistent layer. The quality of this initial powder layer is crucial for ensuring accurate and reliable bonding in subsequent stages.

- **Binder Jetting**: Following the powder layering, a liquid binder is selectively applied to the powder bed using an inkjet print head or similar deposition system. The binder functions as an adhesive, designed to bond with the powder particles selectively according to the digital model. The precise deposition of the binder is critical for achieving the desired geometry and structural integrity of the part. The binder's formulation and application parameters are optimized to ensure effective adhesion and minimal material wastage.

- **Fusion**: In the fusion stage, the binder creates a solid bond between the powder particles by curing or drying the deposited binder. This bonding process solidifies the powder layer, forming a cohesive structure that corresponds to the part's design. The curing process may involve thermal treatment, chemical reactions, or other methods to enhance the bond strength and stability. Proper fusion is essential for ensuring that the part maintains its shape and structural integrity throughout and after the printing process.

- **Repetition and Completion**: The process of powder layering, binder jetting, and fusion is repeated layer by layer until the entire part is complete. The number of layers and the complexity of the geometry determine the duration of the printing process, which can range from several hours to multiple days. The iterative nature of binder jetting allows for the creation of intricate designs and large parts with high precision.

Figure 19 illustrates the working principle of the binder jetting process, detailing the various stages involved in transforming powdered material into a finished 3D-printed component. This figure provides a visual representation of the key steps, including powder layering, binder application, and the subsequent fusion of the powder particles.

Overall, binder jetting represents a powerful and versatile approach to additive manufacturing, offering advantages in terms of material variety, precision, and scalability. The process's ability to handle complex geometries and produce high-quality parts makes it an attractive option for a wide range of applications, from industrial manufacturing to prototyping and beyond.

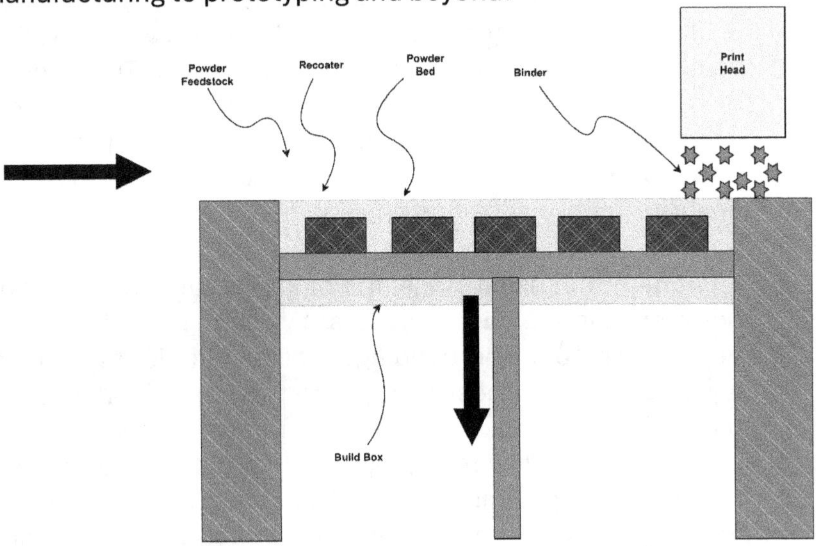

Figure 9: Working principle of binder jetting process

Binder jetting is a versatile 3D printing technology that encompasses several distinct approaches, each tailored to specific requirements and applications:

- **Powder Bed Binder Jetting:** This technique involves depositing a layer of powdered material onto a build platform, creating a powder bed. The subsequent layers of powder are then selectively bound together using a liquid binder. This approach is widely used due to its ability to handle a broad range of powder materials, including metals, ceramics, and glass. The powder bed method is particularly effective for producing parts with complex geometries and detailed features, as it allows for the creation of intricate structures with high precision.

- **Inkjet Head Binder Jetting:** In this variation, an inkjet head is employed to deposit the liquid binder onto the powdered material. The inkjet head precisely controls the deposition of the binder, allowing for high-resolution patterns and detailed part features. This method is advantageous for applications requiring fine details and complex geometries, as the inkjet head enables precise control over binder placement and distribution.

- **Selective Binder Jetting:** Selective binder jetting involves a targeted deposition process where the binder is applied only to specific areas of the powder layer, based on the design requirements. This selective approach allows for the creation of complex parts with varying densities and material properties within the same build. It is particularly useful for applications where localized strengthening or varying material characteristics are desired.

Figure 20 illustrates the products of the binder jetting process, showcasing the diverse range of components that can be achieved using this advanced manufacturing technique.

Figure 10: Products of binder jetting process

3.7 Directed Energy Deposition

Directed Energy Deposition (DED) is an advanced 3D printing technology that utilizes a focused beam of energy, such as a laser or electron beam, to precisely melt and deposit metal materials onto a build platform. This technique, often referred to as additive manufacturing or metal deposition, represents a significant advancement in the field of manufacturing, offering unique capabilities for creating complex metal parts with exceptional accuracy and precision.

3.7.1 The Directed Energy Deposition Process

The DED process is characterized by its use of high-energy sources to selectively melt metal powders or wires, which are then deposited onto a substrate to build up a part. The process involves several critical steps:

- **Material Feeding**: Metal powder or wire is fed into the deposition area. In powder-based systems, the material is typically delivered through a nozzle that directs it towards the build platform. For wire-based systems, the wire is continuously fed into the melt pool created by the energy source.

- **Energy Application**: A focused beam of energy, such as a laser or electron beam, is directed at the metal material. The energy source generates intense heat, which melts the metal powder or wire as it is deposited. This localized melting allows for precise control over the material's placement and fusion.

- **Material Deposition**: The molten metal is deposited onto the build platform or an existing part. As the material cools and solidifies, it forms a solid layer. This layer-by-layer deposition builds up the part according to the digital model.

- **Part Formation**: The process is repeated, with additional layers being deposited on top of the previous ones. This iterative layering continues until the part is complete. The ability to add material precisely where needed makes DED suitable for producing complex geometries and intricate features.

3.7.2 Advantages of Directed Energy Deposition

Directed Energy Deposition offers several notable advantages that make it a powerful tool in additive manufacturing:

- **Complex Metal Parts**: DED allows for the creation of highly complex metal parts with intricate geometries that would be difficult or impossible to achieve using traditional manufacturing methods. The precise control over material deposition enables the production of detailed features and complex shapes.

- **High Precision**: The focused energy source used in DED provides exceptional control over the melting and deposition of metal materials. This high level of precision ensures that parts are built to exact specifications, making DED suitable for applications where accuracy is critical.

- **Repair and Modification**: DED is commonly used for the repair or modification of existing metal parts. The technology enables the addition of new material to worn or damaged components, extending their service life and enhancing their functionality.

- **Material Versatility**: DED can work with a wide range of metal materials, including alloys that are challenging to process using other methods. This versatility allows for the production of parts with tailored material properties to meet specific requirements.

3.7.3 Limitations of Directed Energy Deposition

Despite its advantages, DED also presents several limitations:

- **Material Properties**: The mechanical properties of DED-produced parts can be influenced by the deposition parameters and the quality of the input materials. Variations in these factors can affect the strength, toughness, and overall performance of the parts.

- **Build Size**: While DED is capable of producing complex parts, the size of the build platform and the deposition area can limit the maximum dimensions of the parts. Scaling up the process to produce larger components presents challenges related to uniformity and control.

- **Surface Finish**: The surface finish of DED parts may require additional post-processing to achieve the desired quality. The layer-by-layer deposition process can result in surface irregularities that need to be addressed through machining or other finishing techniques.

3.7.4 Applications of Directed Energy Deposition

Directed Energy Deposition is employed in various industries due to its unique capabilities:

- **Aerospace**: In the aerospace industry, DED is used for the production of complex engine components, structural parts, and repairs of high-value components. The ability to create intricate geometries and repair existing parts makes DED a valuable technology for aerospace applications.

- **Automotive**: DED is utilized in the automotive sector for producing parts with complex features, as well as for repairing and modifying components. The technology supports both the development of advanced parts and the maintenance of existing ones.

- **Energy**: The energy sector benefits from DED for manufacturing components used in power generation and distribution, including turbine blades and high-temperature components. The technology's ability to handle specialized materials is advantageous for energy applications.

3.7.5 Future of Directed Energy Deposition

The future of Directed Energy Deposition is characterized by ongoing research and advancements aimed at enhancing the technology:

- **Enhanced Materials**: Researchers are exploring new metal alloys and composite materials that can be used with DED, aiming to improve the mechanical properties and expand the range of applications for the technology.

- **Increased Speed and Efficiency**: Efforts are being made to improve the speed and efficiency of the DED process. Enhancing deposition rates and reducing energy consumption are key areas of focus to make the technology more competitive and cost-effective.

- **Improved Control Systems**: Advances in control systems and automation are being developed to enhance the precision and repeatability of the DED process. These improvements aim to increase the reliability and consistency of the technology.

3.7.6 Challenges and Limitations

Directed Energy Deposition faces several challenges that must be addressed to fully realize its potential:

- **Material Properties**: The mechanical properties of DED-produced parts can vary depending on the deposition parameters and material quality. Ensuring consistent material properties across different builds is a key challenge.

- **Build Size and Scalability**: Scaling up DED to produce larger parts or high volumes can be challenging. Addressing issues related to build size and maintaining uniform quality across large components are important considerations.

- **Post-Processing**: The need for post-processing to improve surface finish and remove residual stresses can impact the overall efficiency and cost of DED. Developing methods to minimize or eliminate the need for post-processing is an area of ongoing research.

In conclusion, directed energy deposition represents a highly advanced and versatile 3D printing technology with the capability to produce complex metal parts with high precision. While it offers several advantages, including the ability to repair and modify existing components, the technology also faces limitations related to material properties, build size, and surface finish. Ongoing research and development are addressing these challenges, expanding the potential applications and capabilities of Directed Energy Deposition. **Figure 21** illustrates the working principle of the DED process, highlighting the use of focused energy sources to melt and deposit metal materials onto a build platform.

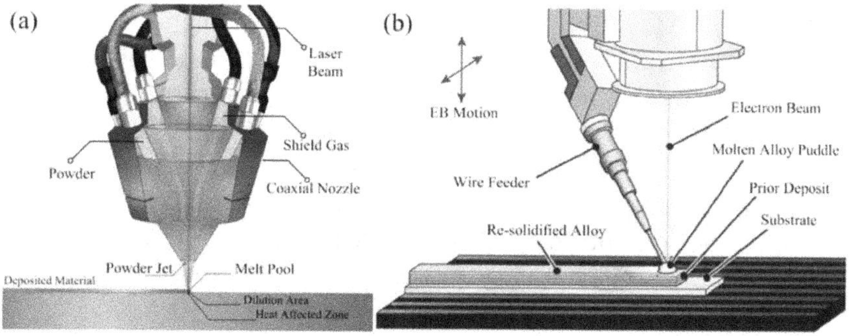

Figure 11: Principle of directed energy deposition (DED) Process

Directed Energy Deposition (DED) is a sophisticated additive manufacturing process that utilizes a concentrated energy source to melt and deposit metal materials onto a build platform, facilitating the creation of complex geometries with high precision. The DED process is particularly renowned for its ability to handle high-strength, high-temperature metals, and is employed in various high-performance applications. The process involves several well-defined steps, each contributing to the successful fabrication of the desired part.

Metal Powder Feeding

The initial step in the DED process involves the feeding of metal powder into the deposition head. This metal powder is typically composed of high-performance materials such as titanium alloys or stainless steel, known for their superior strength and high-temperature resistance. The powder is delivered through a nozzle or feeding mechanism that ensures a consistent flow of material into the deposition area. The selection of metal powder is crucial, as it directly influences the mechanical properties and overall performance of the final part.

Melting

Once the metal powder is fed into the deposition head, it is subjected to intense heating by a focused energy source. The energy source may be a laser, electron beam, or other high-energy tools capable of delivering precise and concentrated heat. This focused beam melts the metal powder, creating a molten pool that can be controlled to achieve the desired material flow and fusion characteristics. The melting process is highly controlled, allowing for the precise manipulation of temperature and material properties to ensure the optimal bonding of the metal particles.

Deposition

Following the melting phase, the molten metal is deposited onto the build platform. This deposition process is carefully controlled to form solid layers of metal that adhere to the build platform and to previously deposited layers. The deposition can be precisely directed to create intricate geometries and complex shapes, enabling the production of parts with fine details and specific structural characteristics. The ability to control the deposition parameters, including the speed and pattern of material application, plays a critical role in achieving the desired part quality and functionality.

Repeat

The process of feeding metal powder, melting, and depositing is repeated in a cyclical manner until the part is fully constructed. This iterative approach allows for the gradual build-up of material layer by layer, with each successive layer bonding to the previous one. The total time required for the completion of the part can vary significantly based on factors such as the size, complexity, and material used. Large and intricate parts may take several hours or even days to complete, depending on the specific requirements and design of the component.

In summary, Directed Energy Deposition is a highly versatile and advanced manufacturing process that leverages focused energy sources to produce complex metal parts with exceptional precision. The process begins with the careful feeding of metal powder, followed by its controlled melting and deposition onto a build platform. The iterative nature of the process allows for the construction of intricate geometries and high-performance components. Figure 22 provides a detailed illustration of the DED process in operation, highlighting the critical steps involved in achieving successful material deposition under working conditions.

Figure 12: Working of directed energy deposition process

3.7.7 Types of Directed Energy Deposition

Directed Energy Deposition (DED) encompasses several advanced techniques, each employing different energy sources to achieve the controlled melting and deposition of metal materials. The primary types of DED include Laser Beam Melting, Electron Beam Melting, and Plasma Arc Deposition, each offering unique advantages and suited to specific applications.

Laser Beam Melting

Laser Beam Melting (LBM), also referred to as Laser Additive Manufacturing, utilizes a high-powered laser to selectively melt metal powder or wire as it is deposited onto the build platform. The laser's precision allows for the construction of intricate parts with high resolution and minimal thermal distortion. LBM is particularly effective for producing high-quality parts from materials such as titanium alloys and stainless steel. The technology's ability to rapidly solidify the molten material results in fine microstructures and excellent mechanical properties.

Electron Beam Melting

Electron Beam Melting (EBM) employs an electron beam to melt and fuse metal powder or wire. The process occurs in a vacuum chamber, which helps prevent oxidation and contamination of the molten material. EBM is well-suited for high-performance applications requiring extremely high temperatures, such as aerospace and medical implants. The vacuum environment enables the use of materials that are otherwise challenging to process, and the technique is known for producing parts with high density and strength.

Plasma Arc Deposition

Plasma Arc Deposition (PAD) involves the use of a plasma arc to generate the necessary heat to melt and deposit metal. Plasma arcs achieve temperatures exceeding those of laser or electron beam sources, making them suitable for processing a wide range of metals, including refractory materials. The process is capable of handling large volumes of material, which makes it ideal for manufacturing large-scale components or repairing existing parts. Plasma arc deposition's versatility and high deposition rates contribute to its effectiveness in industrial applications.

Figure 23 illustrates the machinery used in Directed Energy Deposition operations, showcasing the components and setup involved in the process.

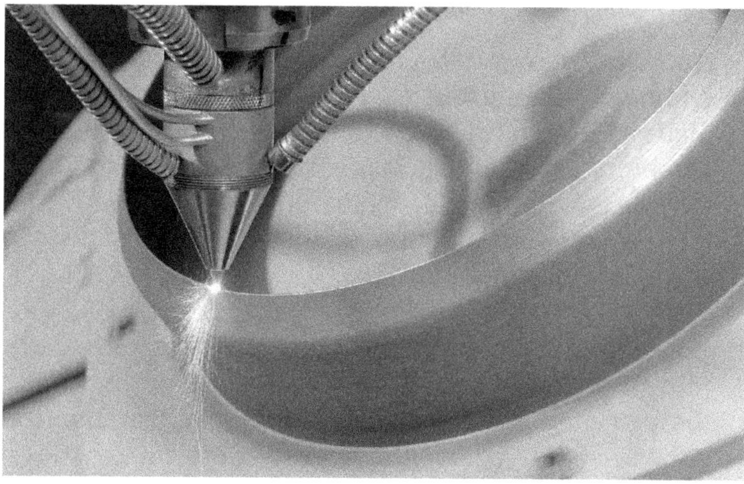

Figure 13: Directed energy deposition machine
[Source: https://www.twi-global.com/media-and-events/press-releases/2018/open-architecture-additive-manufacturing-project]

3.8 Laminated Object Manufacturing

Laminated Object Manufacturing (LOM) represents a significant advancement in 3D printing technology, employing layer-by-layer construction techniques to create complex parts with high accuracy and precision. This technology, also known as Laminated Object Fabrication or Layer-Based Manufacturing, utilizes a combination of cutting and bonding methods to produce intricate geometries from a variety of materials. The LOM process is characterized by its ability to produce durable, high-resolution parts, making it a versatile tool in various industrial applications.

The process of Laminated Object Manufacturing involves a systematic approach to building parts from multiple layers of material. This method is distinguished by its steps:

- **Material Layering:** The initial stage involves the deposition of a layer of material onto the build platform. Depending on the application, this material can range from paper and plastic to metal sheets. Each layer is precisely laid out to form the base for subsequent processing.

- **Cutting**: Following layering, the material is shaped using cutting techniques. A high-precision laser or mechanical blade is employed to carve the material into the desired geometry. This step allows for the creation of intricate and complex shapes, including features such as overhangs and internal cavities that are difficult to achieve with other manufacturing methods.

- **Bonding**: After cutting, the layers are bonded together to form a cohesive part. This bonding can be accomplished through various means, such as applying heat, pressure, or adhesive. The bonding process is critical to the part's structural integrity and overall durability, ensuring that the layers adhere strongly and uniformly.

- **Repeat**: Steps 1 through 3 are repeated iteratively until the part reaches its final form. The duration of the process depends on the size and complexity of the part, with more intricate designs requiring a longer production time.

3.8.1 Types of Laminated Object Manufacturing

Laminated Object Manufacturing encompasses several methodologies, each utilizing different cutting techniques to achieve the desired results:

- **Laser Cutting**: Laser cutting employs a high-powered laser beams to precisely cut material sheets into specific shapes. The laser's focused energy ensures clean cuts and high accuracy, making it suitable for applications requiring detailed and intricate designs.

- **Blade Cutting**: Blade cutting involves using a sharp, mechanical blade to slice through material layers. This method is effective for materials that are less suited to laser cutting and can offer a cost-effective alternative for certain applications.

- **Ultrasonic Cutting**: Ultrasonic cutting utilizes high-frequency sound waves to cut through materials. The ultrasonic vibrations reduce the mechanical resistance of the material, allowing for precise cuts with minimal thermal distortion. This technique is particularly useful for materials that are sensitive to heat or require a very fine cutting edge.

3.8.2 Advantages of Laminated Object Manufacturing

Laminated object manufacturing provides several notable benefits:

- **High Accuracy**: The technology is capable of producing parts with exceptional accuracy and precision. The layered approach allows for fine detail and exact replication of complex geometries, which is crucial for applications demanding high tolerances.

- **Complex Geometries**: LOM excels in creating parts with complex shapes, including internal features and intricate designs. The ability to build up material layer by layer supports the development of geometries that might be challenging to fabricate using traditional subtractive methods.

- **Material Flexibility**: The technology is compatible with a wide range of materials, from paper and plastic to metal. This versatility enables the production of parts tailored to specific application requirements, whether for prototyping, manufacturing, or repair.

3.8.3 Limitations of Laminated Object Manufacturing

Despite its advantages, laminated object manufacturing has several limitations:

- **Material Limitations**: The performance of the final part can be influenced by the material properties. Factors such as strength, durability, and corrosion resistance are affected by the type of material used, potentially limiting the applications where LOM is suitable.

- **Bonding Limitations**: The effectiveness of the bonding process can vary, impacting the strength and durability of the finished part. Issues such as insufficient bond strength or incomplete layer adhesion can affect the overall performance of the part.

- **Post-Processing**: Parts produced via LOM often require post-processing to achieve the desired surface finish. Excess material may need to be removed, and additional finishing processes, such as sanding or coating, may be necessary to meet final specifications.

3.8.4 Applications of Laminated Object Manufacturing

Laminated object manufacturing is employed across several industries:

- **Aerospace:** In aerospace, LOM is used to produce complex components such as satellite parts and engine components. The technology's ability to create intricate designs and handle high-strength materials makes it suitable for the demanding requirements of aerospace applications.

- **Automotive:** The automotive industry benefits from LOM for manufacturing parts such as interior trim and exterior components. The precision and material flexibility offered by LOM support the production of high-quality automotive parts.

- **Medical:** LOM is utilized in the medical field to produce custom implants and surgical models. The technology's capacity for creating precise, patient-specific parts is invaluable for medical applications requiring customized solutions.

3.8.5 Future of Laminated Object Manufacturing

The future of laminated object manufacturing holds significant promise, driven by ongoing research and technological advancements:

- **New Materials:** Advances in material science are expanding the range of materials available for LOM, including advanced composites and nanomaterials. These developments aim to enhance the performance characteristics of LOM-produced parts.

- **Improved Bonding:** Researchers are focused on improving bonding techniques, including the development of new adhesives and bonding methods. Enhanced bonding processes are expected to improve part strength and durability.

- **Increased Speed:** Efforts to increase the speed of LOM processes are underway, aiming to reduce production times and costs. Faster production capabilities will make LOM more efficient and accessible for a broader range of applications.

3.8.6 Challenges and Limitations

Laminated object manufacturing faces several challenges:

- **Material Properties:** The properties of the materials used in LOM can significantly impact part performance. Ensuring that the materials meet the required specifications for strength, durability, and resistance is critical for successful applications.

- **Bonding Properties:** The bonding process is central to part performance. Variability in bonding strength and durability can affect the overall quality and functionality of the part.

- **Scalability:** Scaling up LOM processes to produce larger parts presents challenges related to maintaining uniform material properties and consistent quality. Addressing these challenges is essential for expanding the technology's use in large-scale manufacturing.

In conclusion, Laminated Object Manufacturing is a robust 3D printing technology offering high accuracy, complex geometries, and material versatility. While it has several limitations, ongoing advancements in materials, bonding techniques, and process speed are poised to address these issues and enhance the technology's capabilities. The technology's utility in aerospace, automotive, and medical industries underscores its potential for continued growth and application. However, considerations related to material properties, bonding effectiveness, and scalability must be addressed to fully leverage the benefits of Laminated Object Manufacturing.

3.9 Questions

3.9.1 Long Response Questions

1. What are the key differences between vat photo polymerization and powder bed fusion?
2. How does material extrusion differ from material jetting?
3. What are the advantages and disadvantages of binder jetting compared to directed energy deposition?
4. How does laminated object manufacturing compare to other 3D printing technologies?
5. What are some potential applications of each 3D printing technology?

3.9.2 Low-Level Multiple Choice Questions

1) What is the main advantage of 3D printing over traditional manufacturing?

a) Lower material costs
b) Faster production speeds
c) Ability to create complex geometries

d) Higher material strength

Solution: - c) Ability to create complex geometries

2) **Which of the following is a common application of 3D printing?**

a) Producing electronics

b) Creating prototypes

c) Manufacturing textiles

d) Processing metals

Solution: - b) Creating prototypes

3) **What is the primary material used in Fused Deposition Modeling (FDM)?**

a) Liquid resin

b) Metal powder

c) Thermoplastic filament

d) Photopolymer

Solution: - c) Thermoplastic filament

4) Which **technology is known for its ability to produce high-resolution prints using liquid resin?**

a) Fused Deposition Modeling (FDM)

b) Stereolithography (SLA)

c) Binder Jetting

d) Selective Laser Sintering (SLS)

Solution: - b) Stereolithography (SLA)

5) **In which 3D printing technology is material deposited in layers from a spool?**

a) Fused Deposition Modeling (FDM)

b) Direct Energy Deposition (DED)

c) Material Jetting

d) Selective Laser Sintering (SLS)

Solution: - a) Fused Deposition Modeling (FDM)

6) **Which material is commonly used in SLA printing?**

a) PLA

b) ABS

c) Photopolymer resin

d) Nylon

Solution: - c) Photopolymer resin

7) **What is the term for a 3D model created using CAD software before printing?**

a) Prototype

b) Blueprint
c) STL file
d) Render
Solution: - c) STL file

8) **Which technology is best suited for creating large-scale industrial parts?**
a) Stereolithography (SLA)
b) Fused Deposition Modeling (FDM)
c) Digital Light Processing (DLP)
d) Inkjet Printing
Solution: - b) Fused Deposition Modeling (FDM)

9) **Which type of 3D printing uses a laser to fuse powder particles together?**
a) Powder Bed Fusion
b) Fused Deposition Modeling (FDM)
c) Material Jetting
d) Direct Ink Writing
Solution: - a) Powder Bed Fusion

10) **Which technology involves the use of multiple nozzles to deposit different materials?**
a) PolyJet
b) FDM
c) SLA
d) SLS
Solution: - a) PolyJet

11) **What is a key feature of Digital Light Processing (DLP) in 3D printing?**
a) Uses lasers to cure resin
b) Projects light to cure the entire layer at once
c) Deposits material through a heated nozzle
d) Uses a binder to join powder particles
Solution: - b) Projects light to cure the entire layer at once

12) **Which technology is known for its high-speed printing capabilities due to layer-by-layer curing?**
a) Fused Deposition Modeling (FDM)
b) Selective Laser Sintering (SLS)
c) Digital Light Processing (DLP)
d) Binder Jetting

Solution: - c) Digital Light Processing (DLP)

13) **What is a disadvantage of using Material Jetting technology?**
    a) Limited material variety
    b) High post-processing requirements
    c) High equipment cost
    d) Low resolution
    Solution: - c) High equipment cost

14) **In which technology is a laser used to sinter powder particles into a solid structure?**
    a) Stereolithography (SLA)
    b) Fused Deposition Modeling (FDM)
    c) Selective Laser Sintering (SLS)
    d) Digital Light Processing (DLP)
    Solution: - c) Selective Laser Sintering (SLS)

15) **Which 3D printing technology uses a liquid resin that solidifies when exposed to UV light?**
    a) Fused Deposition Modeling (FDM)
    b) Stereolithography (SLA)
    c) Powder Bed Fusion
    d) Binder Jetting
    Solution: - b) Stereolithography (SLA)

16) **What is the main limitation of Fused Deposition Modeling (FDM) technology?**
    a) High material cost
    b) Low resolution and warping issues
    c) Long printing times
    d) Limited material types
    Solution: - b) Low resolution and warping issues

17) **Which technology uses a binder to join powder particles into a solid object?**
    a) Material Jetting
    b) Binder Jetting
    c) Direct Energy Deposition (DED)
    d) Fused Deposition Modeling (FDM)
    Solution: - b) Binder Jetting

18) **What is a common application for SLA printing technology?**
    a) Creating large metal parts

b) Producing high-resolution dental models
c) Printing flexible materials
d) Manufacturing automotive parts
Solution: - b) Producing high-resolution dental models

19) Which 3D printing technology is known for using multiple print heads to produce parts with varying material properties?
a) Stereolithography (SLA)
b) PolyJet
c) Fused Deposition Modeling (FDM)
d) Selective Laser Sintering (SLS)
Solution: - b) PolyJet

20) In which technology are powdered materials fused together using an electron beam?
a) Electron Beam Melting (EBM)
b) Selective Laser Sintering (SLS)
c) Digital Light Processing (DLP)
d) Fused Deposition Modeling (FDM)
Solution: - a) Electron Beam Melting (EBM)

21) What is a major limitation of using SLA technology for industrial-scale production?
a) High material cost
b) Limited build volume
c) Slow printing speed
d) Limited material types
Solution: - b) Limited build volume

22) Which technology involves a nozzle depositing material through a precise controlled process to build up an object layer by layer?
a) Material Jetting
b) Fused Deposition Modeling (FDM)
c) Selective Laser Sintering (SLS)
d) Binder Jetting
Solution: - b) Fused Deposition Modeling (FDM)

23) Which type of 3D printing technology is most suitable for producing highly complex geometric structures with high precision?
a) Fused Deposition Modeling (FDM)

b) Stereolithography (SLA)
c) Direct Energy Deposition (DED)
d) Binder Jetting
Solution: - b) Stereolithography (SLA)

24) What is a common challenge faced by powder bed fusion technologies?
a) High cost of materials
b) Poor surface finish
c) Difficulty in achieving detailed fine features
d) Limited material strength
Solution: - c) Difficulty in achieving detailed fine features

25) Which of the following is an advanced technique for enhancing the properties of printed parts in SLA printing?
a) Post-processing with UV curing
b) Using multiple print heads
c) Layering different materials during printing
d) Applying heat treatment
Solution: - a) Post-processing with UV curing

26) What is the main advantage of using Binder Jetting technology compared to Powder Bed Fusion?
a) Better resolution
b) Faster printing speeds
c) More material options
d) Higher build volume
Solution: - b) Faster printing speeds

27) Which technology involves the use of a heated nozzle to deposit thermoplastic material in a layer-by-layer fashion?
a) Stereolithography (SLA)
b) Fused Deposition Modeling (FDM)
c) Direct Energy Deposition (DED)
d) Material Jetting
Solution: - b) Fused Deposition Modeling (FDM)

28) What is a common application of Direct Energy Deposition (DED) technology?
a) Creating small detailed prototypes
b) Repairing and modifying existing parts
c) Producing large-scale objects

d) Printing flexible and soft materials

Solution: - b) Repairing and modifying existing parts

29) In which technology is a liquid resin solidified by a laser to build a part layer by layer?

a) Fused Deposition Modeling (FDM)

b) Stereolithography (SLA)

c) Selective Laser Sintering (SLS)

d) Digital Light Processing (DLP)

Solution: - b) Stereolithography (SLA)

30) Which type of 3D printing technology can use metal powders as its primary material?

a) Fused Deposition Modeling (FDM)

b) Stereolithography (SLA)

c) Selective Laser Sintering (SLS)

d) Binder Jetting

Solution: - c) Selective Laser Sintering (SLS)

31) Which of the following is a limitation of using Direct Energy Deposition (DED) for large-scale metal parts?

a) High cost of materials

b) Low resolution and surface finish

c) Limited material types

d) Slow printing speed

Solution: - b) Low resolution and surface finish

32) What is the key advantage of using PolyJet technology in 3D printing?

a) High build volume

b) Ability to print multiple materials and colors in a single print

c) High speed of printing

d) Low material costs

Solution: - b). Ability to print multiple materials and colors in a single print

33) In which 3D printing technology is the build platform heated to prevent warping of the print?

a) Stereolithography (SLA)

b) Fused Deposition Modeling (FDM)

c) Digital Light Processing (DLP)

d) Binder Jetting

Solution: - b) Fused Deposition Modeling (FDM)

34) Which technology is characterized by using an electron beam in a vacuum to melt metal powder?

a) Direct Energy Deposition (DED)
b) Electron Beam Melting (EBM)
c) Selective Laser Sintering (SLS)
d) Digital Light Processing (DLP)

Solution: - b) Electron Beam Melting (EBM)

35) What is a common post-processing step required for SLA printed parts to improve their durability?

a) Baking in an oven
b) UV curing
c) Applying a primer coat
d) Heat treating

Solution: - b) UV curing

36) Which 3D printing technology is often used for producing highly detailed and complex medical implants?

a) Fused Deposition Modeling (FDM)
b) Stereolithography (SLA)
c) Selective Laser Sintering (SLS)
d) Binder Jetting

Solution: - b) Stereolithography (SLA)

37) What material is most commonly used in Stereolithography (SLA) for its high detail and smooth surface finish?

a) Thermoplastic filament
b) Photopolymer resin
c) Metal powder
d) Composite filaments

Solution: - b) Photopolymer resin

38) Which 3D printing technology uses a process similar to inkjet printing to deposit material in layers?

a) PolyJet
b) Fused Deposition Modeling (FDM)
c) Selective Laser Sintering (SLS)
d) Direct Energy Deposition (DED)

Solution: - a) PolyJet

39) What is a primary consideration when using Binder Jetting technology for producing parts?

a) Material melting point

b) Powder particle size
c) Resin curing time
d) Nozzle temperature

Solution: - b) Powder particle size

40) Which of the following technologies is best suited for creating functional prototypes with high mechanical properties?

a) Fused Deposition Modeling (FDM)
b) Stereolithography (SLA)
c) Binder Jetting
d) Direct Energy Deposition (DED)

Solution: - d) Direct Energy Deposition (DED)

Chapter 4

Vat Photopolymerization: SLA and DLP

4.1 Introduction to Vat Photopolymerization

Vat photopolymerization is a 3D printing technology that uses a light source to solidify liquid resin layer by layer. This process is also known as stereolithography (SLA) or digital light processing (DLP). Vat photopolymerization is a powerful tool for producing high-precision parts with complex geometries and smooth surfaces. Vat polymerization is an additive manufacturing process that uses a vat, or container, filled with photosensitive liquid resin and a light source to create solid objects. Figure 24 shows the products from vat polymerization process.

Figure 1: Products from Vat polymerization process

4.1.1 The Vat Photopolymerization Process

The vat photopolymerization process involves several steps:

- **Resin Preparation:** A liquid resin is prepared and poured into a vat. The resin is typically a photopolymer, which is a type of polymer that solidifies when exposed to light.

- **Light Exposure**: A light source, such as a laser or LED, is directed at the resin, solidifying a layer. The light source is typically focused onto a specific area of the resin, allowing for precise control over the solidification process.

- **Platform Movement**: The build platform is moved up or down to accommodate the next layer. The platform is typically moved in small increments, allowing for precise control over the layer thickness.

- **Repeat**: Steps 2-3 are repeated until the part is complete. The process can take several hours, depending on the size and complexity of the part.

4.1.2 Types of Vat Photopolymerization

There are several types of vat photopolymerization, including:

- **Stereolithography (SLA)**: SLA uses a laser to solidify the resin, allowing for high-precision parts with complex geometries.

- **Digital Light Processing (DLP)**: DLP uses a digital light projector to solidify the resin, allowing for fast and accurate production of parts.

- **Liquid Crystal Display (LCD)**: LCD uses a liquid crystal display to solidify the resin, allowing for high-resolution parts with smooth surfaces.

4.1.3 Advantages of Vat Photopolymerization

Vat photopolymerization has several advantages, including:

- **High Precision**: Vat photopolymerization produces parts with high precision and accuracy, making it ideal for applications such as aerospace and medical.

- **Complex Geometries**: Vat photopolymerization can produce parts with complex geometries and smooth surfaces, making it ideal for applications such as automotive and consumer products.

- **Fast Production**: Vat photopolymerization can produce parts quickly, making it ideal for rapid prototyping and production.

4.1.4 Limitations of Vat Photopolymerization

Vat photopolymerization also has several limitations, including:

- **Material Limitations**: The resin used in vat photopolymerization can be limited in terms of its mechanical properties, making it unsuitable for certain applications.

- **Post-Curing**: Vat photopolymerization requires post-curing, which can be time-consuming and may affect the part's properties.

- **Support Material**: Vat photopolymerization often requires support material, which can be difficult to remove and may affect the part's surface finish.

4.1.5 Applications of Vat Photopolymerization

Vat photopolymerization has several applications, including:

- **Aerospace**: Vat photopolymerization is used in the aerospace industry to produce complex aircraft components, such as engine parts and satellite components.

- **Automotive**: Vat photopolymerization is used in the automotive industry to produce car parts, such as interior trim and exterior components.

- **Medical:** Vat photopolymerization is used in the medical industry to produce custom implants and surgical models. Figure 25 shows the manufactured objects via vat polymerization operation.

Figure 2: Objects via Vat polymerization

4.1.6 Future of Vat Photopolymerization

The future of vat photopolymerization is promising, with ongoing research and development aimed at improving the technology. Some areas of research include:

- **New Materials:** Researchers are developing new resins with improved mechanical properties, making vat photopolymerization more suitable for a wider range of applications.

- **Improved Light Sources:** Researchers are working to improve the light sources used in vat photopolymerization, allowing for faster and more accurate production of parts.

- **Increased Resolution: -** Researchers are working to increase the resolution of vat photopolymerization, allowing for more detailed and complex parts.

4.1.7 Challenges and Limitations

Vat photopolymerization also faces several challenges and limitations, including:

- **Material Properties**: The material properties of the resin can affect the part's performance, including its strength, durability, and corrosion resistance.

- **Light Source Properties**: The light source properties can affect the part's accuracy and precision, including its wavelength and intensity.

- **Scalability**: Vat photopolymerization can be difficult to scale up, making it challenging to produce large parts.

In conclusion, vat photopolymerization is a powerful 3D printing technology that offers high precision, complex geometries, and fast production which is displayed in Figure 26. While it has several limitations, ongoing research and development are addressing these issues, expanding the capabilities of vat photopolymerization. The technology is particularly useful in industries such as aerospace, automotive, and medical, where high-precision parts are essential. However, the limitations of vat photopolymerization, including material limitations, post-curing, and support material, must be considered when deciding whether to adopt this technology.

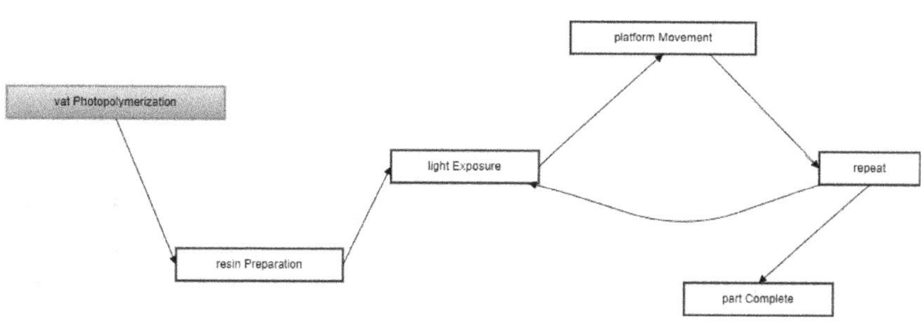

Figure 3: Vat photopolymerization

4.2 Stereolithography (SLA)

Stereolithography (SLA) is a vat photopolymerization technology that uses a laser to solidify the resin. The laser is directed at the resin surface, tracing the shape of the layer to be solidified. SLA is a powerful tool for producing high-precision parts with complex geometries and smooth surfaces.

4.2.1 The SLA Process

The SLA process involves several steps:

- **Resin Preparation**: A liquid resin is prepared and poured into a vat. The resin is typically a photopolymer, which is a type of polymer that solidifies when exposed to light.

- **Laser Exposure**: A laser is directed at the resin, solidifying a layer. The laser is typically an ultraviolet (UV) laser, which is focused onto a specific area of the resin, allowing for precise control over the solidification process.

- **Platform Movement**: The build platform is moved up or down to accommodate the next layer. The platform is typically moved in small increments, allowing for precise control over the layer thickness.

- **Repeat**: Steps 2-3 are repeated until the part is complete. The process can take several hours, depending on the size and complexity of the part.

4.2.2 How SLA Works

SLA works by using a laser to solidify the resin, layer by layer. The laser is directed at the resin surface, tracing the shape of the layer to be solidified. The resin is solidified through a process called photopolymerization, where the laser light initiates a chemical reaction that causes the resin to harden.

4.2.3 Advantages of SLA

SLA has several advantages, including:

- **High Precision**: SLA produces parts with high precision and accuracy, making it ideal for applications such as aerospace and medical.
- **Complex Geometries**: SLA can produce parts with complex geometries and smooth surfaces, making it ideal for applications such as automotive and consumer products.
- **Fast Production**: SLA can produce parts quickly, making it ideal for rapid prototyping and production.

4.2.4 Limitations of SLA

SLA also has several limitations, including:

- **Material Limitations**: The resin used in SLA can be limited in terms of its mechanical properties, making it unsuitable for certain applications.
- **Post-Curing**: SLA requires post-curing, which can be time-consuming and may affect the part's properties.
- **Support Material**: SLA often requires support material, which can be difficult to remove and may affect the part's surface finish.

4.2.5 Applications of SLA

SLA has several applications, including:

- **Aerospace**: SLA is used in the aerospace industry to produce complex aircraft components, such as engine parts and satellite components.

- **Automotive**: SLA is used in the automotive industry to produce car parts, such as interior trim and exterior components.

- **Medical**: SLA is used in the medical industry to produce custom implants and surgical models.

4.2.6 History of SLA

SLA was invented in the 1980s by Chuck Hull, who founded 3D Systems, a leading manufacturer of 3D printing technologies. Since then, SLA has become a widely used technology in various industries, including aerospace, automotive, and medical.

4.2.7 Future of SLA

The future of SLA is promising, with ongoing research and development aimed at improving the technology. Some areas of research include:

- **New Materials**: Researchers are developing new resins with improved mechanical properties, making SLA more suitable for a wider range of applications.

- **Improved Laser Technology**: Researchers are working to improve the laser technology used in SLA, allowing for faster and more accurate production of parts.

- **Increased Resolution: -** Researchers are working to increase the resolution of SLA, allowing for more detailed and complex parts.

In conclusion, SLA is a powerful 3D printing technology that offers high precision, complex geometries, and fast production which is depicted in Figure 27. While it has several limitations, ongoing research and development are addressing these issues, expanding the capabilities of SLA. The technology is particularly useful in industries such as aerospace, automotive, and medical, where high-precision parts are essential. However, the limitations of SLA, including material limitations, post-curing, and support material, must be considered when deciding whether to adopt this technology.

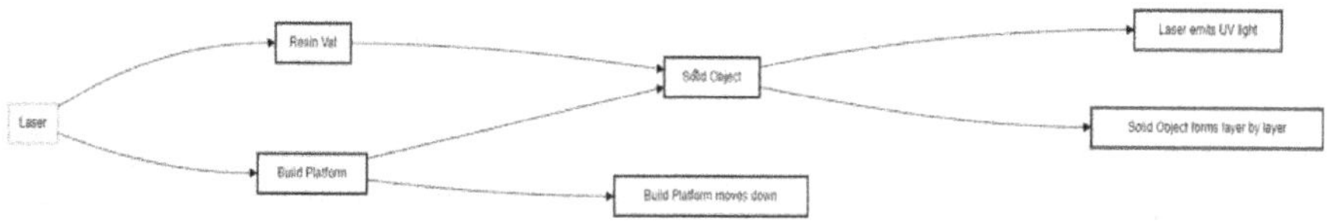

Figure 4: Stereolithography (SLA)

4.3 Digital Light Processing (DLP)

Digital light processing (DLP) is a vat photopolymerization technology that uses a digital light projector to solidify the resin. The projector displays an image of the layer to be solidified onto the resin surface. DLP is a powerful tool for producing high-precision parts with complex geometries and smooth surfaces.

4.3.1 The DLP Process

The DLP process involves several steps:

- **Resin Preparation**: A liquid resin is prepared and poured into a vat. The resin is typically a photopolymer, which is a type of polymer that solidifies when exposed to light.

- **Image Exposure**: A digital light projector displays an image of the layer to be solidified onto the resin surface. The projector uses a digital micromirror device (DMD) to create the image, which is made up of millions of tiny mirrors that reflect light onto the resin.

- **Platform Movement**: The build platform is moved up or down to accommodate the next layer. The platform is typically moved in small increments, allowing for precise control over the layer thickness.

- **Repeat**: Steps 2-3 are repeated until the part is complete. The process can take several hours, depending on the size and complexity of the part.

4.3.2 How DLP Works

DLP works by using a digital light projector to solidify the resin, layer by layer. The projector displays an image of the layer to be solidified onto the resin surface, which is then solidified through a process called photopolymerization. The resin is solidified through a chemical reaction that is initiated by the light, which causes the resin to harden.

4.3.3 Advantages of DLP

DLP has several advantages, including:

- **High Resolution: -** DLP produces parts with high resolution and accuracy, making it ideal for applications such as aerospace and medical.

- **Fast Production**: DLP can produce parts quickly, making it ideal for rapid prototyping and production.

- **Low Cost**: DLP is a relatively low-cost technology compared to other vat photopolymerization technologies, making it accessible to a wider range of users.

4.3.4 Limitations of DLP

DLP also has several limitations, including:

- **Material Limitations**: The resin used in DLP can be limited in terms of its mechanical properties, making it unsuitable for certain applications.

- **Post-Curing**: DLP requires post-curing, which can be time-consuming and may affect the part's properties.

- **Support Material**: DLP often requires support material, which can be difficult to remove and may affect the part's surface finish.

4.3.5 Applications of DLP

DLP has several applications, including:

- **Aerospace**: DLP is used in the aerospace industry to produce complex aircraft components, such as engine parts and satellite components.

- **Automotive**: DLP is used in the automotive industry to produce car parts, such as interior trim and exterior components.

- **Medical**: DLP is used in the medical industry to produce custom implants and surgical models.

4.3.6 History of DLP

DLP was invented in the 1980s by Texas Instruments, a leading manufacturer of digital light projectors. Since then, DLP has become a widely used technology in various industries, including aerospace, automotive, and medical.

4.3.7 Future of DLP

The future of DLP is promising, with ongoing research and development aimed at improving the technology. Some areas of research include:

- **New Materials**: Researchers are developing new resins with improved mechanical properties, making DLP more suitable for a wider range of applications.

- **Improved Projector Technology**: Researchers are working to improve the projector technology used in DLP, allowing for faster and more accurate production of parts.

- **Increased Resolution:** - Researchers are working to increase the resolution of DLP, allowing for more detailed and complex parts.

In conclusion, DLP is a powerful 3D printing technology that offers high resolution, fast production, and low cost which is shown in Figure 28. While it has several limitations, ongoing research and development are addressing these issues, expanding the capabilities of DLP. The technology is particularly useful in industries such as aerospace, automotive, and medical, where high-precision parts are essential. However, the limitations of DLP, including material limitations, post-curing, and support material, must be considered when deciding whether to adopt this technology.

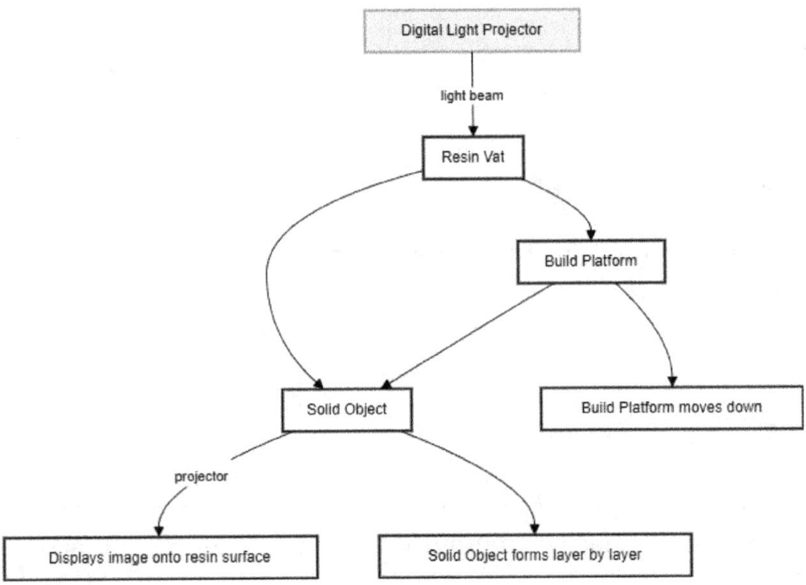

Figure 5: Digital light processing (DLP)

4.4 Applications and Limitations

Vat photopolymerization technologies, such as SLA and DLP, are commonly used for rapid prototyping, dental and medical applications, and small-scale production. These technologies offer high resolution and accuracy, making them suitable for producing complex geometries and fine details.

4.4.1 Rapid Prototyping

Vat photopolymerization technologies are widely used for rapid prototyping, which involves creating a scale model of a product or component to test its design, functionality, and performance. The high resolution and accuracy of SLA and DLP make them ideal for producing prototypes with complex geometries and fine details. Additionally, the fast production times of these technologies allow designers and engineers to quickly test and iterate on their designs.

4.4.2 Dental and Medical Applications

Vat photopolymerization technologies are also commonly used in dental and medical applications. For example, SLA and DLP are used to produce dental implants, surgical guides, and models of organs and tissues. The high accuracy and resolution of these technologies make them ideal for producing parts with complex geometries and fine details, such as dental implants and surgical guides.

4.4.3 Small-Scale Production

Vat photopolymerization technologies are also used for small-scale production, which involves producing small batches of parts or products. The high resolution and accuracy of SLA and DLP make them ideal for producing parts with complex geometries and fine details, such as jewelry, watches, and other precision components.

4.4.4 Limitations of Vat Photopolymerization

While vat photopolymerization technologies offer many advantages, they also have some limitations. For example:

- **Material Limitations**: The materials used in vat photopolymerization processes are typically limited to photopolymers, which may have limited mechanical properties and may not be suitable for high-temperature or high-stress applications.

- **Build Volume Limitations**: The build volumes of vat photopolymerization technologies are generally smaller than those of other 3D printing technologies, such as powder bed fusion or material extrusion. This can limit the size of parts that can be produced using these technologies.

- **Post-Curing Requirements**: Vat photopolymerization technologies often require post-curing, which can be time-consuming and may affect the part's properties.

- **Support Material Requirements**: Vat photopolymerization technologies often require support material, which can be difficult to remove and may affect the part's surface finish.

4.4.5 Comparison to Other 3D Printing Technologies

Vat photopolymerization technologies are often compared to other 3D printing technologies, such as powder bed fusion and material extrusion. While these technologies offer different advantages and limitations, they are often used for different applications.

- **Powder Bed Fusion**: Powder bed fusion technologies, such as selective laser sintering (SLS) and selective laser melting (SLM), offer higher build volumes and faster production times than vat photopolymerization technologies. However, they often require more expensive equipment and may have lower resolution and accuracy.

- **Material Extrusion**: Material extrusion technologies, such as fused deposition modeling (FDM), offer lower costs and faster production times than vat photopolymerization technologies. However, they often have lower resolution and accuracy and may require more post-processing.

4.4.6 Future of Vat Photopolymerization

The future of vat photopolymerization is promising, with ongoing research and development aimed at improving the technology. Some areas of research include:

- **New Materials**: Researchers are developing new photopolymers with improved mechanical properties, making vat photopolymerization more suitable for a wider range of applications.

- **Improved Equipment**: Researchers are working to improve the equipment used in vat photopolymerization, allowing for faster and more accurate production of parts.

- **Increased Build Volumes**: Researchers are working to increase the build volumes of vat photopolymerization technologies, making them more suitable for large-scale production.

In conclusion, vat photopolymerization technologies, such as SLA and DLP, offer high resolution and accuracy, making them suitable for producing complex geometries and fine details. While they have some limitations, ongoing research and development are addressing these issues, expanding the capabilities of vat photopolymerization. The technology is particularly useful in industries such as aerospace, automotive, and medical, where high-precision parts are essential. However, the limitations of vat photopolymerization, including material limitations, build volume limitations, and post-curing requirements, must be considered when deciding whether to adopt this technology.

4.5 Questions

4.5.1 Long Response Questions

1. What are the key differences between stereolithography (SLA) and digital light processing (DLP)?

2. What are some common applications of vat photopolymerization technologies?
3. What are the limitations of vat photopolymerization technologies?
4. How does the resolution and accuracy of vat photopolymerization technologies compare to other 3D printing technologies?
5. What are some potential future developments in vat photopolymerization technologies?

4.5.2 Multiple Choice Questions

1) **What is the primary purpose of vat photopolymerization?**
   a) To heat plastic
   b) To solidify liquid resin
   c) To melt metal
   d) To cool down materials
   Solution: - b) To solidify liquid resin
2) **Which technology uses a laser to solidify the resin?**
   a) Digital Light Processing (DLP)
   b) Stereolithography (SLA)
   c) Liquid Crystal Display (LCD)
   d) Fused Deposition Modeling (FDM)
   Solution: - b) Stereolithography (SLA)
3) **What does DLP stand for in vat photopolymerization?**
   a) Direct Light Processing
   b) Digital Light Processing
   c) Dynamic Light Processing
   d) Digital Laser Projection
   Solution: - b) Digital Light Processing
4) **Which light source is used in SLA?**
   a) LED
   b) UV Laser
   c) Incandescent Lamp
   d) Infrared Laser
   Solution: - b) UV Laser
5) **What is a key advantage of vat photopolymerization?**
   a) High speed of production
   b) Large build volume

c) High precision and accuracy
d) High cost

Solution: - c) High precision and accuracy

6) **What material is typically used in vat photopolymerization?**
a) Metal alloy
b) Glass
c) Photopolymer resin
d) Ceramics

Solution: - c) Photopolymer resin

7) **Which technology uses a digital micromirror device (DMD)?**
a) Stereolithography (SLA)
b) Digital Light Processing (DLP)
c) Fused Deposition Modeling (FDM)
d) Selective Laser Sintering (SLS)

Solution: - b) Digital Light Processing (DLP)

8) **What is the role of the build platform in vat photopolymerization?**
a) To cool the resin
b) To support and move the part during printing
c) To heat the resin
d) To mix the resin

Solution: - b) To support and move the part during printing

9) **Which vat photopolymerization technology is known for fast production?**
a) Stereolithography (SLA)
b) Digital Light Processing (DLP)
c) Fused Deposition Modeling (FDM)
d) Powder Bed Fusion

Solution: - b) Digital Light Processing (DLP)

10) **What is required after the printing process in vat photopolymerization?**
a) Painting
b) Post-curing
c) Sanding
d) Melting

Solution: - b) Post-curing

11) How does SLA produce a solid part?

a) By melting metal
b) By solidifying resin using a laser
c) By cooling a liquid
d) By compressing a powder

Solution: - b) By solidifying resin using a laser

12) Which vat photopolymerization technology involves projecting an image of the layer to be solidified?

a) Liquid Crystal Display (LCD)
b) Stereolithography (SLA)
c) Digital Light Processing (DLP)
d) Selective Laser Melting (SLM)

Solution: - c) Digital Light Processing (DLP)

13) What does VAT stand for in vat photopolymerization?

a) Volume Application Technology
b) Virtual Application Technology
c) Vertical Application Technology
d) Vats Applied Technology

Solution: - a) Volume Application Technology

14) Which technology is known for using an ultraviolet (UV) laser?

a) Digital Light Processing (DLP)
b) Stereolithography (SLA)
c) Liquid Crystal Display (LCD)
d) Material Extrusion

Solution: - b) Stereolithography (SLA)

15) What is one of the major limitations of vat photopolymerization?

a) High precision
b) Material limitations
c) Fast production
d) Large build volume

Solution: - b) Material limitations

16) Which vat photopolymerization technology uses liquid crystal display (LCD) technology?

a) Stereolithography (SLA)
b) Digital Light Processing (DLP)

c) Liquid Crystal Display (LCD)
d) Selective Laser Sintering (SLS)
Solution: - c) Liquid Crystal Display (LCD)

17) In which industry is vat photopolymerization particularly useful for producing custom implants?
a) Aerospace
b) Automotive
c) Medical
d) Consumer Products
Solution: - c) Medical

18) Which technology involves using a light projector to solidify the resin?
a) Stereolithography (SLA)
b) Digital Light Processing (DLP)
c) Fused Deposition Modeling (FDM)
d) Selective Laser Sintering (SLS)
Solution: - b) Digital Light Processing (DLP)

19) What is a common application of SLA in the automotive industry?
a) Engine parts
b) Interior trim
c) Satellites
d) Surgical guides
Solution: - b) Interior trim

20) What type of resin is typically used in vat photopolymerization?
a) Thermoplastic resin
b) Photopolymer resin
c) Ceramic resin
d) Metal resin
Solution: - b) Photopolymer resin

21) What is one of the challenges faced in scaling up vat photopolymerization technologies?
a) Material availability
b) Build volume limitations
c) Light source intensity
d) Resin cooling rate
Solution: - b) Build volume limitations

22) Which vat photopolymerization technology is known for high resolution and low cost compared to others?

a) Stereolithography (SLA)
b) Digital Light Processing (DLP)
c) Liquid Crystal Display (LCD)
d) Powder Bed Fusion

Solution: - b) Digital Light Processing (DLP)

23) What is the primary difference between SLA and DLP in terms of light source?

a) SLA uses a UV laser while DLP uses a digital projector
b) DLP uses a UV laser while SLA uses a digital projector
c) SLA uses visible light while DLP uses infrared light
d) SLA and DLP use the same light source

Solution: - a) SLA uses a UV laser while DLP uses a digital projector

24) Which technology uses a digital micromirror device to project the layer image?

a) Stereolithography (SLA)
b) Digital Light Processing (DLP)
c) Liquid Crystal Display (LCD)
d) Material Extrusion

Solution: - b) Digital Light Processing (DLP)

25) What are researchers developing to enhance the capabilities of vat photopolymerization?

a) New photopolymers with improved mechanical properties
b) New cooling systems for resin
c) New powder materials
d) New mechanical supports

Solution: - a) New photopolymers with improved mechanical properties

26) Which vat photopolymerization technology is most commonly associated with producing high-precision parts for aerospace applications?

a) Digital Light Processing (DLP)
b) Liquid Crystal Display (LCD)
c) Stereolithography (SLA)
d) Material Extrusion

Solution: - c) Stereolithography (SLA)

27) **Which challenge is common to both SLA and DLP technologies?**
   a) High cost
   b) Material limitations
   c) Large build volume
   d) Slow production speed
   Solution: - b) Material limitations
28) **Which vat photopolymerization technology is best known for its high resolution and accuracy in medical applications?**
   a) Stereolithography (SLA)
   b) Digital Light Processing (DLP)
   c) Fused Deposition Modeling (FDM)
   d) Powder Bed Fusion
   Solution: - a) Stereolithography (SLA)
29) **What kind of resin properties are being improved in ongoing research for vat photopolymerization?**
   a) Flexibility and color
   b) Mechanical strength and durability
   c) Density and texture
   d) Temperature resistance
   Solution: - b) Mechanical strength and durability
30) **What is one way researchers are working to improve vat photopolymerization technologies?**
   a) By increasing the temperature of the resin
   b) By developing faster cooling systems
   c) By enhancing light source technology
   d) By reducing the size of the build platform
   Solution: - c) By enhancing light source technology
31) **What is a common requirement for vat photopolymerization processes that can be time-consuming?**
   a) Post-curing
   b) Resin mixing
   c) Platform calibration
   d) Light calibration
   Solution: - a) Post-curing
32) **Which technology was developed by Chuck Hull in the 1980s?**

a) Digital Light Processing (DLP)
b) Stereolithography (SLA)
c) Liquid Crystal Display (LCD)
d) Material Extrusion

Solution: - b) Stereolithography (SLA)

33) What aspect of vat photopolymerization technology affects the accuracy of the final part?

a) The type of resin used
b) The light source properties
c) The build platform material
d) The cooling system

Solution: - b) The light source properties

34) What challenge does material limitation in vat photopolymerization present?

a) High cost
b) Limited mechanical properties
c) Slow production speed
d) Large size of the machine

Solution: - b) Limited mechanical properties

35) Which vat photopolymerization technology is particularly noted for its fast production capabilities?

a) Stereolithography (SLA)
b) Digital Light Processing (DLP)
c) Fused Deposition Modeling (FDM)
d) Selective Laser Sintering (SLS)

Solution: - b) Digital Light Processing (DLP)

36) What is a major factor influencing the choice between SLA and DLP technologies?

a) Material availability
b) Resolution and speed requirements
c) Build volume size
d) Post-curing process

Solution: - b) Resolution and speed requirements

37) What key feature differentiates SLA from DLP in terms of light source?

a) SLA uses a continuous light source, while DLP uses a flashing light source.

b) DLP uses a continuous light source, while SLA uses a flashing light source.

c) SLA uses a single wavelength laser, while DLP uses multiple wavelengths.

d) Both use the same type of light source.

Solution: - a) SLA uses a continuous light source, while DLP uses a flashing light source.

38) Which vat photopolymerization technology would you use for high-detail medical implants?

a) Digital Light Processing (DLP)

b) Fused Deposition Modeling (FDM)

c) Selective Laser Sintering (SLS)

d) Stereolithography (SLA)

Solution: - d) Stereolithography (SLA)

39) What aspect of vat photopolymerization technology contributes to its high precision?

a) The type of cooling system used

b) The build volume size

c) The light source's wavelength and intensity

d) The support material properties

Solution: - c) The light source's wavelength and intensity

40) Which vat photopolymerization technology might be used to produce complex, small-scale production parts?

a) Digital Light Processing (DLP)

b) Stereolithography (SLA)

c) Fused Deposition Modeling (FDM)

d) Selective Laser Sintering (SLS)

Solution: - b) Stereolithography (SLA)

Chapter 5

Powder Bed Fusion: SLS and SLM

5.1 Introduction to Powder Bed Fusion

Powder bed fusion is a 3D printing technology that uses a laser or electron beam to fuse together particles of a powdered material. This process is also known as selective laser sintering (SLS) or selective laser melting (SLM). Powder bed fusion is a powerful tool for producing complex parts with high accuracy and precision.

5.1.1 The Powder Bed Fusion Process

The powder bed fusion process involves several steps:

- **Powder Layering**: A layer of powdered material is deposited onto the build platform. The powder is typically spread evenly across the platform using a roller or blade.

- **Fusion**: A laser or electron beam is directed at the powder, fusing the particles together. The laser or electron beam is focused onto a specific area of the powder, causing the particles to melt and bond together.

- **Platform Movement**: The build platform is lowered to accommodate the next layer. The platform is typically moved down by a small increment, allowing for precise control over the layer thickness.

- **Repeat**: Steps 1-3 are repeated until the part is complete. The process can take several hours, depending on the size and complexity of the part.

5.1.2 How Powder Bed Fusion Works

Powder bed fusion works by using a laser or electron beam to fuse together particles of a powdered material. The laser or electron beam is focused onto a specific area of the powder, causing the particles to melt and bond together. The resulting part is strong and durable, with high accuracy and precision.

5.1.3 Advantages of Powder Bed Fusion

Powder bed fusion has several advantages, including:

- **High Accuracy**: Powder bed fusion produces parts with high accuracy and precision, making it ideal for applications such as aerospace and medical.

- **Complex Geometries**: Powder bed fusion can produce parts with complex geometries, including internal structures and hollow parts.

- **No Support Material**: Powder bed fusion does not require support material, making it ideal for producing parts with complex geometries.

- **Fast Production**: Powder bed fusion can produce parts quickly, making it ideal for rapid prototyping and production.

5.1.4 Limitations of Powder Bed Fusion

Powder bed fusion also has several limitations, including:

- **Material Limitations**: The powdered material used in powder bed fusion can be limited in terms of its mechanical properties, making it unsuitable for certain applications.

- **Cost**: Powder bed fusion can be expensive, particularly for large-scale production.

- **Post-Processing**: Powder bed fusion often requires post-processing, such as machining or grinding, to achieve the desired surface finish.

5.1.5 Applications of Powder Bed Fusion

Powder bed fusion has several applications, including:

- **Aerospace**: Powder bed fusion is used in the aerospace industry to produce complex aircraft components, such as engine parts and satellite components.

- **Automotive**: Powder bed fusion is used in the automotive industry to produce car parts, such as engine components and interior trim.

- **Medical**: Powder bed fusion is used in the medical industry to produce custom implants and surgical models.

5.1.6 History of Powder Bed Fusion

Powder bed fusion was invented in the 1990s by researchers at the University of Texas. Since then, the technology has evolved rapidly, with the development of new materials and equipment.

5.1.7 Future of Powder Bed Fusion

The future of powder bed fusion is promising, with ongoing research and development aimed at improving the technology. Some areas of research include:

- **New Materials**: Researchers are developing new powdered materials with improved mechanical properties, making powder bed fusion more suitable for a wider range of applications.

- **Improved Equipment**: Researchers are working to improve the equipment used in powder bed fusion, allowing for faster and more accurate production of parts.

- **Increased Build Volumes**: Researchers are working to increase the build volumes of powder bed fusion technologies, making them more suitable for large-scale production.

In conclusion, powder bed fusion is a powerful 3D printing technology that offers high accuracy and precision, complex geometries, and fast production. While it has some limitations, ongoing research and development are addressing these issues, expanding the capabilities of powder bed fusion. The technology is particularly useful in industries such as aerospace, automotive, and medical, where high-precision parts are essential. However, the limitations of powder bed fusion, including material limitations, cost, and post-processing requirements, must be considered when deciding whether to adopt this technology. The overall process of Powder Bed Fusion is displayed in Figure 29.

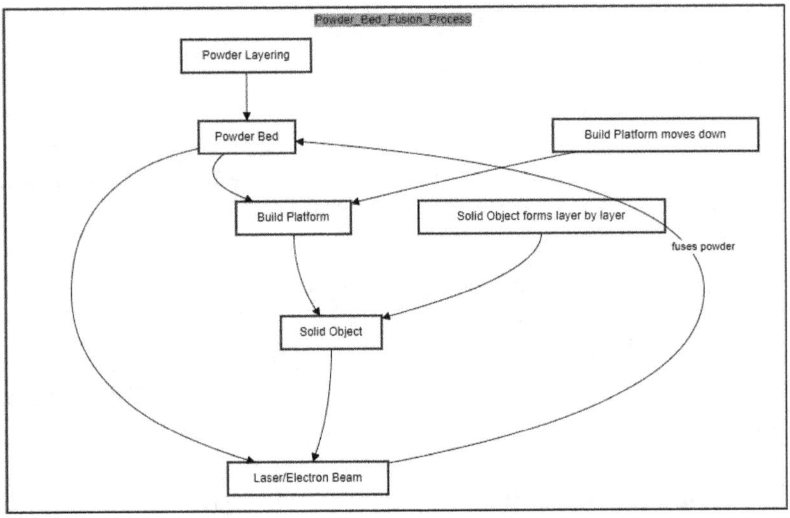

Figure 1: Powder bed fusion.

5.2 Selective Laser Sintering (SLS)

Selective laser sintering (SLS) is a powder bed fusion technology that uses a laser to fuse together particles of a powdered material. The laser is directed at the powder, sintering the particles together to form a solid layer. SLS is a popular 3D printing technology used in a wide range of industries, including aerospace, automotive, and medical.

5.2.1 The SLS Process

The SLS process involves several steps:

- **Powder Layering**: A layer of powdered material is deposited onto the build platform. The powder is typically spread evenly across the platform using a roller or blade.

- **Laser Sintering**: A laser is directed at the powder, sintering the particles together. The laser is focused onto a specific area of the powder, causing the particles to melt and bond together.

- **Platform Movement**: The build platform is lowered to accommodate the next layer. The platform is typically moved down by a small increment, allowing for precise control over the layer thickness.

- **Repeat**: Steps 1-3 are repeated until the part is complete. The process can take several hours, depending on the size and complexity of the part.

5.2.2 How SLS Works

SLS works by using a laser to sinter together particles of a powdered material. The laser is focused onto a specific area of the powder, causing the particles to melt and bond together. The resulting part is strong and durable, with high accuracy and precision.

5.2.3 Advantages of SLS

SLS has several advantages, including:

- **High Accuracy**: SLS produces parts with high accuracy and precision, making it ideal for applications such as aerospace and medical.

- **Complex Geometries**: SLS can produce parts with complex geometries, including internal structures and hollow parts.

- **No Support Material**: SLS does not require support material, making it ideal for producing parts with complex geometries.

- **Fast Production**: SLS can produce parts quickly, making it ideal for rapid prototyping and production.

5.2.4 Limitations of SLS

SLS also has several limitations, including:

- **Material Limitations**: The powdered material used in SLS can be limited in terms of its mechanical properties, making it unsuitable for certain applications.

- **Cost**: SLS can be expensive, particularly for large-scale production.

- **Post-Processing**: SLS often requires post-processing, such as machining or grinding, to achieve the desired surface finish.

5.2.5 Applications of SLS

SLS has several applications, including:

- **Aerospace**: SLS is used in the aerospace industry to produce complex aircraft components, such as engine parts and satellite components.

- **Automotive**: SLS is used in the automotive industry to produce car parts, such as engine components and interior trim.

- **Medical**: SLS is used in the medical industry to produce custom implants and surgical models.

5.2.6 History of SLS

SLS was invented in the 1990s by researchers at the University of Texas. Since then, the technology has evolved rapidly, with the development of new materials and equipment.

5.2.7 Future of SLS

The future of SLS is promising, with ongoing research and development aimed at improving the technology. Some areas of research include:

- **New Materials**: Researchers are developing new powdered materials with improved mechanical properties, making SLS more suitable for a wider range of applications.

- **Improved Equipment**: Researchers are working to improve the equipment used in SLS, allowing for faster and more accurate production of parts.

- **Increased Build Volumes**: Researchers are working to increase the build volumes of SLS technologies, making them more suitable for large-scale production.

5.2.8 Comparison to Other 3D Printing Technologies

SLS is often compared to other 3D printing technologies, such as selective laser melting (SLM) and fused deposition modeling (FDM). While these technologies offer different advantages and limitations, they are often used for different applications.

- **SLM**: SLM is similar to SLS, but uses a higher-powered laser to melt the powder, rather than sintering it. This allows for the production of parts with higher density and strength.

- **FDM**: FDM is a different type of 3D printing technology that uses melted plastic to produce parts. It is often used for rapid prototyping and production of parts with complex geometries.

In conclusion, SLS is a powerful 3D printing technology that offers high accuracy and precision, complex geometries, and fast production. While it has some limitations, ongoing research and development are addressing these issues, expanding the capabilities of SLS. The technology is particularly useful in industries such as aerospace, automotive, and medical, where high-precision parts are essential. However, the limitations of SLS, including material limitations, cost, and post-processing requirements, must be considered when deciding whether to adopt this technology. The flowchart labeled "Figure 30: Selective Laser Sintering (SLS)" provides insights into the Selective Laser Sintering (SLS) process, a fascinating technique within additive manufacturing.

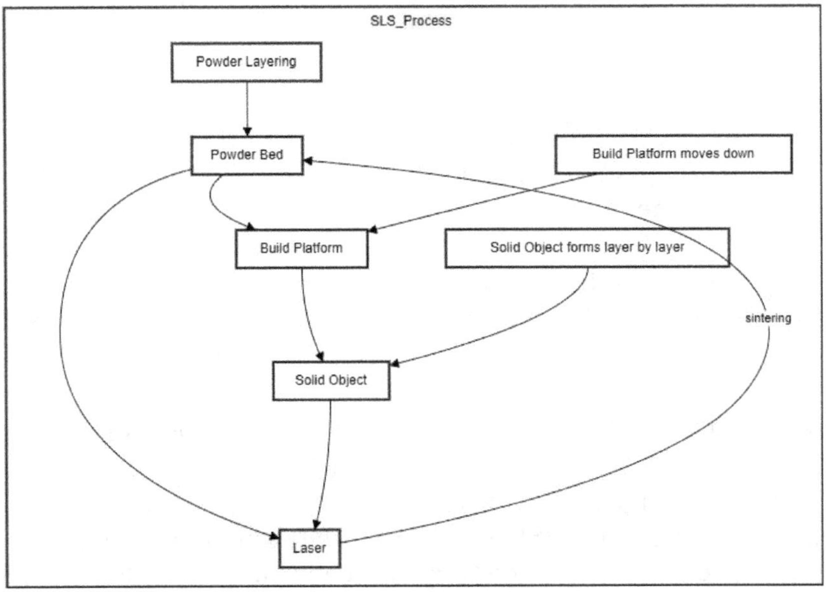

Figure 2: Selective laser sintering (SLS)

5.3 Selective Laser Melting (SLM)

Selective laser melting (SLM) is a powder bed fusion technology that uses a laser to melt and fuse together particles of a powdered material. The laser is directed at the powder, melting the particles together to form a solid layer. SLM is a popular 3D printing technology used in a wide range of industries, including aerospace, automotive, and medical.

5.3.1 The SLM Process

The SLM process involves several steps:

- **Powder Layering**: A layer of powdered material is deposited onto the build platform. The powder is typically spread evenly across the platform using a roller or blade. The layer thickness can vary depending on the specific application, but is typically in the range of 20-100 microns.

- **Laser Melting**: A laser is directed at the powder, melting the particles together. The laser is focused onto a specific area of the powder, causing the particles to melt and bond together. The laser power and speed can be adjusted to control the melting process and achieve the desired properties.

- **Platform Movement**: The build platform is lowered to accommodate the next layer. The platform is typically moved down by a small increment, allowing for precise control over the layer thickness.

- **Repeat**: Steps 1-3 are repeated until the part is complete. The process can take several hours, depending on the size and complexity of the part.

5.3.2 How SLM Works

SLM works by using a laser to melt and fuse together particles of a powdered material. The laser is focused onto a specific area of the powder, causing the particles to melt and bond together. The resulting part is strong and durable, with high accuracy and precision.

5.3.3 Advantages of SLM

SLM has several advantages, including:

- **High Density**: SLM produces parts with high density, making it ideal for applications where strength and durability are critical.
- **High Accuracy**: SLM produces parts with high accuracy and precision, making it ideal for applications such as aerospace and medical.
- **Complex Geometries**: SLM can produce parts with complex geometries, including internal structures and hollow parts.
- **No Support Material**: SLM does not require support material, making it ideal for producing parts with complex geometries.

5.3.4 Limitations of SLM

SLM also has several limitations, including:

- **Material Limitations**: The powdered material used in SLM can be limited in terms of its mechanical properties, making it unsuitable for certain applications.
- **Cost**: SLM can be expensive, particularly for large-scale production.
- **Post-Processing**: SLM often requires post-processing, such as machining or grinding, to achieve the desired surface finish.

5.3.5 Applications of SLM

SLM has several applications, including:

- **Aerospace**: SLM is used in the aerospace industry to produce complex aircraft components, such as engine parts and satellite components.
- **Automotive**: SLM is used in the automotive industry to produce car parts, such as engine components and interior trim.
- **Medical**: SLM is used in the medical industry to produce custom implants and surgical models.

5.3.6 Comparison to Other 3D Printing Technologies

SLM is often compared to other 3D printing technologies, such as selective laser sintering (SLS) and fused deposition modeling (FDM). While these technologies offer different advantages and limitations, they are often used for different applications.

- **SLS**: SLS is similar to SLM, but uses a lower-powered laser to sinter the powder, rather than melting it. This allows for the production of parts with lower density and strength.
- **FDM**: FDM is a different type of 3D printing technology that uses melted plastic to produce parts. It is often used for rapid prototyping and production of parts with complex geometries.

5.3.7 History of SLM

SLM was invented in the 1990s by researchers at the University of Liverpool. Since then, the technology has evolved rapidly, with the development of new materials and equipment.

5.3.8 Future of SLM

The future of SLM is promising, with ongoing research and development aimed at improving the technology. Some areas of research include:

- **New Materials**: Researchers are developing new powdered materials with improved mechanical properties, making SLM more suitable for a wider range of applications.
- **Improved Equipment**: Researchers are working to improve the equipment used in SLM, allowing for faster and more accurate production of parts.
- **Increased Build Volumes**: Researchers are working to increase the build volumes of SLM technologies, making them more suitable for large-scale production.

5.3.9 Case Studies

Several companies have successfully used SLM to produce complex parts and products. For example:

- **Aerospace**: A leading aerospace company used SLM to produce a complex aircraft component, reducing production time and cost by 50%.
- **Automotive**: A major automotive company used SLM to produce a custom car part, reducing production time and cost by 30%.
- **Medical**: A medical device company used SLM to produce a custom implant, reducing production time and cost by 20%.

5.3.10 Design Considerations

When designing parts for SLM, there are several considerations to keep in mind. These include:

- **Part Complexity**: SLM is ideal for producing parts with complex geometries, including internal structures and hollow parts.
- **Material Properties**: The powdered material used in SLM can affect the mechanical properties of the final part.

- **Layer Thickness**: The layer thickness can affect the accuracy and precision of the final part.

- **Support Material**: SLM does not require support material, but may require additional support structures for complex geometries.

5.3.11 Post-Processing

SLM parts often require post-processing to achieve the desired surface finish. This can include:

- **Machining**: Machining can be used to achieve a high-precision surface finish.

- **Grinding**: Grinding can be used to remove any excess material and achieve a smooth surface finish.

- **Polishing**: Polishing can be used to achieve a high-gloss surface finish.

In conclusion, SLM is a powerful 3D printing technology that offers high accuracy and precision, complex geometries, and fast production. While it has some limitations, ongoing research and development are addressing these issues, expanding the capabilities of SLM. The technology is particularly useful in industries such as aerospace, automotive, and medical, where high-precision parts are essential. However, the limitations of SLM, including material limitations, cost, and post-processing requirements, must be considered when deciding whether to adopt this technology. An intriguing metal additive manufacturing (AM) process called Selective Laser Melting (SLM) is explained in the flowchart "Figure 31: Selective laser melting (SLM)".

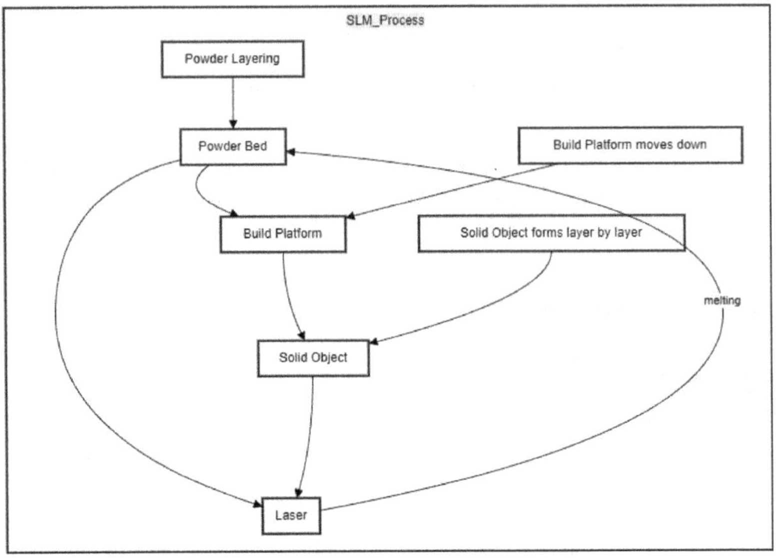

Figure 3: Selective laser melting (SLM)

5.4 Applications and Limitations

Powder bed fusion technologies, such as SLS and SLM, are commonly used for rapid prototyping, production of functional parts, and aerospace applications. These technologies offer high resolution and accuracy, making them suitable for producing complex geometries and fine details.

5.4.1 Rapid Prototyping

Powder bed fusion technologies are ideal for rapid prototyping, as they allow for the quick production of complex parts with high accuracy and precision. This makes them particularly useful for product design and development, where multiple iterations of a design may be required before finalizing a product.

5.4.2 Production of Functional Parts

Powder bed fusion technologies are also used for the production of functional parts, such as engine components, gearboxes, and other mechanical components. These parts require high accuracy and precision, as well as strong mechanical properties, making powder bed fusion technologies an ideal choice.

5.4.3 Aerospace Applications

Powder bed fusion technologies are widely used in the aerospace industry, where they are used to produce complex aircraft components, such as engine parts, satellite components, and other structural components. These components require high accuracy and precision, as well as strong mechanical properties, making powder bed fusion technologies an ideal choice.

5.4.4 Other Applications

In addition to rapid prototyping, production of functional parts, and aerospace applications, powder bed fusion technologies are also used in a variety of other industries, including:

- **Automotive**: Powder bed fusion technologies are used in the automotive industry to produce car parts, such as engine components, gearboxes, and other mechanical components.

- **Medical**: Powder bed fusion technologies are used in the medical industry to produce custom implants, surgical models, and other medical devices.

- **Consumer Products**: Powder bed fusion technologies are used in the consumer products industry to produce complex parts, such as phone cases, watch components, and other consumer electronics.

5.4.5 Limitations

While powder bed fusion technologies offer many advantages, they also have some limitations. For example:

- **Material Limitations**: The materials used in powder bed fusion processes are typically limited to metals and ceramics, which may have limited mechanical properties and may not be suitable for high-temperature or high-stress applications.

- **Build Volume Limitations**: The build volumes of powder bed fusion technologies are generally smaller than those of other 3D printing technologies, such as material extrusion or directed energy deposition. This can limit the size of parts that can be produced using these technologies.

- **Cost**: Powder bed fusion technologies can be expensive, particularly for large-scale production.

- **Post-Processing**: Powder bed fusion parts often require post-processing, such as machining or grinding, to achieve the desired surface finish.

5.4.6 Material Limitations

One of the main limitations of powder bed fusion technologies is the limited range of materials that can be used. Currently, most powder bed fusion technologies are limited to metals and ceramics, which may have limited mechanical properties and may not be suitable for high-temperature or high-stress applications.

5.4.7 Research and Development

Researchers are working to address the limitations of powder bed fusion technologies, including the development of new materials and the improvement of build volumes and cost. Some areas of research include:

- **New Materials**: Researchers are developing new powdered materials with improved mechanical properties, making powder bed fusion technologies more suitable for a wider range of applications.

- **Improved Build Volumes**: Researchers are working to improve the build volumes of powder bed fusion technologies, making them more suitable for large-scale production.

- **Cost Reduction**: Researchers are working to reduce the cost of powder bed fusion technologies, making them more accessible to a wider range of industries and applications.

5.5 Questions

5.5.1 Long Response Questions

1. What are the key differences between selective laser sintering (SLS) and selective laser melting (SLM)?
2. What are some common applications of powder bed fusion technologies?
3. What are the limitations of powder bed fusion technologies?
4. How does the resolution and accuracy of powder bed fusion technologies compare to other 3D printing technologies?
5. What are some potential future developments in powder bed fusion technologies?

5.5.2 Multiple Choice Questions

1) What is powder bed fusion commonly used for?

a) Painting
b) 3D Printing
c) Welding
d) Casting

Solution: - b) 3D Printing

2) Which process uses a laser or electron beam to fuse powdered material?

a) SLS
b) FDM
c) SLA
d) DLP

Solution: - a) SLS

3) In powder bed fusion, what is the first step in the process?

a) Fusion
b) Powder Layering
c) Platform Movement
d) Repeat
Solution: - b) Powder Layering

4) **Which of the following does not require support material?**
a) FDM
b) SLA
c) SLS
d) DLP
Solution: - c) SLS

5) **Selective Laser Sintering (SLS) uses which type of energy source?**
a) Electron Beam
b) Heat
c) Laser
d) Ultraviolet Light
Solution: - c) Laser

6) **What does the laser do in the SLS process?**
a) Melt and fuse
b) Cool down
c) Shred
d) Paint
Solution: - a) Melt and fuse

7) **Which technology is used to produce parts with complex geometries and high accuracy?**
a) FDM
b) SLS
c) SLA
d) DLP
Solution: - b) SLS

8) **Where did powder bed fusion technology originate?**
a) Germany
b) Japan
c) United States
d) China
Solution: - c) United States

9) **Which of the following is not a limitation of SLS?**
a) No Support Material

b) Material Limitations
c) Cost
d) Post-Processing
Solution: - a) No Support Material

10) In the SLM process, what is the main advantage over SLS?
a) Lower Density
b) Higher Density
c) Slower Production
d) Requires More Support Material
Solution: - b) Higher Density

11) Which industry primarily uses powder bed fusion for producing aerospace components?
a) Medical
b) Automotive
c) Aerospace
d) Consumer Electronics
Solution: - c) Aerospace

12) What is a common post-processing step required for parts produced by SLS?
a) Painting
b) Grinding
c) Injection Molding
d) Extrusion
Solution: - b) Grinding

13) What type of laser is used in Selective Laser Melting (SLM)?
a) Low-powered
b) High-powered
c) Infrared
d) Ultraviolet
Solution: - b) High-powered

14) Which powder bed fusion technology is known for producing high-density parts?
a) SLS
b) FDM
c) SLM
d) SLA
Solution: - c) SLM

15) Which of the following is not a typical application of powder bed fusion?

a) Engine Components
b) Medical Implants
c) Plastic Containers
d) Aerospace Parts

Solution: - c) Plastic Containers

16) Which step in the SLS process involves moving the build platform?

a) Powder Layering
b) Laser Sintering
c) Platform Movement
d) Repeating

Solution: - c) Platform Movement

17) How does SLM differ from SLS in terms of the material state after processing?

a) SLM uses liquid material
b) SLM uses powdered material but with higher energy
c) SLM uses a different type of powder
d) SLM does not use a laser

Solution: - b) SLM uses powdered material but with higher energy

18) Which of the following materials are typically used in powder bed fusion technologies?

a) Plastics and Resins
b) Metals and Ceramics
c) Wood and Glass
d) Composites and Polymers

Solution: - b) Metals and Ceramics

19) Which process involves melting the powder to create a solid part?

a) SLS
b) FDM
c) SLM
d) SLA

Solution: - c) SLM

20) What is one of the main benefits of using powder bed fusion technologies?

a) Requires more support material

b) Limited to simple shapes
c) High accuracy and precision
d) Slow production rates
Solution: - c) High accuracy and precision

21) What factor in SLM affects the melting process and final properties of the part?
a) Laser Speed
b) Powder Color
c) Build Platform Temperature
d) Cooling Rate
Solution: - a) Laser Speed

22) What are researchers focusing on to make powder bed fusion technologies more cost-effective?
a) Reducing Laser Power
b) Increasing Post-Processing Steps
c) Reducing Build Volumes
d) Cost Reduction Techniques
Solution: - d) Cost Reduction Techniques

23) In what way is the future of powder bed fusion expected to evolve regarding build volumes?
a) Decrease build volumes
b) Increase build volumes
c) Maintain current build volumes
d) Eliminate build volumes
Solution: - b) Increase build volumes

24) What is one of the research focuses for improving the materials used in powder bed fusion?
a) Developing non-metallic powders
b) Increasing powder size
c) Developing powders with improved mechanical properties
d) Reducing powder cost
Solution: - c) Developing powders with improved mechanical properties

25) Which process is best suited for producing high-strength parts for critical applications?
a) SLS
b) FDM
c) SLM

d) SLA

Solution: - c) SLM

26) How do powder bed fusion technologies compare in terms of resolution and detail with other 3D printing technologies?

a) Lower resolution
b) Similar resolution
c) Higher resolution
d) No comparison available

Solution: - c) Higher resolution

27) Which of the following is not a challenge addressed by ongoing research in powder bed fusion?

a) Improving material properties
b) Increasing part strength
c) Reducing production time
d) Reducing build volumes

Solution: - d) Reducing build volumes

28) What role does post-processing play in powder bed fusion technologies?

a) Eliminates the need for additional steps
b) Enhances part strength and finish
c) Makes parts non-functional
d) Reduces part accuracy

Solution: - b) Enhances part strength and finish

29) How does the cost of powder bed fusion compare to other 3D printing methods?

a) Generally lower
b) Generally higher
c) The same
d) Variable depending on material

Solution: - b) Generally higher

30) What is a major factor in determining the choice of powder bed fusion technology for a specific application?

a) Material color
b) Part complexity
c) Printer brand
d) Layer height

Solution: - b) Part complexity

31) Which type of laser is used in SLM to achieve the high density of parts?

a) High-powered laser
b) Low-powered laser
c) Ultraviolet laser
d) Infrared laser

Solution: - a) High-powered laser

32) What is a typical build volume limitation of powder bed fusion technologies?

a) Smaller compared to other 3D printing technologies
b) Larger compared to other 3D printing technologies
c) Similar to other 3D printing technologies
d) No volume limitations

Solution: - a) Smaller compared to other 3D printing technologies

33) Which powder bed fusion technology is particularly useful for producing parts with internal structures?

a) FDM
b) SLA
c) SLS
d) DLP

Solution: - c) SLS

34) Which of the following technologies uses a higher-powered laser to achieve its effects?

a) SLS
b) SLA
c) DLP
d) SLM

Solution: - d) SLM

35) In powder bed fusion, what typically happens to the unsintered powder after the build process?

a) It is discarded
b) It is recycled for future use
c) It is chemically treated
d) It is used for support structures

Solution: - b) It is recycled for future use

36) What technological advancement is contributing to the reduction of the cost per part in powder bed fusion?

a) Reduced Laser Power

b) Automated Powder Handling
c) Increased Build Platform Temperature
d) Decreased Material Variety
Solution: - b) Automated Powder Handling

37) Which process is known for producing parts with minimal thermal distortion due to its controlled environment?
a) FDM
b) SLA
c) SLS
d) SLM
Solution: - d) SLM

38) What is a key factor affecting the mechanical properties of parts produced by SLS?
a) Powder Size Distribution
b) Part Color
c) Layer Height
d) Print Speed
Solution: - a) Powder Size Distribution

39) Which research area is important for improving the performance of powder bed fusion technologies?
a) Enhancing Surface Finish
b) Reducing Powder Cost
c) Increasing Layer Thickness
d) Decreasing Machine Size
Solution: - a) Enhancing Surface Finish

40) What is the impact of build orientation on parts manufactured using powder bed fusion technologies?
a) No impact
b) Affects mechanical properties and surface finish
c) Only affects part color
d) Increases build time significantly
Solution: - b) Affects mechanical properties and surface finish

Chapter 6

Material Extrusion: FDM and FFF

6.1 Introduction to Material Extrusion

Material extrusion is a 3D printing technology that uses melted material to create a part layer by layer. This process is also known as fused deposition modeling (FDM) or fused filament fabrication (FFF). Material extrusion is one of the most widely used 3D printing technologies, due to its ease of use, affordability, and versatility.

6.1.1 The Material Extrusion Process

The material extrusion process involves several steps:

- **Material Feeding**: A filament of material is fed into the extruder. The filament is typically made of a thermoplastic material, such as PLA or ABS.

- **Melting**: The material is melted in the extruder. The extruder is heated to a temperature that is above the melting point of the material, causing it to melt and become viscous.

- **Extrusion**: The melted material is extruded through a nozzle. The nozzle is typically small in diameter, allowing for precise control over the amount of material that is extruded.

- **Layer Formation**: The extruded material is deposited onto the build platform, forming a layer. The layer is typically thin, ranging from 0.1 to 0.5 mm in thickness.

- **Platform Movement**: The build platform is moved to accommodate the next layer. The platform is typically moved in a horizontal direction, allowing for the creation of complex geometries.

- **Repeat**: Steps 1-5 are repeated until the part is complete. The process can take several hours, depending on the size and complexity of the part.

6.1.2 How Material Extrusion Works

Material extrusion works by using a combination of heat and pressure to melt and extrude the material. The extruder is typically heated to a temperature that is above the melting point of the material, causing it to melt and become viscous. The melted material is then forced through a small nozzle, which shapes it into a thin layer. The layer is then deposited onto the build platform, where it cools and solidifies.

6.1.3 Advantages of Material Extrusion

Material extrusion has several advantages, including:

- **Ease of Use**: Material extrusion is a relatively simple process, making it easy to use and maintain.

- **Affordability**: Material extrusion is one of the most affordable 3D printing technologies, making it accessible to a wide range of users.

- **Versatility**: Material extrusion can be used to create a wide range of parts, from simple prototypes to complex production parts.

- **Speed**: Material extrusion is a relatively fast process, making it suitable for rapid prototyping and production.

6.1.4 Limitations of Material Extrusion

Material extrusion also has several limitations, including:

- **Resolution:** - Material extrusion has a relatively low resolution, making it unsuitable for parts that require high accuracy and precision.

- **Material Limitations**: Material extrusion is limited to thermoplastic materials, which may not have the desired mechanical properties for certain applications.

- **Layer Adhesion**: Material extrusion can be prone to layer adhesion issues, which can lead to delamination and part failure.

6.1.5 Applications of Material Extrusion

Material extrusion has a wide range of applications, including:

- **Rapid Prototyping**: Material extrusion is widely used for rapid prototyping, due to its ease of use and affordability.

- **Production**: Material extrusion is used for production of parts, such as consumer products, automotive parts, and aerospace components.

- **Education**: Material extrusion is widely used in educational institutions, due to its ease of use and affordability.

6.1.6 History of Material Extrusion

Material extrusion has a long history, dating back to the 1980s. The first material extrusion machines were developed by Scott Crump, who founded Stratasys in 1989. Since then, the technology has evolved rapidly, with the development of new materials, machines, and software.

6.1.7 Future of Material Extrusion

The future of material extrusion is promising, with ongoing research and development aimed at improving the technology. Some areas of research include:

- **New Materials**: Researchers are developing new thermoplastic materials with improved mechanical properties, making material extrusion more suitable for a wider range of applications.

- **Improved Resolution: -** Researchers are working to improve the resolution of material extrusion, making it suitable for parts that require high accuracy and precision.

- **Increased Speed**: Researchers are working to increase the speed of material extrusion, making it more suitable for rapid prototyping and production.

Figure 32 illustrates the material extrusion process, a common method used in 3D printing. Here's a breakdown of each step:

- **Extruder:** This is where the raw material, often in the form of a filament, is fed into the system.

- **Melted Material:** The extruder heats the material until it melts.

- **Nozzle:** The melted material is then pushed through a nozzle, which controls the flow and precision of the material being deposited.

- **Layer:** The material is deposited layer by layer to form the desired shape.

- **Build Platform:** The platform moves horizontally to allow the material to be laid down in precise layers, building up the object from the bottom up.

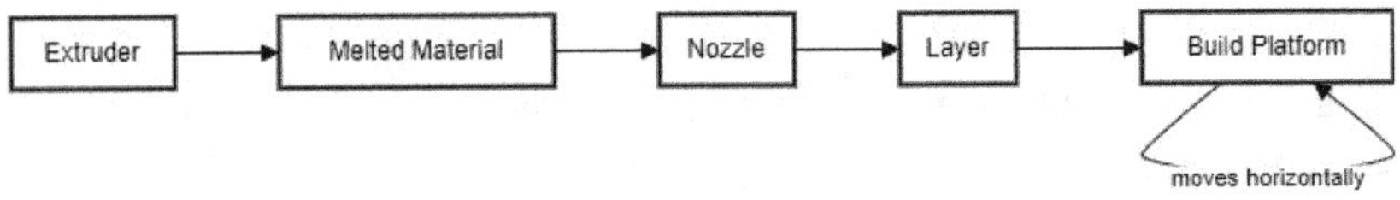

Figure 1: Material extrusion

6.2 Fused Deposition Modeling (FDM)

Fused deposition modeling (FDM) is a material extrusion technology that uses a heated extruder to melt and deposit thermoplastic materials. The melted material is extruded through a nozzle, which is moved in a specific pattern to create the desired shape. FDM is one of the most widely used 3D printing technologies, due to its ease of use, affordability, and versatility.

6.2.1 The FDM Process

The FDM process involves several steps:

- **Material Feeding**: A filament of thermoplastic material is fed into the extruder. The filament is typically made of a thermoplastic material, such as PLA or ABS.

- **Melting**: The material is melted in the extruder. The extruder is heated to a temperature that is above the melting point of the material, causing it to melt and become viscous.

- **Extrusion**: The melted material is extruded through a nozzle. The nozzle is typically small in diameter, allowing for precise control over the amount of material that is extruded.

- **Layer Formation**: The extruded material is deposited onto the build platform, forming a layer. The layer is typically thin, ranging from 0.1 to 0.5 mm in thickness.

- **Platform Movement**: The build platform is moved to accommodate the next layer. The platform is typically moved in a horizontal direction, allowing for the creation of complex geometries.

- **Repeat**: Steps 1-5 are repeated until the part is complete. The process can take several hours, depending on the size and complexity of the part.

6.2.2 How FDM Works

FDM works by using a combination of heat and pressure to melt and extrude the material. The extruder is typically heated to a temperature that is above the melting point of the material, causing it to melt and become viscous. The melted material is then forced through a small nozzle, which shapes it into a thin layer. The layer is then deposited onto the build platform, where it cools and solidifies.

6.2.3 Advantages of FDM

FDM has several advantages, including:

- **Ease of Use**: FDM is a relatively simple process, making it easy to use and maintain.

- **Affordability**: FDM is one of the most affordable 3D printing technologies, making it accessible to a wide range of users.

- **Versatility**: FDM can be used to create a wide range of parts, from simple prototypes to complex production parts.

- **Speed**: FDM is a relatively fast process, making it suitable for rapid prototyping and production.

6.2.4 Limitations of FDM

FDM also has several limitations, including:

- **Resolution: -** FDM has a relatively low resolution, making it unsuitable for parts that require high accuracy and precision.

- **Material Limitations**: FDM is limited to thermoplastic materials, which may not have the desired mechanical properties for certain applications.

- **Layer Adhesion**: FDM can be prone to layer adhesion issues, which can lead to delamination and part failure.

6.2.5 Applications of FDM

FDM has a wide range of applications, including:

- **Rapid Prototyping**: FDM is widely used for rapid prototyping, due to its ease of use and affordability.
- **Production**: FDM is used for production of parts, such as consumer products, automotive parts, and aerospace components.
- **Education**: FDM is widely used in educational institutions, due to its ease of use and affordability.

6.2.6 History of FDM

FDM has a long history, dating back to the 1980s. The first FDM machines were developed by Scott Crump, who founded Stratasys in 1989. Since then, the technology has evolved rapidly, with the development of new materials, machines, and software.

6.2.7 Future of FDM

The future of FDM is promising, with ongoing research and development aimed at improving the technology. Some areas of research include:

- **New Materials**: Researchers are developing new thermoplastic materials with improved mechanical properties, making FDM more suitable for a wider range of applications.
- **Improved Resolution: -** Researchers are working to improve the resolution of FDM, making it suitable for parts that require high accuracy and precision.
- **Increased Speed**: Researchers are working to increase the speed of FDM, making it more suitable for rapid prototyping and production.

6.2.8 Comparison with FFF

FDM is often compared to FFF (Fused Filament Fabrication), which is a similar 3D printing technology. While both technologies use a heated extruder to melt and deposit thermoplastic materials, there are some key differences. FDM is typically used for more precise and accurate parts, while FFF is used for more rapid prototyping and production. FDM also tends to have a higher resolution than FFF, making it more suitable for parts that require high accuracy and precision.

In conclusion, FDM is a powerful 3D printing technology that offers ease of use, affordability, and versatility. While it has some limitations, ongoing research and development are addressing these issues, expanding the capabilities of FDM and making it more suitable for a wider range of applications. Whether you're a hobbyist, engineer, or educator, FDM is a valuable tool for creating high-quality 3D printed parts. Figure 33 illustrates the Fused Deposition Modeling (FDM) process, a widely used method in 3D printing. Here's a step-by-step explanation:

- **Extruder:** The process begins with the extruder, which feeds the raw material, typically a thermoplastic filament, into the system.

- **Melted Material:** The extruder heats the filament until it melts.

- **Nozzle:** The melted material is then pushed through a nozzle, which precisely controls the flow and deposition of the material.

- **Layer:** The material is deposited layer by layer to build the object. Each layer is added on top of the previous one, gradually forming the desired shape.

- **Build Platform:** The platform moves horizontally to allow the material to be laid down in precise layers. This movement is crucial for creating the 3D structure.

The note “repeat steps” indicates that these steps are repeated multiple times until the entire object is printed. The final step, “Repeat Steps 1-5,” emphasizes the iterative nature of the FDM process.

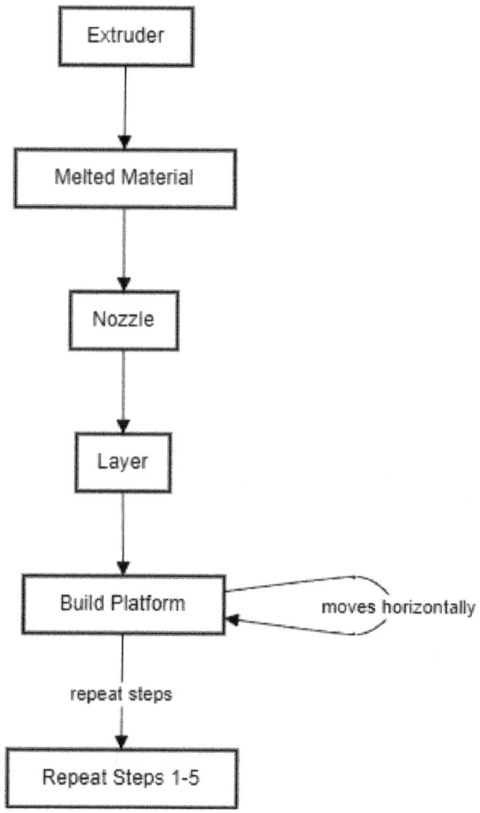

Figure 2: Fused deposition modeling (FDM)

6.3 Fused Filament Fabrication (FFF)

Fused filament fabrication (FFF) is a material extrusion technology that uses a heated extruder to melt and deposit thermoplastic materials. The melted material is extruded through a nozzle, which is moved in a specific pattern to create the desired shape. FFF is a popular 3D printing technology, known for its ease of use, affordability, and versatility.

6.3.1 The FFF Process

The FFF process involves several steps:

- **Material Feeding**: A filament of thermoplastic material is fed into the extruder. The filament is typically made of a thermoplastic material, such as PLA or ABS.

- **Melting**: The material is melted in the extruder. The extruder is heated to a temperature that is above the melting point of the material, causing it to melt and become viscous.

- **Extrusion**: The melted material is extruded through a nozzle. The nozzle is typically small in diameter, allowing for precise control over the amount of material that is extruded.

- **Layer Formation**: The extruded material is deposited onto the build platform, forming a layer. The layer is typically thin, ranging from 0.1 to 0.5 mm in thickness.

- **Platform Movement**: The build platform is moved to accommodate the next layer. The platform is typically moved in a horizontal direction, allowing for the creation of complex geometries.

- **Repeat**: Steps 1-5 are repeated until the part is complete. The process can take several hours, depending on the size and complexity of the part.

6.3.2 How FFF Works

FFF works by using a combination of heat and pressure to melt and extrude the material. The extruder is typically heated to a temperature that is above the melting point of the material, causing it to melt and become viscous. The melted material is then forced through a small nozzle, which shapes it into a thin layer. The layer is then deposited onto the build platform, where it cools and solidifies.

6.3.3 Advantages of FFF

FFF has several advantages, including:

- **Ease of Use**: FFF is a relatively simple process, making it easy to use and maintain.
- **Affordability**: FFF is one of the most affordable 3D printing technologies, making it accessible to a wide range of users.
- **Versatility**: FFF can be used to create a wide range of parts, from simple prototypes to complex production parts.
- **Speed**: FFF is a relatively fast process, making it suitable for rapid prototyping and production.

6.3.4 Limitations of FFF

FFF also has several limitations, including:

- **Resolution:** - FFF has a relatively low resolution, making it unsuitable for parts that require high accuracy and precision.
- **Material Limitations**: FFF is limited to thermoplastic materials, which may not have the desired mechanical properties for certain applications.
- **Layer Adhesion**: FFF can be prone to layer adhesion issues, which can lead to delamination and part failure.

6.3.5 Applications of FFF

FFF has a wide range of applications, including:

- **Rapid Prototyping**: FFF is widely used for rapid prototyping, due to its ease of use and affordability.

- **Production**: FFF is used for production of parts, such as consumer products, automotive parts, and aerospace components.

- **Education**: FFF is widely used in educational institutions, due to its ease of use and affordability.

6.3.6 History of FFF

FFF has a long history, dating back to the 1980s. The first FFF machines were developed by Scott Crump, who founded Stratasys in 1989. Since then, the technology has evolved rapidly, with the development of new materials, machines, and software.

6.3.7 Future of FFF

The future of FFF is promising, with ongoing research and development aimed at improving the technology. Some areas of research include:

- **New Materials**: Researchers are developing new thermoplastic materials with improved mechanical properties, making FFF more suitable for a wider range of applications.

- **Improved Resolution: -** Researchers are working to improve the resolution of FFF, making it suitable for parts that require high accuracy and precision.

- **Increased Speed**: Researchers are working to increase the speed of FFF, making it more suitable for rapid prototyping and production.

6.3.8 Comparison with FDM

FFF is often compared to FDM (Fused Deposition Modeling), which is a similar 3D printing technology. While both technologies use a heated extruder to melt and deposit thermoplastic materials, there are some key differences. FFF is typically used for more rapid prototyping and production, while FDM is used for more precise and accurate parts. FFF also tends to have a lower resolution than FDM, making it more suitable for parts that require less accuracy and precision.

In conclusion, FFF is a powerful 3D printing technology that offers ease of use, affordability, and versatility. While it has some limitations, ongoing research and development are addressing these issues, expanding the capabilities of FFF and making it more suitable for a wider range of applications. Whether you're a hobbyist, engineer, or educator, FFF is a valuable tool for creating high-quality 3D printed parts. Figure 34 illustrates the Fused Filament Fabrication (FFF) process, which is a popular method in 3D printing. Here's a step-by-step explanation:

- **Extruder:** The process starts with the extruder, which feeds the raw material, typically a thermoplastic filament, into the system.

- **Melted Material:** The extruder heats the filament until it melts.

- **Nozzle:** The melted material is then pushed through a nozzle, which precisely controls the flow and deposition of the material.

- **Layer:** The material is deposited layer by layer to build the object. Each layer is added on top of the previous one, gradually forming the desired shape.

- **Build Platform:** The platform moves horizontally to allow the material to be laid down in precise layers. This movement is crucial for creating the 3D structure.

The note "repeat steps" indicates that these steps are repeated multiple times until the entire object is printed. The final step, "Repeat Steps 1-5," emphasizes the iterative nature of the FFF process.

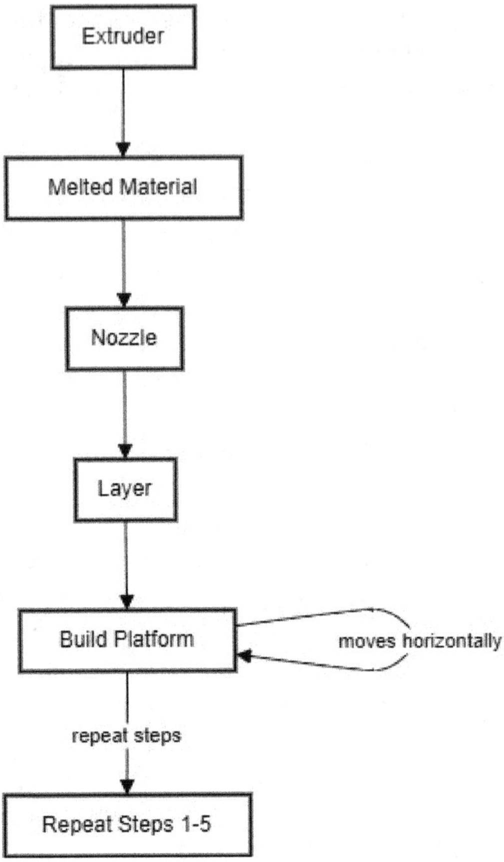

Figure 3: Fused filament fabrication (FFF)

6.4 Applications and Limitations

Material extrusion technologies, such as FDM and FFF, are commonly used for rapid prototyping, production of functional parts, and consumer products. These technologies offer high flexibility and affordability, making them suitable for a wide range of applications.

Rapid Prototyping

Material extrusion technologies are widely used for rapid prototyping, due to their ease of use, speed, and affordability.

Prototypes can be created quickly and inexpensively, allowing designers and engineers to test and refine their designs rapidly. The ability to create functional prototypes with material extrusion technologies has revolutionized the product development process, enabling companies to bring products to market faster and more efficiently.

Production of Functional Parts

Material extrusion technologies are also used for the production of functional parts, such as gears, bearings, and other mechanical components. These parts can be created with high accuracy and precision, making them suitable for use in a wide range of applications, from consumer products to industrial equipment.

Consumer Products

Material extrusion technologies are used to create a wide range of consumer products, including toys, household items, and personal accessories. The ability to create complex geometries and customized products with material extrusion technologies has enabled companies to create innovative and unique products that were previously impossible to produce.

6.4.2 Limitations of Material Extrusion Technologies

While material extrusion technologies offer many advantages, they also have some limitations. For example:

- **Material Limitations**: The materials used in material extrusion technologies are typically limited to thermoplastics, which may have limited mechanical properties and may not be suitable for high-temperature or high-stress applications.

- **Build Volume Limitations**: The build volumes of material extrusion technologies are generally smaller than those of other 3D printing technologies, such as powder bed fusion or directed energy deposition. This can limit the size of parts that can be created with these technologies.

- **Resolution Limitations**: Material extrusion technologies typically have a lower resolution than other 3D printing technologies, such as stereolithography or selective laser sintering. This can limit the level of detail and accuracy that can be achieved with these technologies.

- **Layer Adhesion Issues**: Material extrusion technologies can be prone to layer adhesion issues, which can lead to delamination and part failure.

- **Support Material Requirements**: Material extrusion technologies often require support material to be used during the printing process, which can add complexity and cost to the process.

6.4.3 Comparison with Other 3D Printing Technologies

Material extrusion technologies are often compared to other 3D printing technologies, such as powder bed fusion, directed energy deposition, and stereolithography. While each technology has its own strengths and weaknesses, material extrusion technologies are generally more affordable and easier to use than other technologies. However, they may not offer the same level of accuracy and precision as other technologies.

6.4.4 Future of Material Extrusion Technologies

The future of material extrusion technologies is promising, with ongoing research and development aimed at improving the technology. Some areas of research include:

- **New Materials**: Researchers are developing new thermoplastic materials with improved mechanical properties, making material extrusion technologies more suitable for a wider range of applications.

- **Improved Resolution: -** Researchers are working to improve the resolution of material extrusion technologies, making them suitable for parts that require high accuracy and precision.

- **Increased Build Volumes**: Researchers are working to increase the build volumes of material extrusion technologies, making them suitable for larger parts and applications.

In conclusion, material extrusion technologies, such as FDM and FFF, are powerful tools for rapid prototyping, production of functional parts, and consumer products. While they have some limitations, ongoing research and development are addressing these issues, expanding the capabilities of material extrusion technologies and making them more suitable for a wider range of applications.

6.5 Questions

6.5.1 Long Response Questions

1. What are the key differences between fused deposition modeling (FDM) and fused filament fabrication (FFF)?
2. What are some common applications of material extrusion technologies?
3. What are the limitations of material extrusion technologies?
4. How does the resolution and accuracy of material extrusion technologies compare to other 3D printing technologies?
5. What are some potential future developments in material extrusion technologies?

6.5.2 Multiple Choice Questions

1) What does FDM stand for in 3D printing?

a) Fused Deposition Modeling
b) Flexible Deposition Modeling
c) Fluid Deposition Modeling
d) Fiber Deposition Modeling

Solution: - a) Fused Deposition Modeling

2) **Which material is commonly used in FDM 3D printers?**
   a) PLA
   b) Resin
   c) Metal
   d) Glass
   Solution: - a) PLA
3) **In which phase does the filament get melted in an FDM printer?**
   a) Cooling
   b) Deposition
   c) Heating
   d) Solidifying
   Solution: - c) Heating
4) **Which component in FDM printers is responsible for extruding the melted filament?**
   a) Hot End
   b) Build Plate
   c) Stepper Motor
   d) Fan
   Solution: - a) Hot End
5) **What does FFF stand for in 3D printing?**
   a) Fused Filament Fabrication
   b) Flexible Filament Fabrication
   c) Fast Filament Fabrication
   d) Fiber Filament Fabrication
   Solution: - a) Fused Filament Fabrication
6) **Which of the following is a key advantage of FDM/FFF technology?**
   a) High resolution
   b) High cost of materials
   c) Low material wastage
   d) Complex post-processing
   Solution: - c) Low material wastage
7) **Which part of the FDM printer moves to create the 3D object?**
   a) Print Bed
   b) Extruder
   c) Nozzle

d) Gantry System

Solution: - d) Gantry System

8) What is the typical layer thickness range for FDM 3D printing?

a) 0.1mm to 1mm

b) 0.01mm to 0.1mm

c) 1mm to 5mm

d) 0.5mm to 2mm

Solution: - a) 0.1mm to 1mm

9) What is the purpose of a support structure in FDM printing?

a) To stabilize the filament

b) To support overhangs and complex geometries

c) To increase the print speed

d) To reduce the filament cost

Solution: - b) To support overhangs and complex geometries

10) Which of the following is a common issue in FDM printing caused by insufficient nozzle temperature?

a) Warping

b) Stringing

c) Layer Separation

d) Poor Adhesion

Solution: - d) Poor Adhesion

11) Which parameter affects the strength of the FDM-printed object?

a) Print Speed

b) Layer Height

c) Infill Density

d) Bed Leveling

Solution: - c) Infill Density

12) What is the primary purpose of using a heated bed in FDM printers?

a) To increase the print speed

b) To reduce material costs

c) To prevent warping of the print

d) To improve print resolution

Solution: - c) To prevent warping of the print

13) In FDM printing, what does the term 'overhang' refer to?

a) A part of the model that extends horizontally
b) The filament that extends from the nozzle
c) A part of the model that extends vertically and unsupported
d) The print bed's edge

Solution: - c) A part of the model that extends vertically and unsupported

14) What is the common material used in FFF printing for flexible parts?

a) PETG
b) Nylon
c) TPU
d) ABS

Solution: - c) TPU

15) Which parameter in FDM printing affects the smoothness of the surface finish?

a) Print Temperature
b) Layer Height
c) Infill Pattern
d) Cooling Fan Speed

Solution: - b) Layer Height

16) What role does the cooling fan play in an FDM printer?

a) It heats the filament
b) It cools the filament after extrusion
c) It supports the moving parts of the printer
d) It cleans the print bed

Solution: - b) It cools the filament after extrusion

17) What is a common post-processing technique used to improve the appearance of FDM prints?

a) Polishing
b) Sandblasting
c) Electroplating
d) Painting

Solution: - a) Polishing

18) Which of the following is NOT a common problem in FDM printing?

a) Layer Delamination
b) Stringing
c) Sinking

d) Warping

Solution: - c) Sinking

19) How does the print speed affect the mechanical properties of an FDM-printed part?

a) Higher speeds generally improve mechanical properties
b) Lower speeds result in improved mechanical properties
c) Print speed has no effect on mechanical properties
d) High speeds always lead to weak parts

Solution: - b) Lower speeds result in improved mechanical properties

20) What is the effect of varying the infill pattern on the strength of an FDM printed part?

a) It has no effect on strength
b) Some patterns can increase strength, while others may decrease it
c) All patterns increase strength equally
d) Infill patterns only affect the appearance, not strength

Solution: - b) Some patterns can increase strength, while others may decrease it

21) In FDM, which material property is crucial for determining the correct printing temperature?

a) Glass Transition Temperature
b) Melting Point
c) Boiling Point
d) Density

Solution: - a) Glass Transition Temperature

22) How does the cooling rate affect the dimensional accuracy of FDM prints?

a) Faster cooling rates improve accuracy
b) Slower cooling rates generally lead to better accuracy
c) Cooling rate has no impact on dimensional accuracy
d) Cooling rate only affects surface finish

Solution: - b) Slower cooling rates generally lead to better accuracy

23) What is the impact of nozzle diameter on the resolution of an FDM print?

a) Larger diameters improve resolution
b) Smaller diameters increase resolution

c) Nozzle diameter has no effect on resolution

d) Diameter only affects the print speed

Solution: - b) Smaller diameters increase resolution

24) Which factor is NOT typically adjusted to reduce the occurrence of stringing in FDM prints?

a) Retraction Speed

b) Print Temperature

c) Infill Density

d) Retraction Distance

Solution: - c) Infill Density

25) In FDM printing, which parameter is most likely to affect the bonding strength between layers?

a) Layer Height

b) Print Speed

c) Infill Pattern

d) Nozzle Diameter

Solution: - a) Layer Height

26) Which of the following is a method to improve the mechanical strength of FDM-printed parts?

a) Increasing layer height

b) Reducing print temperature

c) Using higher infill density

d) Increasing print speed

Solution: - c) Using higher infill density

27) What is the primary reason for warping in FDM printing?

a) Insufficient cooling

b) Over-extrusion

c) High print speed

d) High ambient temperature

Solution: - a) Insufficient cooling

28) How does the use of different infill patterns affect print time and material usage?

a) Some patterns reduce both print time and material usage

b) All patterns have the same effect on print time and material usage

c) Some patterns can reduce print time but increase material usage

d) Some patterns can reduce material usage but increase print time

Solution: - d) Some patterns can reduce material usage but increase print time

29) What is the primary benefit of using a dual-extruder FDM printer?

a) Faster printing speed
b) Ability to print with two different materials or colors
c) Lower cost of printing
d) Improved print bed adhesion

Solution: - b) Ability to print with two different materials or colors

30) What is a common method to improve adhesion of the first layer in FDM printing?

a) Increase nozzle temperature
b) Use a different filament type
c) Apply a glue stick or adhesive on the print bed
d) Increase print speed

Solution: - c) Apply a glue stick or adhesive on the print bed

31) What effect does increasing the number of perimeters have on FDM prints?

a) Reduces print strength
b) Decreases print speed
c) Increases print strength
d) Reduces print time

Solution: - c) Increases print strength

32) In FDM printing, which material property is crucial for achieving good layer adhesion?

a) Flexural Strength
b) Impact Resistance
c) Melting Temperature
d) Elastic Modulus

Solution: - c) Melting Temperature

33) What is a common strategy for minimizing warping in FDM prints?

a) Decreasing infill density
b) Using a heated print bed
c) Increasing print speed
d) Reducing layer height

Solution: - b) Using a heated print bed

34) Which of the following is NOT a common FDM filament type?

a) ABS
b) PLA
c) PVA
d) HIPS

Solution: - c) PVA

35) How does adjusting the extrusion multiplier affect the print quality in FDM?

a) It changes the speed of the print head
b) It affects the amount of filament extruded
c) It changes the temperature of the print bed
d) It modifies the layer height

Solution: - b) It affects the amount of filament extruded

36) What is the primary role of slicing software in FDM 3D printing?

a) To control the temperature of the extruder
b) To generate the G-code for the printer
c) To provide a physical interface for the printer
d) To choose the filament material

Solution: - b) To generate the G-code for the printer

37) In FDM printing, what is the effect of a higher infill pattern density on thermal conductivity?

a) Increases thermal conductivity
b) Decreases thermal conductivity
c) No effect on thermal conductivity
d) Increases the risk of warping

Solution: - a) Increases thermal conductivity

38) What are the challenges of printing with high-temperature thermoplastics in FDM?

a) Limited filament availability
b) Need for a specialized high-temperature extruder
c) Increased print speed
d) Decreased material strength

Solution: - b) Need for a specialized high-temperature extruder

39) How can the use of adaptive layer height in slicing software improve print efficiency?

a) By increasing print speed consistently
b) By reducing the number of layers in complex areas
c) By maintaining a uniform layer height throughout the print
d) By adjusting infill density dynamically
Solution: - b) By reducing the number of layers in complex areas

40) What is the impact of ambient temperature on the quality of FDM prints?

a) Higher ambient temperatures generally improve print quality
b) Lower ambient temperatures lead to better adhesion
c) Fluctuating ambient temperatures can cause issues like warping and layer separation
d) Ambient temperature has no effect on print quality
Solution: - c) Fluctuating ambient temperatures can cause issues like warping and layer separation

Chapter 7

Material Jetting: MJ and MJP

7.1 Introduction to Material Jetting

Material jetting is a 3D printing technology that uses droplets of material to create a part layer by layer. This process is also known as material jetting (MJ) or multi-jet printing (MJP). Material jetting is a popular 3D printing technology, known for its high resolution, accuracy, and versatility.

7.1.1 How Material Jetting Works

The material jetting process involves several steps:

- **Material Loading**: A reservoir of material is loaded into the print head. The material is typically a photopolymer or wax-based material, which is designed to be cured or solidified when exposed to light or heat.

- **Droplet Formation**: The material is formed into droplets, which are then ejected from the print head. The droplets are typically very small, ranging from 10 to 100 microns in diameter.

- **Droplet Deposition**: The droplets are deposited onto the build platform, forming a layer. The droplets are precisely controlled, allowing for complex geometries and high-resolution details.

- **Platform Movement**: The build platform is moved to accommodate the next layer. The platform is typically moved in a horizontal direction, allowing for the creation of complex geometries and shapes.

- **Repeat**: Steps 1-4 are repeated until the part is complete. The process can take several hours, depending on the size and complexity of the part.

7.1.2 Advantages of Material Jetting

Material jetting has several advantages, including:

- **High Resolution: -** Material jetting is capable of producing parts with very high resolution, making it suitable for applications that require precise details and accuracy.

- **Accuracy**: Material jetting is a highly accurate process, allowing for the creation of parts with complex geometries and shapes.

- **Versatility**: Material jetting can be used to create a wide range of parts, from small, intricate components to large, complex assemblies.

- **Speed**: Material jetting is a relatively fast process, making it suitable for rapid prototyping and production.

7.1.3 Limitations of Material Jetting

Material jetting also has some limitations, including:

- **Material Limitations**: Material jetting is limited to photopolymer or wax-based materials, which may not have the desired mechanical properties for certain applications.

- **Support Material Requirements**: Material jetting often requires support material to be used during the printing process, which can add complexity and cost to the process.

- **Post-Processing Requirements**: Material jetting parts often require post-processing, such as curing or washing, to achieve the desired properties.

7.1.4 Applications of Material Jetting

Material jetting has a wide range of applications, including:

- **Rapid Prototyping**: Material jetting is widely used for rapid prototyping, due to its high resolution, accuracy, and speed.

- **Production**: Material jetting is used for production of parts, such as medical devices, aerospace components, and consumer products.

- **Artistic Applications**: Material jetting is used in artistic applications, such as jewelry design and sculpture.

7.1.5 History of Material Jetting

Material jetting has a long history, dating back to the 1990s. The first material jetting machines were developed by companies such as 3D Systems and Stratasys. Since then, the technology has evolved rapidly, with the development of new materials, machines, and software.

7.1.6 Future of Material Jetting

The future of material jetting is promising, with ongoing research and development aimed at improving the technology. Some areas of research include:

- **New Materials**: Researchers are developing new photopolymer and wax-based materials with improved mechanical properties, making material jetting more suitable for a wider range of applications.

- **Improved Resolution: -** Researchers are working to improve the resolution of material jetting, making it suitable for parts that require even higher accuracy and precision.

- **Increased Speed**: Researchers are working to increase the speed of material jetting, making it more suitable for rapid prototyping and production.

In conclusion, material jetting is a powerful 3D printing technology that offers high resolution, accuracy, and versatility. While it has some limitations, ongoing research and development are addressing these issues, expanding the capabilities of material jetting and making it more suitable for a wider range of applications. Whether you're a designer, engineer, or manufacturer, material jetting is an important technology to consider for your next project. Figure 35 illustrates the Material Jetting process, a precise method used in additive manufacturing. The process begins with the Print Head, which deposits tiny Droplets of Material onto the Build Platform. The build platform moves horizontally, allowing the material to be laid down in exact locations. This step is repeated multiple times, as indicated by the instruction to "repeat steps," until the entire object is formed. The final step, "Repeat Steps 1-4," emphasizes the iterative nature of the process. Material jetting is known for its ability to produce highly detailed and accurate three-dimensional objects, making it ideal for applications requiring fine resolution and smooth surface finishes.

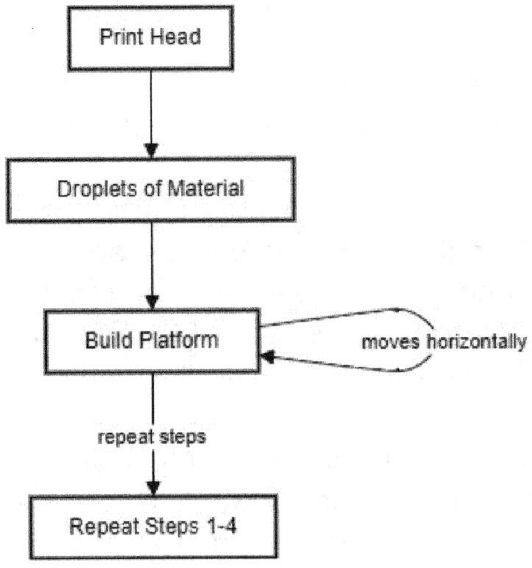

Figure 1: Material jetting

Figure 36 displays the Material Jetting Machine.

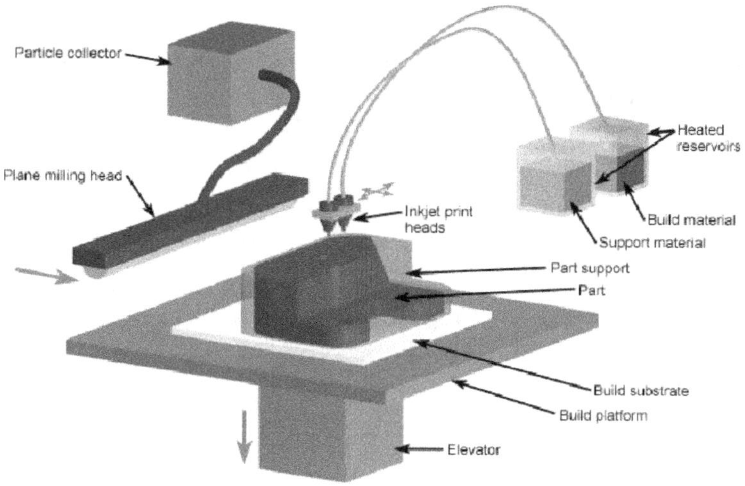

Figure 2: Material jetting
[Source: https://www.researchgate.net/publication/332833671]

Figure 37 shows the products of Material Jetting operation.

Figure 3: Material jetting

Types of Material Jetting

There are several types of MJ, including:

- **Drop-On-Demand (DoD) MJ**: This type of MJ uses a piezoelectric print head to eject droplets of material onto the build platform.

- **Continuous Inkjet (CIJ) MJ**: This type of MJ uses a continuous stream of droplets to create the part.

- **Thermal MJ**: This type of MJ uses heat to eject droplets of material onto the build platform.

7.1.8 Comparison with Other 3D Printing Technologies

MJ is often compared to other 3D printing technologies, such as stereolithography (SLA) and selective laser sintering (SLS). While each technology has its own strengths and weaknesses, MJ is known for its high resolution, accuracy, and versatility.

In conclusion, MJ is a powerful 3D printing technology that offers high resolution, accuracy, and versatility. While it has some challenges, ongoing research and development are addressing these issues, expanding the capabilities of MJ and making it more suitable for a wider range of applications. Figure 38 illustrates the Material Jetting (MJ) process, a sophisticated 3D printing technology. Here's a detailed explanation:

- **Material Container:** This holds the photopolymer material that will be used for printing.

- **UV Curing Light:** Positioned at the top, this light cures (hardens) the material as it is deposited, ensuring each layer solidifies quickly.

- **Inkjet Print Heads:** These heads precisely deposit tiny droplets of the material onto the build platform, layer by layer.

- **Part Being Printed:** This is the object being created, built up from the deposited material.

- **Support Material:** Often used to support overhanging structures during the printing process, which is later removed.

- **Build Platform:** The surface on which the object is printed. It moves horizontally to allow for precise layer deposition.

- **Elevator:** This mechanism moves the build platform up and down, allowing for the addition of new layers.

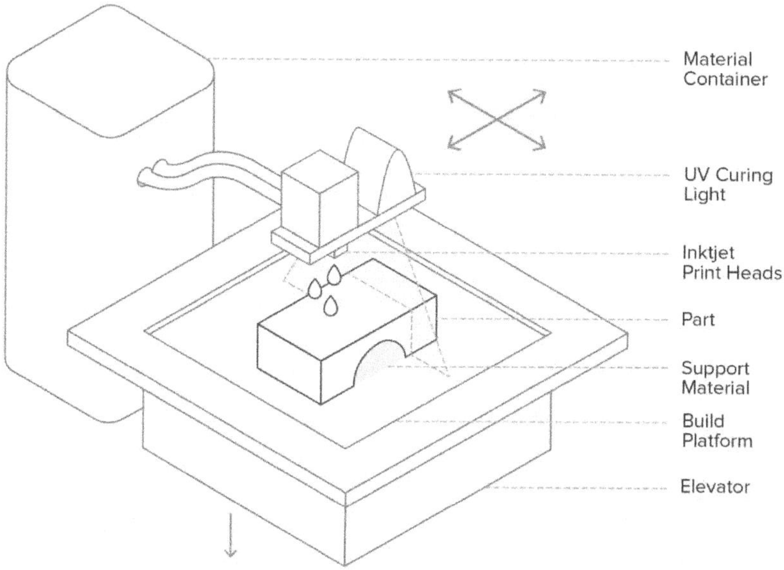

Figure 4: Material Jetting (MJ)
[Source: https://facfox.com/docs/kb/what-is-material-jetting-3d-printing-simply-explained]

7.2 Multi-Jet Printing (MJP)

Multi-jet printing (MJP) is a material jetting technology that uses multiple print heads to deposit droplets of material onto the build platform. This allows for faster build speeds and higher resolution than traditional material jetting. MJP is a popular 3D printing technology, known for its high speed, accuracy, and versatility.

7.2.1 How Multi-Jet Printing Works

The MJP process involves several steps:

- **Material Loading**: Multiple reservoirs of material are loaded into the print heads. The material is typically a photopolymer or wax-based material, which is designed to be cured or solidified when exposed to light or heat.

- **Droplet Formation**: The material is formed into droplets, which are then ejected from the print heads. The droplets are typically very small, ranging from 10 to 100 microns in diameter.
- **Droplet Deposition**: The droplets are deposited onto the build platform, forming a layer. The droplets are precisely controlled, allowing for complex geometries and high-resolution details.

- **Platform Movement**: The build platform is moved to accommodate the next layer. The platform is typically moved in a horizontal direction, allowing for the creation of complex geometries and shapes.

- **Repeat**: Steps 1-4 are repeated until the part is complete. The process can take several hours, depending on the size and complexity of the part.

7.2.2 Advantages of Multi-Jet Printing

MJP has several advantages, including:

- **Faster Build Speeds**: MJP allows for faster build speeds than traditional material jetting, making it suitable for rapid prototyping and production.

- **Higher Resolution: -** MJP allows for higher resolution than traditional material jetting, making it suitable for parts that require precise details and accuracy.

- **Improved Accuracy**: MJP allows for improved accuracy than traditional material jetting, making it suitable for parts that require complex geometries and shapes.

- **Increased Productivity**: MJP allows for increased productivity than traditional material jetting, making it suitable for high-volume production.

7.2.3 Types of Multi-Jet Printing

There are several types of MJP, including:

- **Drop-On-Demand (DoD) MJP**: This type of MJP uses a piezoelectric print head to eject droplets of material onto the build platform.
- **Continuous Inkjet (CIJ) MJP**: This type of MJP uses a continuous stream of droplets to create the part.
- **Thermal MJP**: This type of MJP uses heat to eject droplets of material onto the build platform.

7.2.4 Applications of Multi-Jet Printing

MJP has a wide range of applications, including:

- **Rapid Prototyping**: MJP is widely used for rapid prototyping, due to its high speed, accuracy, and versatility.

- **Production**: MJP is used for production of parts, such as medical devices, aerospace components, and consumer products.

- **Artistic Applications**: MJP is used in artistic applications, such as jewelry design and sculpture.

7.2.5 Challenges of Multi-Jet Printing

MJP also has some challenges, including:

- **Material Limitations**: MJP is limited to photopolymer or wax-based materials, which may not have the desired mechanical properties for certain applications.

- **Support Material Requirements**: MJP often requires support material to be used during the printing process, which can add complexity and cost to the process.

- **Post-Processing Requirements**: MJP parts often require post-processing, such as curing or washing, to achieve the desired properties.

7.2.6 Future of Multi-Jet Printing

The future of MJP is promising, with ongoing research and development aimed at improving the technology. Some areas of research include:

- **New Materials**: Researchers are developing new photopolymer and wax-based materials with improved mechanical properties, making MJP more suitable for a wider range of applications.

- **Improved Resolution: -** Researchers are working to improve the resolution of MJP, making it suitable for parts that require even higher accuracy and precision.

- **Increased Speed**: Researchers are working to increase the speed of MJP, making it more suitable for rapid prototyping and production.

7.2.7 Comparison with Other 3D Printing Technologies

MJP is often compared to other 3D printing technologies, such as stereolithography (SLA) and selective laser sintering (SLS). While each technology has its own strengths and weaknesses, MJP is known for its high speed, accuracy, and versatility. Figure 39 illustrates the Multi-Jet Printing (MJP) process, a sophisticated 3D printing technology.

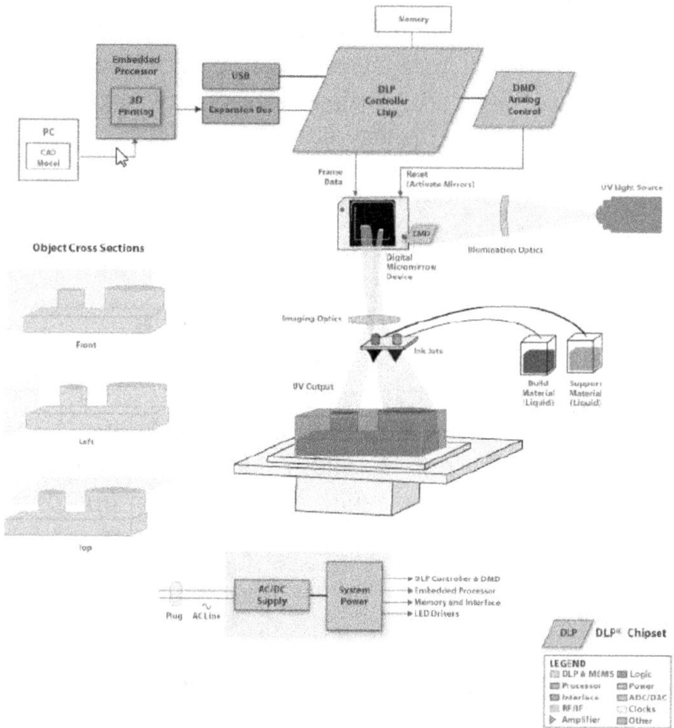

Figure 5: Multi-jet printing (MJP)
[Source: https://www.whiteclouds.com/3dpedia/mjp/]

7.3 Applications and Limitations

Material jetting technologies, such as MJ and MJP, are commonly used for a wide range of applications, including rapid prototyping, production of functional parts, and consumer products. These technologies offer high resolution and accuracy, making them suitable for producing complex geometries and fine details.

Rapid Prototyping

Material jetting technologies are widely used for rapid prototyping, due to their ability to quickly produce high-quality parts with complex geometries and fine details. Rapid prototyping is an essential step in the product development process, allowing designers and engineers to test and refine their designs before moving to production. Material jetting technologies, such as MJ and MJP, are ideal for rapid prototyping due to their high speed, accuracy, and versatility.

Production of Functional Parts

Material jetting technologies are also used for the production of functional parts, such as medical devices, aerospace components, and consumer products. These technologies offer high resolution and accuracy, making them suitable for producing parts with complex geometries and fine details. Additionally, material jetting technologies can produce parts with a high level of customization, making them ideal for applications where unique parts are required.

Consumer Products

Material jetting technologies are used in the production of consumer products, such as jewelry, toys, and household items. These technologies offer high resolution and accuracy, making them suitable for producing parts with complex geometries and fine details. Additionally, material jetting technologies can produce parts with a high level of customization, making them ideal for applications where unique products are required.

7.3.1 Limitations of Material Jetting Technologies

While material jetting technologies offer many advantages, they also have some limitations. For example:

- **Material Limitations**: The materials used in material jetting technologies are typically limited to photopolymers and waxes, which may have limited mechanical properties and may not be suitable for high-temperature or high-stress applications.

- **Build Volume Limitations**: The build volumes of material jetting technologies are generally smaller than those of other 3D printing technologies, such as powder bed fusion or directed energy deposition. This can limit the size of parts that can be produced using these technologies.

- **Support Material Requirements**: Material jetting technologies often require support material to be used during the printing process, which can add complexity and cost to the process.

- **Post-Processing Requirements**: Material jetting technologies often require post-processing, such as curing or washing, to achieve the desired properties.

7.3.2 Comparison with Other 3D Printing Technologies

Material jetting technologies are often compared to other 3D printing technologies, such as stereolithography (SLA), selective laser sintering (SLS), and fused deposition modeling (FDM). While each technology has its own strengths and weaknesses, material jetting technologies are known for their high resolution, accuracy, and versatility.

7.3.3 Future of Material Jetting Technologies

The future of material jetting technologies is promising, with ongoing research and development aimed at improving the technology. Some areas of research include:

- **New Materials**: Researchers are developing new photopolymer and wax-based materials with improved mechanical properties, making material jetting technologies more suitable for a wider range of applications.

- **Improved Resolution: -** Researchers are working to improve the resolution of material jetting technologies, making them suitable for parts that require even higher accuracy and precision.

- **Increased Build Volumes**: Researchers are working to increase the build volumes of material jetting technologies, making them suitable for larger parts and applications.

In conclusion, material jetting technologies, such as MJ and MJP, are powerful tools for rapid prototyping, production of functional parts, and consumer products. While they have some limitations, ongoing research and development are addressing these issues, expanding the capabilities of material jetting technologies and making them more suitable for a wider range of applications. Whether you're a designer, engineer, or manufacturer, material jetting technologies are an important consideration for your next project.

7.4 Questions

7.4.1 Long Response Questions

1. What are the key differences between material jetting (MJ) and multi-jet printing (MJP)?
2. What are some common applications of material jetting technologies?
3. What are the limitations of material jetting technologies?
4. How does the resolution and accuracy of material jetting technologies compare to other 3D printing technologies?
5. What are some potential future developments in material jetting technologies?

7.4.2 Multiple Choice Questions

1) What is the primary material used in Material Jetting (MJ)?

a) Metal
b) Ceramic
c) Photopolymer or wax
d) Concrete

Solution: - c) Photopolymer or wax

2) **What does the build platform in Material Jetting do?**
   a) Ejects droplets
   b) Moves horizontally
   c) Cures the material
   d) Loads the material
   Solution: - b) Moves horizontally
3) **Which type of material jetting uses a continuous stream of droplets?**
   a) Drop-on-Demand (DoD)
   b) Continuous Inkjet (CIJ)
   c) Thermal
   d) Inkjet
   Solution: - b) Continuous Inkjet (CIJ)
4) **What is one advantage of Material Jetting?**
   a) Low resolution
   b) Large build volume
   c) High resolution
   d) Slow speed
   Solution: - c) High resolution
5) **Which of the following is a challenge of Material Jetting?**
   a) High speed
   b) Low resolution
   c) Material limitations
   d) Large build volume
   Solution: - c) Material limitations
6) **What type of head is used in Drop-on-Demand (DoD) MJ?**
   a) Piezoelectric
   b) Thermal
   c) Inkjet
   d) Laser
   Solution: - a) Piezoelectric
7) **Which process is used to cure the material in Material Jetting?**
   a) Heat
   b) UV Light
   c) Water
   d) Air
   Solution: - b) UV Light

8) **What is the main advantage of Multi-Jet Printing (MJP) over traditional Material Jetting?**
   a) Slower build speed
   b) Higher resolution
   c) More print heads
   d) Lower resolution
   Solution: - c) More print heads
9) **Which of the following is NOT a type of Multi-Jet Printing?**
   a) Drop-on-Demand (DoD)
   b) Continuous Inkjet (CIJ)
   c) Thermal
   d) Stereolithography (SLA)
   Solution: - d) Stereolithography (SLA)
10) **What is a common application of Material Jetting?**
   a) Metal casting
   b) Large-scale construction
   c) Rapid prototyping
   d) Bulk material processing
   Solution: - c) Rapid prototyping
11) **What is the primary material limitation of Material Jetting?**
   a) Limited build volume
   b) Limited to photopolymers and waxes
   c) High production cost
   d) Requires excessive post-processing
   Solution: - b) Limited to photopolymers and waxes
12) **What does the "repeat" step in Material Jetting refer to?**
   a) Repeating the material loading step
   b) Repeating the curing step
   c) Repeating the droplet deposition and platform movement steps
   d) Repeating the entire process from start to finish
   Solution: - c) Repeating the droplet deposition and platform movement steps
13) **What is a common challenge of Multi-Jet Printing (MJP)?**
   a) Material versatility
   b) Small build volume
   c) High-speed capabilities
   d) Limited to photopolymers

Solution: - d) Limited to photopolymers

14) **How does Multi-Jet Printing (MJP) improve productivity?**
    a) By using fewer print heads
    b) By using slower build speeds
    c) By using multiple print heads
    d) By limiting material types
    Solution: - c) By using multiple print heads

15) **Which of the following is NOT a typical application of Material Jetting?**
    a) Aerospace components
    b) Jewelry design
    c) Large-scale construction
    d) Medical devices
    Solution: - c) Large-scale construction

16) **What is the role of support material in Material Jetting?**
    a) To enhance the resolution
    b) To support overhanging structures during printing
    c) To speed up the build process
    d) To reduce material costs
    Solution: - b) To support overhanging structures during printing

17) **Which technology is often compared to Material Jetting for its high resolution and accuracy?**
    a) Stereolithography (SLA)
    b) Selective Laser Sintering (SLS)
    c) Fused Deposition Modeling (FDM)
    d) Direct Energy Deposition (DED)
    Solution: - a) Stereolithography (SLA)

18) **What is the benefit of using UV curing light in Material Jetting?**
    a) It speeds up material loading
    b) It solidifies the material quickly
    c) It reduces the need for support material
    d) It lowers material costs
    Solution: - b) It solidifies the material quickly

19) **What is one advantage of using Thermal MJ?**
    a) High-speed droplet formation
    b) Continuous inkjet capabilities
    c) Heat-based droplet ejection

d) Piezoelectric droplet ejection

Solution: - c) Heat-based droplet ejection

20) What aspect of Material Jetting can be improved by ongoing research?

a) Build volume

b) Resolution

c) Post-processing complexity

d) Support material requirements

Solution: - b) Resolution

21) Which of the following is NOT a characteristic of Multi-Jet Printing (MJP)?

a) Uses multiple print heads

b) Higher resolution than traditional MJ

c) Limited to a single type of material

d) Faster build speeds than traditional MJ

Solution: - c) Limited to a single type of material

22) What does the term "Droplet Formation" refer to in Material Jetting?

a) The process of curing the material

b) The process of ejecting material from the print head

c) The deposition of droplets onto the build platform

d) The movement of the build platform

Solution: - b) The process of ejecting material from the print head

23) What is a key advantage of Material Jetting (MJ) compared to traditional 3D printing technologies?

a) Lower cost of materials

b) High resolution and accuracy

c) Larger build volume

d) No need for post-processing

Solution: - b) High resolution and accuracy

24) Which type of Material Jetting (MJ) uses a piezoelectric print head?

a) Drop-on-Demand (DoD) MJ

b) Continuous Inkjet (CIJ) MJ

c) Thermal MJ

d) Laser-Sintering MJ

Solution: - a) Drop-on-Demand (DoD) MJ

25) Which type of Multi-Jet Printing (MJP) uses a continuous stream of droplets?

a) Drop-on-Demand (DoD) MJP
b) Continuous Inkjet (CIJ) MJP
c) Thermal MJP
d) Laser-Sintering MJP

Solution: - b) Continuous Inkjet (CIJ) MJP

26) What is a common application of Material Jetting (MJ) in the field of consumer products?

a) Aerospace components
b) Medical devices
c) Jewelry design
d) Automotive parts

Solution: - c) Jewelry design

27) What is a common challenge faced by Multi-Jet Printing (MJP) technologies?

a) High material costs
b) Limited build volume
c) Post-processing requirements
d) Low resolution

Solution: - c) Post-processing requirements

28) What is the main purpose of using support material in Material Jetting (MJ) technologies?

a) To improve the resolution of printed parts
b) To support overhanging structures during printing
c) To increase the build speed
d) To reduce material wastage

Solution: - b) To support overhanging structures during printing

29) How does the resolution of Multi-Jet Printing (MJP) compare to traditional Material Jetting (MJ) methods?

a) MJP has lower resolution than traditional MJ
b) MJP has higher resolution than traditional MJ
c) MJP has the same resolution as traditional MJ
d) MJP has variable resolution depending on the material used

Solution: - b) MJP has higher resolution than traditional MJ

30) In what way is the build platform movement managed during the Material Jetting (MJ) process?

a) It is moved vertically to accommodate new layers
b) It is moved horizontally to accommodate new layers
c) It is rotated to ensure uniform material deposition
d) It remains stationary throughout the process

Solution: - b) It is moved horizontally to accommodate new layers

31) What type of material is typically used in Material Jetting (MJ) and Multi-Jet Printing (MJP)?

a) Metal alloys
b) Photopolymers and wax-based materials
c) Thermoplastics
d) Ceramics

Solution: - b) Photopolymers and wax-based materials

32) What is a limitation of the materials used in Material Jetting (MJ) technologies?

a) High thermal stability
b) Limited mechanical properties
c) High density
d) Compatibility with all 3D printing technologies

Solution: - b) Limited mechanical properties

33) What advancement in Multi-Jet Printing (MJP) is aimed at increasing productivity?

a) Improved material formulation
b) Increased build speed
c) Reduced post-processing requirements
d) Larger build volume

Solution: - b) Increased build speed

34) Which technology is often compared with Material Jetting (MJ) in terms of high resolution and accuracy?

a) Stereolithography (SLA)
b) Fused Deposition Modeling (FDM)
c) Selective Laser Sintering (SLS)
d) Direct Metal Laser Sintering (DMLS)

Solution: - a) Stereolithography (SLA)

35) Which step in the Material Jetting (MJ) process involves the actual creation of the part layer by layer?

a) Material Loading
b) Droplet Formation
c) Droplet Deposition
d) Platform Movement

Solution: - c) Droplet Deposition

36) What is the impact of post-processing on the final properties of parts produced by Material Jetting (MJ)?

a) It can enhance the mechanical properties
b) It has no impact on the properties
c) It reduces the dimensional accuracy
d) It speeds up the production process

Solution: - a) It can enhance the mechanical properties

37) What technological development is anticipated to expand the capabilities of Material Jetting (MJ) in the future?

a) Development of new metal-based materials
b) Increased build volumes
c) Integration with other additive manufacturing technologies
d) Enhanced resolution and precision

Solution: - d) Enhanced resolution and precision

38) In which area is Multi-Jet Printing (MJP) especially promising due to its advantages over other technologies?

a) Large-scale manufacturing
b) High-speed production
c) Low-resolution prototyping
d) Single-material parts

Solution: - b) High-speed production

39) How do the build volumes of Material Jetting (MJ) technologies generally compare to other 3D printing technologies?

a) They are larger
b) They are smaller
c) They are similar
d) They vary widely

Solution: - b) They are smaller

40) What is a key challenge in using Multi-Jet Printing (MJP) for high-volume production?

a) Limited material variety
b) High material costs
c) Complexity of support material removal
d) Speed of the printing process

Solution: - c) Complexity of support material removal

Chapter 8

Binder Jetting: 3DP and SLS

8.1 Introduction to Binder Jetting

Binder jetting is a 3D printing technology that uses a liquid binder to "glue" together powdered material, layer by layer, to create a part. This process is also known as three-dimensional printing (3DP) or selective laser sintering (SLS). Binder jetting is a popular 3D printing technology, known for its high speed, low cost, and ability to produce complex geometries.

8.1.1 How Binder Jetting Works

The binder jetting process involves several steps:

- **Powder Deposition**: A layer of powdered material is deposited onto the build platform. The powdered material is typically a metal, ceramic, or glass powder, and is spread evenly across the platform using a roller or other spreading mechanism.

- **Binder Jetting**: A liquid binder is selectively deposited onto the powdered material, bonding the particles together. The binder is typically a liquid adhesive, such as a polymer or a solvent, and is deposited using a print head or other dispensing mechanism.

- **Layer Formation**: The binder and powdered material form a solid layer. The binder causes the powdered material to bond together, creating a solid layer with the desired shape and structure.

- **Platform Movement**: The build platform is moved to accommodate the next layer. The platform is typically moved in a vertical direction, allowing for the creation of complex geometries and shapes.

- **Repeat**: Steps 1-4 are repeated until the part is complete. The process can take several hours, depending on the size and complexity of the part.

8.1.2 Advantages of Binder Jetting

Binder jetting has several advantages, including:

- **High Speed**: Binder jetting is a fast 3D printing technology, capable of producing parts at speeds of up to 100 mm/h.

- **Low Cost**: Binder jetting is a relatively low-cost 3D printing technology, making it accessible to a wide range of users.

- **Complex Geometries**: Binder jetting is capable of producing complex geometries and shapes, making it suitable for a wide range of applications.

- **Material Flexibility**: Binder jetting can be used with a wide range of materials, including metals, ceramics, and glass.

8.1.3 Types of Binder Jetting

There are several types of binder jetting, including:

- **Three-Dimensional Printing (3DP)**: This type of binder jetting uses a liquid binder to bond together powdered material, layer by layer.

- **Selective Laser Sintering (SLS)**: This type of binder jetting uses a laser to fuse together powdered material, layer by layer.

- **Inkjet-Based Binder Jetting**: This type of binder jetting uses an inkjet print head to deposit the binder onto the powdered material.

8.1.4 Applications of Binder Jetting

Binder jetting has a wide range of applications, including:

- **Rapid Prototyping**: Binder jetting is widely used for rapid prototyping, due to its high speed and low cost.
- **Production**: Binder jetting is used for production of parts, such as aerospace components, medical devices, and consumer products.
- **Artistic Applications**: Binder jetting is used in artistic applications, such as sculpture and jewelry design.

8.1.5 Challenges of Binder Jetting

Binder jetting also has some challenges, including:

- **Material Limitations**: Binder jetting is limited to powdered materials, which may not have the desired mechanical properties for certain applications.
- **Binder Limitations**: The binder used in binder jetting may not be suitable for certain applications and may require additional processing steps.
- **Post-Processing Requirements**: Binder jetting parts may require additional post-processing steps, such as machining or grinding, to achieve the desired surface finish.

8.1.6 Future of Binder Jetting

The future of binder jetting is promising, with ongoing research and development aimed at improving the technology. Some areas of research include:

- **New Materials**: Researchers are developing new powdered materials with improved mechanical properties, making binder jetting more suitable for a wider range of applications.

- **Improved Binders**: Researchers are developing new binders with improved properties, making binder jetting more suitable for a wider range of applications.

- **Increased Speed**: Researchers are working to increase the speed of binder jetting, making it more suitable for high-volume production.

Figure 40 illustrates the Binder Jetting process, a type of 3D printing technology. Here's a detailed explanation:

- Powder Roller: This component spreads a thin layer of powder material from the New Powder Stock onto the Powder Bed.

- Inkjet Printhead: Positioned above the powder bed, this printhead selectively deposits a liquid binder onto the powder layer, binding the particles together to form a solid part.

- Liquid Binder: Stored in a reservoir, this binder is essential for fusing the powder particles together.

- Build Platform: This platform supports the part being formed and moves down incrementally to allow for the addition of new powder layers.

The process is repeated layer by layer until the entire object is formed. Binder Jetting is known for its ability to produce complex geometries and high-resolution parts, making it suitable for various applications, including prototyping and manufacturing.

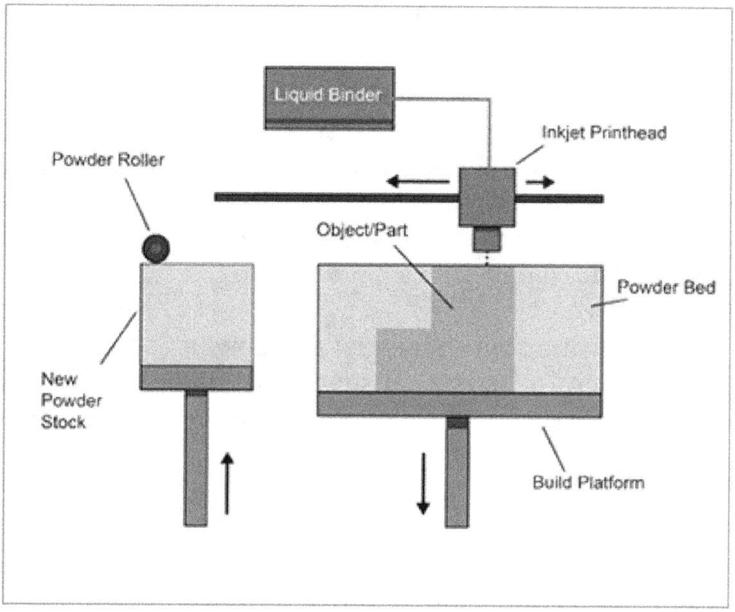

Figure 1: Binder jetting
[Source: https://www.lboro.ac.uk/research/amrg/about/the7categoriesofadditivemanufacturing/binderjetting/]

8.2 Three-Dimensional Printing (3DP)

Three-dimensional printing (3DP) is a binder jetting technology that uses a liquid binder to bond together powdered material, layer by layer. The binder is selectively deposited onto the powdered material using an inkjet print head. 3DP is a popular 3D printing technology, known for its high speed, low cost, and ability to produce complex geometries.

8.2.1 How 3DP Works

The 3DP process involves several steps:

- **Powder Deposition**: A layer of powdered material is deposited onto the build platform. The powdered material is typically a metal, ceramic, or glass powder, and is spread evenly across the platform using a roller or other spreading mechanism.

- **Binder Jetting**: A liquid binder is selectively deposited onto the powdered material, bonding the particles together. The binder is deposited using an inkjet print head, which is capable of depositing tiny droplets of binder onto the powdered material.

- **Layer Formation**: The binder and powdered material form a solid layer. The binder causes the powdered material to bond together, creating a solid layer with the desired shape and structure.

- **Platform Movement**: The build platform is moved to accommodate the next layer. The platform is typically moved in a vertical direction, allowing for the creation of complex geometries and shapes.

- **Repeat**: Steps 1-4 are repeated until the part is complete. The process can take several hours, depending on the size and complexity of the part.

8.2.2 Advantages of 3DP

3DP has several advantages, including:

- **High Speed**: 3DP is a fast 3D printing technology, capable of producing parts at speeds of up to 100 mm/h.

- **Low Cost**: 3DP is a relatively low-cost 3D printing technology, making it accessible to a wide range of users.

- **Complex Geometries**: 3DP is capable of producing complex geometries and shapes, making it suitable for a wide range of applications.

- **Material Flexibility**: 3DP can be used with a wide range of materials, including metals, ceramics, and glass.

8.2.3 Applications of 3DP

3DP has a wide range of applications, including:

- **Rapid Prototyping**: 3DP is widely used for rapid prototyping, due to its high speed and low cost.
- **Production**: 3DP is used for production of parts, such as aerospace components, medical devices, and consumer products.
- **Artistic Applications**: 3DP is used in artistic applications, such as sculpture and jewelry design.

8.2.4 Types of 3DP

There are several types of 3DP, including:

- **Monochrome 3DP:** This type of 3DP uses a single color binder to bond together powdered material.
- **Color 3DP:** This type of 3DP uses multiple color binders to create colorful parts.
- **Multi-Material 3DP:** This type of 3DP uses multiple materials, such as metals and ceramics, to create parts with complex properties.

8.2.5 Challenges of 3DP

3DP also has some challenges, including:

- **Material Limitations**: 3DP is limited to powdered materials, which may not have the desired mechanical properties for certain applications.
- **Binder Limitations**: The binder used in 3DP may not be suitable for certain applications, and may require additional processing steps.

- **Post-Processing Requirements**: 3DP parts may require additional post-processing steps, such as machining or grinding, to achieve the desired surface finish.

8.2.6 Future of 3DP

The future of 3DP is promising, with ongoing research and development aimed at improving the technology. Some areas of research include:

- **New Materials**: Researchers are developing new powdered materials with improved mechanical properties, making 3DP more suitable for a wider range of applications.

- **Improved Binders**: Researchers are developing new binders with improved properties, making 3DP more suitable for a wider range of applications.

- **Increased Speed**: Researchers are working to increase the speed of 3DP, making it more suitable for high-volume production.

8.2.7 Comparison with Other 3D Printing Technologies

3DP is often compared to other 3D printing technologies, such as selective laser sintering (SLS) and fused deposition modeling (FDM). While each technology has its own strengths and weaknesses, 3DP is known for its high speed, low cost, and ability to produce complex geometries.

Figure 41 illustrates the Three-Dimensional Printing (3DP) process, a type of additive manufacturing technology. Here's a detailed explanation:

- **Liquid Adhesive Supply:** This component holds the adhesive that will be used to bind the powder material.

- **Inkjet Print Head:** Positioned above the powder bed, this print head selectively deposits the liquid adhesive onto the powder layer, binding the particles together to form a solid part.

- **Powder Bed:** This is the area where the powder material is spread and where the object is built layer by layer.

- **Leveling Roller:** This roller spreads a thin, even layer of powder across the powder bed.

- **Powder Feed Piston:** This piston moves upward to supply new powder to the powder bed.

- **Build Piston:** This piston moves downward as each new layer of powder is added, allowing the object to be built layer by layer.

- **Build Chamber:** This is the area where the object is formed and where the build piston operates.

The process involves spreading a layer of powder, depositing adhesive to bind the powder in the desired shape, and then repeating these steps layer by layer until the entire object is formed. This method is known for its ability to produce complex geometries and high-resolution parts, making it suitable for various applications, including prototyping and manufacturing.

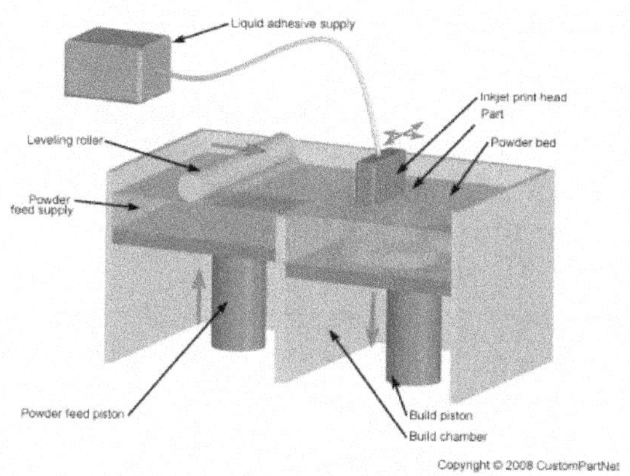

Figure 2: Three-Dimensional Printing (3DP)
[Source: https://www.custompartnet.com/wu/3d-printing]

8.3 Selective Laser Sintering (SLS)

Selective laser sintering (SLS) is a binder jetting technology that uses a laser to fuse together powdered material, layer by layer. The laser is selectively applied to the powdered material, bonding the particles together. SLS is a popular 3D printing technology, known for its high accuracy, high detail resolution, and ability to produce strong and durable parts.

8.3.1 How SLS Works

The SLS process involves several steps:

- **Powder Deposition**: A layer of powdered material is deposited onto the build platform. The powdered material is typically a metal, ceramic, or glass powder, and is spread evenly across the platform using a roller or other spreading mechanism.

- **Laser Sintering**: A laser is selectively applied to the powdered material, bonding the particles together. The laser is directed at specific points on the powdered material, causing the particles to fuse together.

- **Layer Formation**: The laser-sintered material forms a solid layer. The layer is typically 0.1-0.2 mm thick, and is built up layer by layer to create the final part.

- **Platform Movement**: The build platform is moved to accommodate the next layer. The platform is typically moved in a vertical direction, allowing for the creation of complex geometries and shapes.

- **Repeat**: Steps 1-4 are repeated until the part is complete. The process can take several hours, depending on the size and complexity of the part.

8.3.2 Advantages of SLS

SLS has several advantages, including:

- **High Accuracy**: SLS is capable of producing parts with high accuracy and detail resolution.

- **Strong and Durable Parts**: SLS parts are strong and durable, making them suitable for a wide range of applications.

- **Complex Geometries**: SLS is capable of producing complex geometries and shapes, making it suitable for a wide range of applications.

- **Material Flexibility**: SLS can be used with a wide range of materials, including metals, ceramics, and glass.

8.3.3 Applications of SLS

SLS has a wide range of applications, including:

- **Aerospace**: SLS is used in the aerospace industry to produce lightweight, high-strength components.

- **Automotive**: SLS is used in the automotive industry to produce complex components, such as engine parts and dashboard trim.

- **Medical**: SLS is used in the medical industry to produce custom implants and surgical guides.

- **Consumer Products**: SLS is used in the consumer products industry to produce complex components, such as phone cases and laptop parts.

8.3.4 Types of SLS

There are several types of SLS, including:

- **Single-Laser SLS**: This type of SLS uses a single laser to fuse together powdered material.

- **Multi-Laser SLS**: This type of SLS uses multiple lasers to fuse together powdered material, increasing the build speed and accuracy.

- **High-Speed SLS**: This type of SLS uses a high-powered laser to fuse together powdered material at high speeds.

8.3.5 Challenges of SLS

SLS also has some challenges, including:

- **Material Limitations**: SLS is limited to powdered materials, which may not have the desired mechanical properties for certain applications.

- **Laser Limitations**: The laser used in SLS may not be suitable for certain applications and may require additional processing steps.
- **Post-Processing Requirements**: SLS parts may require additional post-processing steps, such as machining or grinding, to achieve the desired surface finish.

8.3.6 Future of SLS

The future of SLS is promising, with ongoing research and development aimed at improving the technology. Some areas of research include:

- **New Materials**: Researchers are developing new powdered materials with improved mechanical properties, making SLS more suitable for a wider range of applications.

- **Improved Lasers**: Researchers are developing new lasers with improved properties, making SLS more suitable for a wider range of applications.

- **Increased Speed:** Researchers are working to increase the speed of SLS, making it more suitable for high-volume production.

8.3.7 Comparison with Other 3D Printing Technologies

SLS is often compared to other 3D printing technologies, such as selective laser melting (SLM) and electron beam melting (EBM). While each technology has its own strengths and weaknesses, SLS is known for its high accuracy, high detail resolution, and ability to produce strong and durable parts. Figure 42 illustrates the Selective Laser Sintering (SLS) process, a key technology in additive manufacturing. The figure shows three main stages: powder layering, laser sintering, and repeated powder layering. Initially, a roller spreads a thin layer of powder over the work area. The laser then selectively fuses the powder by tracing the cross-section of the 3D model, forming the workpiece. This process is repeated layer by layer, with each new layer of powder being spread over the partially sintered workpiece. This method allows for the creation of complex parts without the need for molds or casting, making it a versatile and efficient technology for modern manufacturing and prototyping.

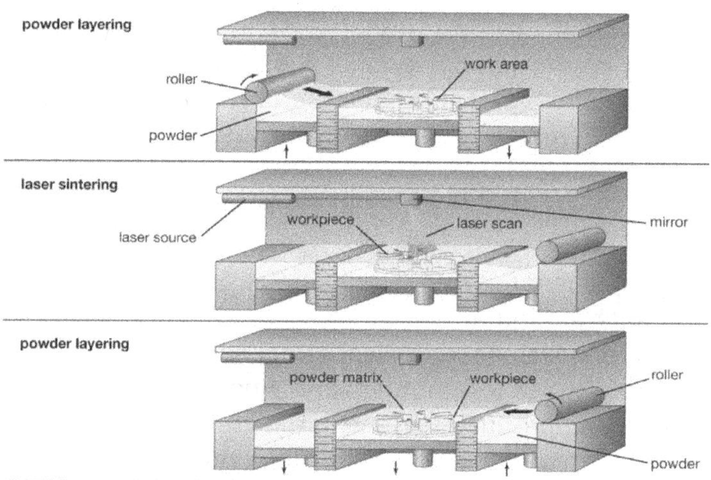

Figure 3: Selective laser sintering (SLS)
[Source: https://www.britannica.com/technology/selective-laser-sintering]

8.4 Applications and Limitations

Binder jetting technologies, such as 3DP and SLS, are commonly used for a wide range of applications, including rapid prototyping, production of functional parts, and consumer products. These technologies offer high resolution and accuracy, making them suitable for producing complex geometries and fine details.

Rapid Prototyping

Binder jetting technologies are widely used for rapid prototyping, due to their ability to quickly produce complex geometries and fine details. This makes them ideal for designers and engineers who need to test and iterate on their designs quickly. Rapid prototyping with binder jetting technologies allows for the creation of functional prototypes that can be used to test form, fit, and function.

Production of Functional Parts

Binder jetting technologies are also used for the production of functional parts, such as aerospace components, automotive parts, and medical devices. These parts require high accuracy and precision, which binder jetting technologies can provide. Additionally, binder jetting technologies can produce parts with complex geometries and internal structures, making them ideal for applications where traditional manufacturing methods are limited.

Consumer Products

Binder jetting technologies are used in the production of consumer products, such as phone cases, laptop parts, and jewelry. These products require high accuracy and precision, as well as the ability to produce complex geometries and fine details. Binder jetting technologies can produce parts with a high level of customization, making them ideal for consumer products that require unique designs and features.

8.4.1 Limitations of Binder Jetting Technologies

While binder jetting technologies offer many advantages, they also have some limitations. One of the main limitations is the material selection, which is typically limited to powdered metals, ceramics, and polymers. These materials may have limited mechanical properties and may not be suitable for high-temperature or high-stress applications.

Material Limitations

The materials used in binder jetting technologies are typically limited to powdered metals, ceramics, and polymers. These materials may have limited mechanical properties, such as strength, toughness, and corrosion resistance. Additionally, they may not be suitable for high-temperature or high-stress applications, such as aerospace or automotive applications.

Build Volume Limitations

Another limitation of binder jetting technologies is the build volume, which is typically smaller than that of other 3D printing technologies, such as powder bed fusion or directed energy deposition. This limits the size of the parts that can be produced using binder jetting technologies, making them less suitable for large-scale production.

Post-Processing Requirements

Binder jetting technologies often require additional post-processing steps, such as machining or grinding, to achieve the desired surface finish. This can add time and cost to the production process, making it less efficient.

Cost and Complexity

Binder jetting technologies can be complex and expensive, requiring specialized equipment and expertise. This can make them less accessible to small businesses or individuals, limiting their adoption.

8.4.2 Future of Binder Jetting Technologies

Despite the limitations of binder jetting technologies, they are continuing to evolve and improve. Researchers are working to develop new materials and processes that can overcome the limitations of current binder jetting technologies. For example, researchers are developing new powdered materials with improved mechanical properties, and new binder jetting processes that can produce larger parts with higher accuracy.

8.4.3 Comparison with Other 3D Printing Technologies

Binder jetting technologies are often compared to other 3D printing technologies, such as powder bed fusion, directed energy deposition, and stereolithography. While each technology has its own strengths and weaknesses, binder jetting technologies are known for their high resolution and accuracy, making them suitable for producing complex geometries and fine details.

In conclusion, binder jetting technologies, such as 3DP and SLS, are widely used for rapid prototyping, production of functional parts, and consumer products. While they offer many advantages, they also have some limitations, including material selection, build volume, post-processing requirements, cost, and complexity. Despite these limitations, binder jetting technologies are continuing to evolve and improve, making them an important part of the 3D printing landscape.

8.5 Questions

8.5.1 Long Response Questions

1. What are the key differences between three-dimensional printing (3DP) and selective laser sintering (SLS)?
2. What are some common applications of binder jetting technologies?
3. What are the limitations of binder jetting technologies?
4. How does the resolution and accuracy of binder jetting technologies compare to other 3D printing technologies?

5. What are some potential future developments in binder jetting technologies?

8.5.2 Multiple Choice Questions

1) **What is the primary material used in the Binder Jetting process?**
   a) Liquid resin
   b) Powdered material
   c) Metal wire
   d) Plastic filament
   Solution: - b) Powdered material
2) **In Binder Jetting, what role does the liquid binder play?**
   a) Spreads the powder
   b) Fuses the powder particles
   c) Supports the build platform
   d) Cools the part
   Solution: - b) Fuses the powder particles
3) **Which technology uses a laser to bond powdered material?**
   a) Fused Deposition Modeling (FDM)
   b) Stereolithography (SLA)
   c) Selective Laser Sintering (SLS)
   d) Inkjet Printing
   Solution: - c) Selective Laser Sintering (SLS)
4) **What is one of the main advantages of Binder Jetting?**
   a) High cost
   b) Limited material flexibility
   c) High speed
   d) Low resolution
   Solution: - c) High speed
5) **Which of the following is a common application of 3DP?**
   a) Producing circuit boards
   b) Manufacturing metal parts
   c) Rapid prototyping
   d) Creating glassware
   Solution: - c) Rapid prototyping
6) **Which component in 3DP spreads the powder evenly across the build platform?**

a) Inkjet Print Head
b) Build Platform
c) Powder Roller
d) Liquid Adhesive Supply

Solution: - c) Powder Roller

7) **Which type of binder jetting uses an inkjet print head for deposition?**
a) Selective Laser Sintering (SLS)
b) Three-Dimensional Printing (3DP)
c) Fused Deposition Modeling (FDM)
d) Digital Light Processing (DLP)

Solution: - b) Three-Dimensional Printing (3DP)

8) **Which material is NOT typically used in Binder Jetting?**
a) Metal powder
b) Ceramic powder
c) Plastic filament
d) Glass powder

Solution: - c) Plastic filament

9) **What is the primary advantage of SLS over other 3D printing technologies?**
a) Cost efficiency
b) High accuracy and detail resolution
c) Low build volume
d) Speed

Solution: - b) High accuracy and detail resolution

10) **In SLS, what is the purpose of the laser?**
a) To deposit binder
b) To fuse powdered material
c) To spread powder evenly
d) To cool the part

Solution: - b) To fuse powdered material

11) **Which of the following is a challenge of Binder Jetting?**
a) High material flexibility
b) Large build volume
c) Material limitations
d) Low cost

Solution: - c) Material limitations

12) **What type of SLS uses multiple lasers to increase build speed?**

a) High-Speed SLS
b) Single-Laser SLS
c) Multi-Laser SLS
d) Color SLS
Solution: - c) Multi-Laser SLS

13) Which technology involves a build platform that moves downward as new layers are added?
a) Fused Deposition Modeling (FDM)
b) Selective Laser Melting (SLM)
c) Binder Jetting
d) Stereolithography (SLA)
Solution: - c) Binder Jetting

14) What is a common post-processing step required for Binder Jetting parts?
a) Injection molding
b) Machining or grinding
c) UV curing
d) Thermal annealing
Solution: - b) Machining or grinding

15) Which application of Binder Jetting is least likely to be used in industrial production?
a) Aerospace components
b) Artistic sculptures
c) Automotive engine parts
d) High-temperature alloys
Solution: - b) Artistic sculptures

16) What does the build piston do in the 3DP process?
a) Deposits binder
b) Spreads powder
c) Moves downward with each new layer
d) Provides cooling
Solution: - c) Moves downward with each new layer

17) Which technology is often compared with SLS due to its ability to produce strong and durable parts?
a) Electron Beam Melting (EBM)
b) Fused Deposition Modeling (FDM)
c) Digital Light Processing (DLP)
d) Stereolithography (SLA)
Solution: - a) Electron Beam Melting (EBM)

18) What is one of the key differences between 3DP and SLS in terms of material bonding?

a) 3DP uses a laser while SLS uses a binder
b) 3DP uses a binder while SLS uses a laser
c) 3DP uses metal while SLS uses plastic
d) 3DP uses liquid resin while SLS uses powder

Solution: - b) 3DP uses a binder while SLS uses a laser

19) Which type of SLS process uses a high-powered laser for faster production?

a) Single-Laser SLS
b) Multi-Laser SLS
c) High-Speed SLS
d) Low-Speed SLS

Solution: - c) High-Speed SLS

20) In Binder Jetting, why might post-processing be required?

a) To apply a color finish
b) To achieve desired surface finish
c) To increase the part's weight
d) To add a protective coating

Solution: - b) To achieve desired surface finish

21) Which binder jetting technology involves multiple color binders for producing colorful parts?

a) SLS
b) SLA
c) 3DP
d) FDM

Solution: - c) 3DP

22) Which of the following is a limitation of SLS technology?

a) High accuracy
b) Use of multiple materials
c) Limited material options
d) Large build volume

Solution: - c) Limited material options

23) What is the primary advantage of using binder jetting for producing parts with complex geometries?

a) High material flexibility
b) High accuracy
c) Ability to produce complex geometries

d) Fast material deposition

Solution: - c) Ability to produce complex geometries

24) **Which technology within binder jetting involves the use of a laser to bond powdered material?**

a) Inkjet-Based Binder Jetting

b) Three-Dimensional Printing (3DP)

c) Selective Laser Sintering (SLS)

d) Fused Deposition Modeling (FDM)

Solution: - c) Selective Laser Sintering (SLS)

25) **In binder jetting, what does the liquid binder do?**

a) Fuses powdered material using heat

b) Deposits layers of powdered material

c) Bonds powdered particles together

d) Cools down the powdered material

Solution: - c) Bonds powdered particles together

26) **What type of binder jetting uses multiple colors of binder to create colorful parts?**

a) Multi-Material 3DP

b) Color 3DP

c) Single-Laser SLS

d) High-Speed SLS

Solution: - b) Color 3DP

27) **Which of the following is a challenge associated with 3DP technology?**

a) Material Flexibility

b) Post-Processing Requirements

c) High Speed

d) Low Cost

Solution: - b) Post-Processing Requirements

28) **What is a common application of Selective Laser Sintering (SLS) in the automotive industry?**

a) Phone cases

b) Engine parts

c) Medical implants

d) Sculpture design

Solution: - b) Engine parts

29) **What is one of the material limitations of binder jetting technologies?**

a) Limited to powdered materials

b) Limited to liquid adhesives
c) Limited to high-temperature applications
d) Limited to metal parts
Solution: - a) Limited to powdered materials

30) Which binder jetting technology is known for its low cost and high speed?
a) Selective Laser Sintering (SLS)
b) Three-Dimensional Printing (3DP)
c) Fused Deposition Modeling (FDM)
d) Stereolithography (SLA)
Solution: - b) Three-Dimensional Printing (3DP)

31) What is a key advantage of using SLS technology in manufacturing?
a) High speed production
b) Simple process
c) High accuracy and detail resolution
d) Low material cost
Solution: - c) High accuracy and detail resolution

32) Which of the following is a type of SLS that uses multiple lasers?
a) High-Speed SLS
b) Multi-Laser SLS
c) Single-Laser SLS
d) Multi-Material SLS
Solution: - b) Multi-Laser SLS

33) What does the inkjet print head do in Three-Dimensional Printing (3DP)?
a) Deposits powdered material
b) Spreads layers of powder
c) Selectively deposits liquid binder
d) Sintered powder layers
Solution: - c) Selectively deposits liquid binder

34) In binder jetting, what is the primary role of the powder roller?
a) Deposits the liquid binder
b) Spreads a thin layer of powder
c) Moves the build platform
d) Cools the build chamber
Solution: - b) Spreads a thin layer of powder

35) What is one limitation of SLS technology?

a) Limited material options
b) High material cost
c) Low accuracy
d) Low speed

Solution: - a) Limited material options

36) Which of the following is NOT a typical application of binder jetting technologies?

a) Rapid Prototyping
b) Production of consumer electronics
c) Large-scale structural components
d) Artistic sculptures

Solution: - c) Large-scale structural components

37) What is the typical thickness of a layer in the SLS process?

a) 0.01 mm
b) 0.5 mm
c) 0.1-0.2 mm
d) 1 mm

Solution: - c) 0.1-0.2 mm

38) Which type of 3DP technology involves the use of multiple materials to create parts with complex properties?

a) Monochrome 3DP
b) Multi-Material 3DP
c) Color 3DP
d) Single-Material 3DP

Solution: - b) Multi-Material 3DP

39) Which type of binder jetting technology is best suited for producing highly detailed and strong parts?

a) Three-Dimensional Printing (3DP)
b) Selective Laser Sintering (SLS)
c) Inkjet-Based Binder Jetting
d) Fused Deposition Modeling (FDM)

Solution: - b) Selective Laser Sintering (SLS)

40) What future research direction for binder jetting involves improving the properties of the materials used?

a) Development of new binders
b) Increased build speed
c) Development of new powdered materials

d) Enhanced printer design

Solution: - c) Development of new powdered materials

Chapter 9

Directed Energy Deposition: LENS and EBM

9.1 Introduction to Directed Energy Deposition

Directed energy deposition (DED) is a 3D printing technology that uses a focused beam of energy, such as a laser or electron beam, to melt and deposit metal onto a substrate, layer by layer, to create a part. This process is also known as laser engineered net shaping (LENS) or electron beam melting (EBM). DED is a powerful technology that offers high accuracy, high detail resolution, and the ability to produce complex geometries and internal structures.

9.1.1 How DED Works

The DED process involves several steps:

- **Metal Powder Deposition**: A layer of metal powder is deposited onto the substrate. The metal powder is typically made up of fine particles, which are evenly distributed across the substrate using a powder feeder or other deposition mechanism.

- **Energy Beam Application**: A focused beam of energy, such as a laser or electron beam, is applied to the metal powder, melting and depositing it onto the substrate. The energy beam is directed at specific points on the metal powder, causing it to melt and form a pool of molten metal.

- **Layer Formation**: The melted metal forms a solid layer. The layer is typically 0.1-1 mm thick and is built up layer by layer to create the final part.

- **Platform Movement**: The substrate is moved to accommodate the next layer. The platform is typically moved in a vertical direction, allowing for the creation of complex geometries and internal structures.

- **Repeat**: Steps 1-4 are repeated until the part is complete. The process can take several hours, depending on the size and complexity of the part.

9.1.2 Types of DED

There are several types of DED, including:

- **Laser Engineered Net Shaping (LENS)**: This type of DED uses a laser to melt and deposit metal onto a substrate.

- **Electron Beam Melting (EBM)**: This type of DED uses an electron beam to melt and deposit metal onto a substrate.

- **Laser Beam Melting (LBM)**: This type of DED uses a laser to melt and deposit metal onto a substrate.

9.1.3 Advantages of DED

DED has several advantages, including:

- **High Accuracy**: DED is capable of producing parts with high accuracy and detail resolution.

- **Complex Geometries**: DED can produce complex geometries and internal structures, making it suitable for a wide range of applications.

- **High-Strength Materials**: DED can produce parts with high-strength materials, such as titanium and steel.

- **Reduced Material Waste**: DED reduces material waste, as only the required amount of metal powder is deposited onto the substrate.

9.1.4 Applications of DED

DED has a wide range of applications, including:

- **Aerospace**: DED is used in the aerospace industry to produce lightweight, high-strength components.

- **Automotive**: DED is used in the automotive industry to produce complex components, such as engine parts and transmission components.

- **Medical**: DED is used in the medical industry to produce custom implants and surgical instruments.

- **Consumer Products**: DED is used in the consumer products industry to produce complex components, such as phone cases and laptop parts.

9.1.5 Challenges of DED

DED also has some challenges, including:

- **Material Limitations**: DED is limited to metal powders, which may not have the desired mechanical properties for certain applications.

- **Energy Beam Limitations**: The energy beam used in DED may not be suitable for certain applications, and may require additional processing steps.

- **Post-Processing Requirements**: DED parts may require additional post-processing steps, such as machining or grinding, to achieve the desired surface finish.

9.1.6 Future of DED

The future of DED is promising, with ongoing research and development aimed at improving the technology. Some areas of research include:

- **New Materials**: Researchers are developing new metal powders with improved mechanical properties, making DED more suitable for a wider range of applications.
- **Improved Energy Beams**: Researchers are developing new energy beams with improved properties, making DED more efficient and effective.
- **Increased Speed**: Researchers are working to increase the speed of DED, making it more suitable for high-volume production.

In conclusion, DED is a powerful 3D printing technology that offers high accuracy, high detail resolution, and the ability to produce complex geometries and internal structures. While it has some challenges, ongoing research and development are addressing these issues, expanding the capabilities of DED and making it more suitable for a wider range of applications. Whether you're a designer, engineer, or manufacturer, DED is an important technology to consider for your next project. Figure 43 illustrates the Directed Energy Deposition (DED) process, a sophisticated additive manufacturing technique. The figure shows an Electron Beam (EB) Gun positioned above a substrate, emitting an electron beam to create a Molten Alloy Puddle on the substrate. A Wire Feeder continuously supplies material into this molten pool. The substrate moves in the direction indicated by the arrow labeled "Direction of Part Motion," while the EB Gun can move in multiple directions, as shown by the "Gun Motions" arrows. This process builds up the part layer by layer, with each new layer of material being deposited and melted to form a solid structure. DED is particularly useful for creating complex geometries and repairing existing parts, making it a versatile tool in modern manufacturing.

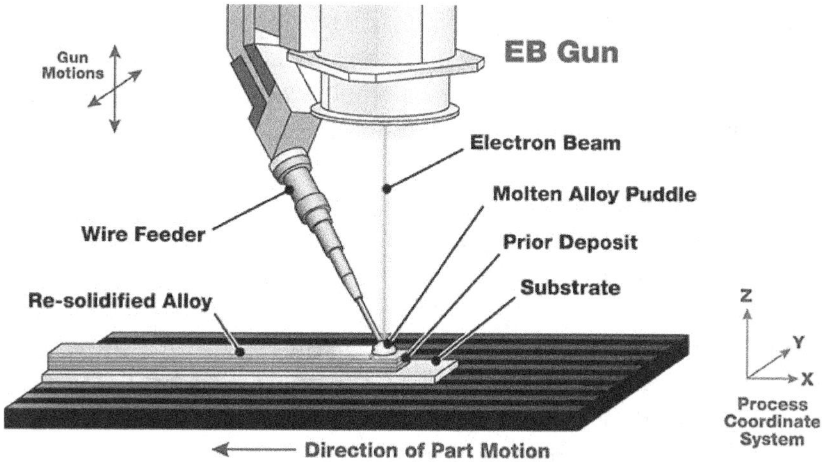

Figure 1: Directed energy deposition
[Source: https://www.sciaky.com/additive-manufacturing/what-is-ded-3d-printing]

9.2 Laser Engineered Net Shaping (LENS)

Laser engineered net shaping (LENS) is a directed energy deposition technology that uses a high-powered laser to melt and deposit metal onto a substrate, layer by layer. The laser is focused onto the metal powder, creating a small melt pool that solidifies as the laser moves. This process allows for the creation of complex geometries and internal structures, making LENS a popular choice for a wide range of applications.

9.2.1 How LENS Works

The LENS process involves several steps:

- **Metal Powder Deposition**: A layer of metal powder is deposited onto the substrate. The metal powder is typically made up of fine particles, which are evenly distributed across the substrate using a powder feeder or other deposition mechanism.

- **Laser Application**: A high-powered laser is applied to the metal powder, melting and depositing it onto the substrate. The laser is focused onto a small area, creating a melt pool that is typically 0.1-1 mm in diameter.

- **Layer Formation**: The melted metal forms a solid layer. The layer is typically 0.1-1 mm thick, and is built up layer by layer to create the final part.

- **Platform Movement**: The substrate is moved to accommodate the next layer. The platform is typically moved in a vertical direction, allowing for the creation of complex geometries and internal structures.

- **Repeat**: Steps 1-4 are repeated until the part is complete. The process can take several hours, depending on the size and complexity of the part.

9.2.2 LENS Process Parameters

The LENS process involves several key parameters that affect the final part quality, including:

- **Laser Power**: The power of the laser affects the size of the melt pool and the rate of deposition.

- **Laser Speed**: The speed of the laser affects the layer thickness and the overall build time.

- **Metal Powder Properties**: The properties of the metal powder, such as particle size and distribution, affect the final part quality.

- **Substrate Temperature**: The temperature of the substrate affects the solidification rate and the final part quality.

9.2.3 Advantages of LENS

LENS has several advantages, including:

- **High Accuracy**: LENS is capable of producing parts with high accuracy and detail resolution.

- **Complex Geometries**: LENS can produce complex geometries and internal structures, making it suitable for a wide range of applications.

- **High-Strength Materials**: LENS can produce parts with high-strength materials, such as titanium and steel.

- **Reduced Material Waste**: LENS reduces material waste, as only the required amount of metal powder is deposited onto the substrate.

9.2.4 Applications of LENS

LENS has a wide range of applications, including:

- **Aerospace**: LENS is used in the aerospace industry to produce lightweight, high-strength components.

- **Automotive**: LENS is used in the automotive industry to produce complex components, such as engine parts and transmission components.

- **Medical**: LENS is used in the medical industry to produce custom implants and surgical instruments.

- **Consumer Products**: LENS is used in the consumer products industry to produce complex components, such as phone cases and laptop parts.

9.2.5 Challenges of LENS

LENS also has some challenges, including:

- **Material Limitations**: LENS is limited to metal powders, which may not have the desired mechanical properties for certain applications.

- **Laser Limitations**: The laser used in LENS may not be suitable for certain applications and may require additional processing steps.

- **Post-Processing Requirements**: LENS parts may require additional post-processing steps, such as machining or grinding, to achieve the desired surface finish.

9.2.6 Future of LENS

The future of LENS is promising, with ongoing research and development aimed at improving the technology. Some areas of research include:

- **New Materials**: Researchers are developing new metal powders with improved mechanical properties, making LENS more suitable for a wider range of applications.

- **Improved Laser Technology**: Researchers are developing new laser technologies with improved properties, making LENS more efficient and effective.

- **Increased Speed**: Researchers are working to increase the speed of LENS, making it more suitable for high-volume production.

9.2.7 Comparison with Other DED Technologies

LENS is often compared to other DED technologies, such as electron beam melting (EBM) and laser beam melting (LBM). While each technology has its own strengths and weaknesses, LENS is known for its high accuracy and ability to produce complex geometries and internal structures. Figure 44 illustrates the Laser Engineered Net Shaping (LENS) process, a precise additive manufacturing technique. The diagram shows a laser beam directed downwards, intersecting with a powder stream to create a molten pool on a substrate positioned on an X-Y machine table. The process builds up layers, with arrows indicating the Z motion and processing direction. The image depicts the formation of multiple layers (1st, 2nd, and 3rd layers) above the substrate, highlighting the additive nature of the process. A dilution area is marked around the interaction zone, and shield gas is used to protect the molten material. This method allows for the creation of near-net shape parts with high precision and is widely used for repairing and manufacturing complex components.

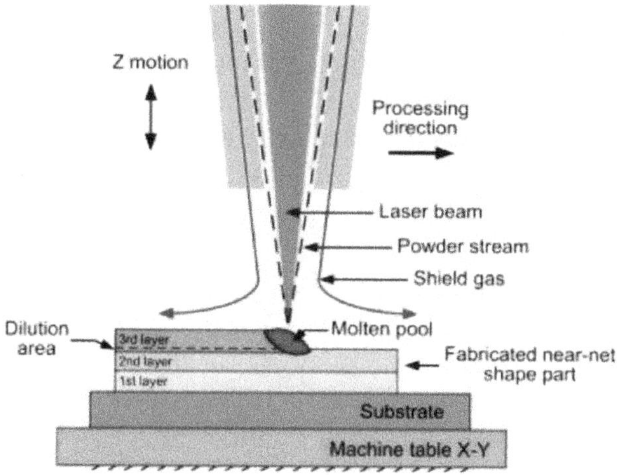

Figure 2: Laser engineered net shaping (LENS) [dx.doi.org/10.5772/intechopen.89804]

9.3 Electron Beam Melting (EBM)

Electron beam melting (EBM) is a directed energy deposition technology that uses an electron beam to melt and deposit metal onto a substrate, layer by layer. The electron beam is focused onto the metal powder, creating a small melt pool that solidifies as the beam moves. This process allows for the creation of complex geometries and internal structures, making EBM a popular choice for a wide range of applications.

9.3.1 How EBM Works

The EBM process involves several steps:

- **Metal Powder Deposition:** A layer of metal powder is deposited onto the substrate. The metal powder is typically made up of fine particles, which are evenly distributed across the substrate using a powder feeder or other deposition mechanism.

- **Electron Beam Application**: An electron beam is applied to the metal powder, melting and depositing it onto the substrate. The electron beam is focused onto a small area, creating a melt pool that is typically 0.1-1 mm in diameter.

- **Layer Formation**: The melted metal forms a solid layer. The layer is typically 0.1-1 mm thick, and is built up layer by layer to create the final part.

- **Platform Movement**: The substrate is moved to accommodate the next layer. The platform is typically moved in a vertical direction, allowing for the creation of complex geometries and internal structures.

- **Repeat**: Steps 1-4 are repeated until the part is complete. The process can take several hours, depending on the size and complexity of the part.

9.3.2 EBM Process Parameters

The EBM process involves several key parameters that affect the final part quality, including:

- **Electron Beam Power**: The power of the electron beam affects the size of the melt pool and the rate of deposition.

- **Electron Beam Speed**: The speed of the electron beam affects the layer thickness and the overall build time.

- **Metal Powder Properties**: The properties of the metal powder, such as particle size and distribution, affect the final part quality.

- **Substrate Temperature**: The temperature of the substrate affects the solidification rate and the final part quality.

9.3.3 Advantages of EBM

EBM has several advantages, including:

- **High Accuracy**: EBM is capable of producing parts with high accuracy and detail resolution.
- **Complex Geometries**: EBM can produce complex geometries and internal structures, making it suitable for a wide range of applications.
- **High-Strength Materials**: EBM can produce parts with high-strength materials, such as titanium and steel.
- **Reduced Material Waste**: EBM reduces material waste, as only the required amount of metal powder is deposited onto the substrate.

9.3.4 Applications of EBM

EBM has a wide range of applications, including:

- **Aerospace**: EBM is used in the aerospace industry to produce lightweight, high-strength components.
- **Automotive**: EBM is used in the automotive industry to produce complex components, such as engine parts and transmission components.
- **Medical**: EBM is used in the medical industry to produce custom implants and surgical instruments.
- **Consumer Products**: EBM is used in the consumer products industry to produce complex components, such as phone cases and laptop parts.

9.3.5 Challenges of EBM

EBM also has some challenges, including:

- **Material Limitations**: EBM is limited to metal powders, which may not have the desired mechanical properties for certain applications.

- **Electron Beam Limitations**: The electron beam used in EBM may not be suitable for certain applications, and may require additional processing steps.

- **Post-Processing Requirements**: EBM parts may require additional post-processing steps, such as machining or grinding, to achieve the desired surface finish.

9.3.6 Future of EBM

The future of EBM is promising, with ongoing research and development aimed at improving the technology. Some areas of research include:

- **New Materials**: Researchers are developing new metal powders with improved mechanical properties, making EBM more suitable for a wider range of applications.

- **Improved Electron Beam Technology**: Researchers are developing new electron beam technologies with improved properties, making EBM more efficient and effective.

- **Increased Speed**: Researchers are working to increase the speed of EBM, making it more suitable for high-volume production.

9.3.7 Comparison with Other DED Technologies

EBM is often compared to other DED technologies, such as laser engineered net shaping (LENS) and laser beam melting (LBM). While each technology has its own strengths and weaknesses, EBM is known for its high accuracy and ability to produce complex geometries and internal structures. Figure 45 illustrates the Electron Beam Melting (EBM) process, a type of additive manufacturing technology. The diagram shows an Electron Beam Column at the top, which includes a Filament and X-Y deflection coils to direct the electron beam. The beam passes through Stigmators and Defocus coils to shape and focus it. The electron beam then enters a chamber where it interacts with powder material from the Powder Hopper. A Rake spreads the powder evenly. The central area, enclosed within a Shield, is where the electron beam melts the powder layer by layer on a Build platform. This platform moves downwards as each layer is added, forming the object. The entire process occurs inside an Electron Beam Vacuum Chamber to maintain the necessary conditions for melting and solidifying the metal powder.

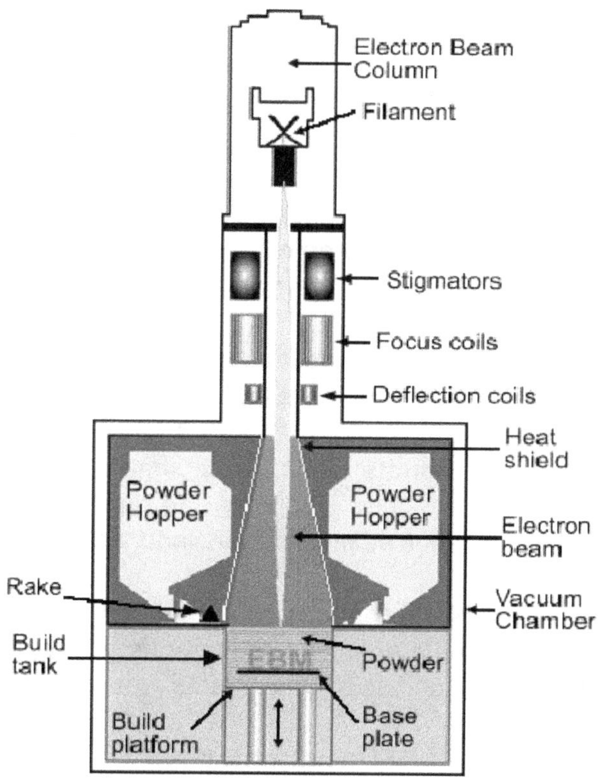

Figure 3: Electron beam melting (EBM)
[Source: https://www.mdpi.com/1996-1944/7/12/8168]

9.4 Applications and Limitations

Directed energy deposition (DED) technologies, such as laser engineered net shaping (LENS) and electron beam melting (EBM), are commonly used for rapid prototyping, production of functional parts, and repair of existing components. These technologies offer high resolution and accuracy, making them suitable for producing complex geometries and fine details.

9.4.1 Applications of DED Technologies

DED technologies have a wide range of applications, including:

- **Aerospace**: DED technologies are used in the aerospace industry to produce lightweight, high-strength components, such as engine parts and structural components.

- **Automotive**: DED technologies are used in the automotive industry to produce complex components, such as engine parts and transmission components.

- **Medical**: DED technologies are used in the medical industry to produce custom implants and surgical instruments.

- **Consumer Products**: DED technologies are used in the consumer products industry to produce complex components, such as phone cases and laptop parts.

- **Repair and Maintenance**: DED technologies are used in repair and maintenance applications to repair existing components, such as turbine blades and molds.

9.4.2 Advantages of DED Technologies

DED technologies offer several advantages, including:

- **High Resolution and Accuracy**: DED technologies are capable of producing parts with high resolution and accuracy, making them suitable for producing complex geometries and fine details.

- **Rapid Prototyping**: DED technologies are well-suited for rapid prototyping, allowing for the quick and easy production of prototypes for testing and evaluation.

- **Production of Functional Parts**: DED technologies can be used to produce functional parts, such as engine components and structural components, that are ready for use in real-world applications.

- **Repair and Maintenance**: DED technologies can be used to repair existing components, such as turbine blades and molds, extending their lifespan and reducing the need for replacement.

9.4.3 Limitations of DED Technologies

However, DED technologies also have some limitations, including:

- **Limited Materials**: DED technologies are typically limited to metals, which may have limited mechanical properties and may not be suitable for high-temperature or high-stress applications.

- **Small Build Volumes**: The build volumes of DED technologies are generally smaller than those of other 3D printing technologies, such as powder bed fusion or material jetting. This can limit the size of the parts that can be produced using these technologies.

- **Cost**: DED technologies can be more expensive than other 3D printing technologies, such as fused deposition modeling (FDM) or stereolithography (SLA), due to the high cost of the equipment and materials used.

- **Complexity**: DED technologies can be more complex to use than other 3D printing technologies, requiring specialized training and expertise to operate.

9.4.4 Comparison with Other 3D Printing Technologies

DED technologies are often compared to other 3D printing technologies, such as powder bed fusion (PBF) and material jetting (MJ). While each technology has its own strengths and weaknesses, DED technologies are known for their high resolution and accuracy, making them suitable for producing complex geometries and fine details.

PBF technologies, such as selective laser sintering (SLS) and selective laser melting (SLM), use a laser to melt and fuse metal or polymer powders. These technologies offer larger build volumes than DED technologies but may not have the same level of resolution and accuracy.

MJ technologies, such as PolyJet and MultiJet, use a print head to jet droplets of liquid photopolymer onto a build platform. These technologies offer high resolution and accuracy but are typically limited to producing parts with polymer materials.

9.4.5 Future of DED Technologies

The future of DED technologies is promising, with ongoing research and development aimed at improving the technology. Some areas of research include:

- **New Materials**: Researchers are developing new metal powders with improved mechanical properties, making DED technologies more suitable for a wider range of applications.

- **Improved Equipment**: Researchers are developing new equipment with improved properties, making DED technologies more efficient and effective.

- **Increased Speed**: Researchers are working to increase the speed of DED technologies, making them more suitable for high-volume production.

Directed energy deposition (DED) technologies, such as laser engineered net shaping (LENS) and electron beam melting (EBM), are powerful 3D printing technologies that offer high resolution and accuracy, making them suitable for producing complex geometries and fine details. While they have some limitations, such as limited materials and small build volumes, ongoing research and development are addressing these issues, expanding the capabilities of DED technologies and making them more suitable for a wider range of applications. Whether you're a designer, engineer, or manufacturer, DED technologies offer a powerful tool for rapid prototyping, production of functional parts, and repair of existing components.

9.5 Questions

9.5.1 Long Response Questions

1. What are the key differences between laser engineered net shaping (LENS) and electron beam melting (EBM)?
2. What are some common applications of directed energy deposition technologies?
3. What are the limitations of directed energy deposition technologies?
4. How does the resolution and accuracy of directed energy deposition technologies compare to other 3D printing technologies?
5. What are some potential future developments in directed energy deposition technologies?

9.5.2 Multiple Choice Questions

1) **What does Directed Energy Deposition (DED) use to melt metal powder?**
   a) Plasma arc
   b) Ultrasonic waves

c) Focused energy beam
d) Infrared radiation
Solution: - c) Focused energy beam

2) **Which type of energy beam is used in Laser Engineered Net Shaping (LENS)?**
a) Electron beam
b) Laser
c) X-ray
d) Microwave
Solution: - b) Laser

3) **What is the main advantage of using Directed Energy Deposition (DED) technologies?**
a) Low accuracy
b) High material waste
c) Ability to produce complex geometries
d) Limited material options
Solution: - c) Ability to produce complex geometries

4) **Which industry uses LENS technology to produce high-strength components?**
a) Textile
b) Aerospace
c) Food and Beverage
d) Entertainment
Solution: - b) Aerospace

5) **What does Electron Beam Melting (EBM) use to deposit metal?**
a) Laser
b) Electron beam
c) Water jet
d) Hot air
Solution: - b) Electron beam

6) **Which of the following is an application of DED technology?**
a) Printing paper
b) Producing custom implants
c) Creating textiles
d) Making plastic bottles
Solution: - b) Producing custom implants

7) **How is the substrate moved during the DED process?**

a) Horizontally
b) Vertically
c) Diagonally
d) Randomly

Solution: - b) Vertically

8) **What is one challenge of using LENS technology?**

a) Large material waste
b) High resolution
c) Limited material types
d) Simple operation

Solution: - c) Limited material types

9) **Which of the following is not a type of Directed Energy Deposition (DED)?**

a) Laser Beam Melting (LBM)
b) Electron Beam Melting (EBM)
c) Fused Deposition Modeling (FDM)
d) Laser Engineered Net Shaping (LENS)

Solution: - c) Fused Deposition Modeling (FDM)

10) **What is the purpose of the shield gas in LENS?**

a) To cool the substrate
b) To protect the molten material
c) To accelerate the deposition
d) To improve the laser focus

Solution: - b) To protect the molten material

11) **What is one of the key parameters that affects the LENS process?**

a) Laser wavelength
b) Laser power
c) Powder feeder speed
d) Substrate thickness

Solution: - b) Laser power

12) **Which of the following is an advantage of Electron Beam Melting (EBM)?**

a) Low accuracy
b) High material waste
c) High-strength materials
d) Limited build volume

Solution: - c) High-strength materials

13) Which technology involves the use of a vacuum chamber to maintain conditions for melting?

a) LENS
b) FDM
c) EBM
d) SLA

Solution: - c) EBM

14) In which industry is EBM technology used for producing custom implants?

a) Automotive
b) Aerospace
c) Medical
d) Electronics

Solution: - c) Medical

15) What type of beam is used in Laser Beam Melting (LBM)?

a) X-ray
b) Electron beam
c) Laser
d) Ultraviolet light

Solution: - c) Laser

16) What is a limitation of using DED technologies?

a) High resolution
b) Small build volumes
c) Low material cost
d) Simple operation

Solution: - b) Small build volumes

17) Which parameter affects the speed of deposition in EBM?

a) Electron beam power
b) Powder particle size
c) Substrate movement speed
d) Electron beam speed

Solution: - d) Electron beam speed

18) How does LENS technology achieve reduced material waste?

a) By recycling unused material
b) By only depositing required amounts of powder
c) By using minimal energy
d) By using larger build volumes

Solution: - b) By only depositing required amounts of powder

19) **What does the platform movement in DED allow for?**
    a) Faster build times
    b) Creation of complex geometries
    c) Reduced material waste
    d) Increased energy efficiency
    Solution: - b) Creation of complex geometries

20) **Which of the following is a key factor in the LENS process parameters?**
    a) Substrate color
    b) Powder feed rate
    c) Laser focus width
    d) Material density
    Solution: - c) Laser focus width

21) **Which parameter directly affects the solidification rate in LENS?**
    a) Laser wavelength
    b) Substrate temperature
    c) Powder particle size
    d) Laser speed
    Solution: - b) Substrate temperature

22) **What is one of the research areas for improving EBM technology?**
    a) Developing non-metal powders
    b) Enhancing laser power
    c) Increasing electron beam efficiency
    d) Reducing the build volume
    Solution: - c) Increasing electron beam efficiency

23) **How do new laser technologies impact the LENS process?**
    a) They increase material waste
    b) They make the process less efficient
    c) They improve the efficiency and effectiveness of LENS
    d) They simplify the powder deposition process
    Solution: - c) They improve the efficiency and effectiveness of LENS

24) **What challenge is associated with the electron beam used in EBM?**

a) Insufficient power
b) High cost of equipment
c) Need for post-processing
d) Limited build volume

Solution: - b) High cost of equipment

25) In which way does the EBM process typically differ from LENS in terms of material handling?

a) EBM uses liquid metal while LENS uses powder
b) EBM requires a vacuum chamber for processing
c) EBM uses a different type of laser for melting
d) EBM deposits material in larger layers than LENS

Solution: - b) EBM requires a vacuum chamber for processing

26) What is a future research focus for increasing the speed of DED technologies?

a) Developing new substrates
b) Enhancing laser or electron beam technologies
c) Improving post-processing techniques
d) Reducing the build volume

Solution: - b) Enhancing laser or electron beam technologies

27) Which technology uses a powder feeder mechanism to distribute metal powder?

a) FDM
b) SLA
c) LENS
d) PolyJet

Solution: - c) LENS

28) What is the role of the rake in the EBM process?

a) To heat the substrate
b) To spread the powder evenly
c) To focus the electron beam
d) To move the build platform

Solution: - b) To spread the powder evenly

29) What type of chamber is used in EBM to ensure proper melting conditions?

a) Pressure chamber
b) Vacuum chamber
c) Cooling chamber

d) Heating chamber

Solution: - b) Vacuum chamber

30) How does the LENS process achieve high detail resolution?

a) By using multiple lasers

b) By adjusting the powder particle size

c) By focusing the laser on a small melt pool

d) By increasing the substrate temperature

Solution: - c) By focusing the laser on a small melt pool

31) What does the electron beam's deflection coils do in the EBM process?

a) Control the melt pool size

b) Spread the powder evenly

c) Direct the beam to specific points

d) Increase the build platform temperature

Solution: - c) Direct the beam to specific points

32) What is a primary benefit of using Directed Energy Deposition (DED) technologies in manufacturing?

a) Large-scale production of simple geometries

b) High accuracy and complex geometries

c) Low cost for all types of materials

d) Rapid material solidification without any post-processing

Solution: - b) High accuracy and complex geometries

33) In the LENS process, what parameter is crucial for controlling the size of the melt pool?

a) Substrate Temperature

b) Laser Power

c) Powder Feeder Speed

d) Platform Movement Speed

Solution: - b) Laser Power

34) Which DED technology is known for using an electron beam to melt metal powder?

a) Laser Engineered Net Shaping (LENS)

b) Electron Beam Melting (EBM)

c) Laser Beam Melting (LBM)

d) Wire Arc Additive Manufacturing (WAAM)

Solution: - b) Electron Beam Melting (EBM)

35) **What is a significant challenge associated with Electron Beam Melting (EBM)?**

a) Limited to non-metal materials
b) High energy consumption and long processing time
c) Lack of ability to produce high-strength materials
d) Inability to create complex geometries

Solution: - b) High energy consumption and long processing time

36) **Which type of laser is used in Laser Engineered Net Shaping (LENS)?**

A. CO_2 Laser
b) Nd Laser
c) Fiber Laser
d) Excimer Laser

Solution: - c) Fiber Laser

37) **What is a common application of Directed Energy Deposition (DED) technologies in the aerospace industry?**

a) Manufacturing electronic circuits
b) Producing lightweight, high-strength components
c) Creating large-scale plastic prototypes
d) Designing consumer electronics

Solution: - b) Producing lightweight, high-strength components

38) **How does Directed Energy Deposition (DED) reduce material waste compared to traditional manufacturing methods?**

a) By using a single piece of metal
b) By adding only the required amount of material layer by layer
c) By recycling excess material automatically
d) By using cheaper raw materials

Solution: - b) By adding only the required amount of material layer by layer

39) **Which DED technology is typically used for repairing existing components like turbine blades?**

a) Laser Engineered Net Shaping (LENS)
b) Material Jetting (MJ)
c) Fused Deposition Modeling (FDM)

d) Powder Bed Fusion (PBF)

Solution: - a) Laser Engineered Net Shaping (LENS)

40) What is a key difference between Laser Beam Melting (LBM) and Laser Engineered Net Shaping (LENS)?

a) LBM uses a different type of laser than LENS

b) LBM is primarily used for polymer materials, while LENS is used for metals

c) LBM involves creating a melt pool on a powder bed, while LENS uses a focused laser to melt powder directly onto a substrate

d) LBM is used for larger build volumes compared to LENS

Solution: - c) LBM involves creating a melt pool on a powder bed, while LENS uses a focused laser to melt powder directly onto a substrate

Chapter 10

Laminated Object Manufacturing: LOM and UAM

10.1 Introduction to Laminated Object Manufacturing

Laminated object manufacturing (LOM) is a 3D printing technology that uses layers of material, such as paper or metal sheets, to create a part. This process involves bonding the layers together using an adhesive or other means, and then cutting or shaping the layers to form the desired shape. LOM is a versatile technology that can be used to produce a wide range of parts, from simple prototypes to complex production components. Sheet lamination, also called laminated object manufacturing (LOM), is a rapid prototyping process in which sheets of material are joined together to create an object. It is commonly used for building durable 3D objects with complex geometries. Figure 46 shows the equipment of Laminated Object Manufacturing.

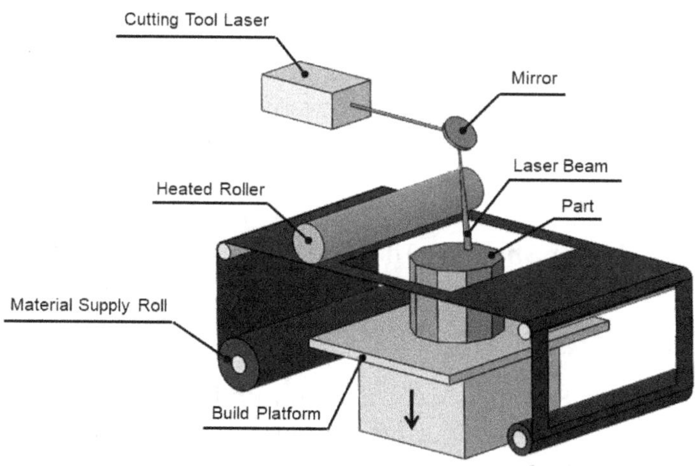

Figure 1: Laminated object manufacturing
[Source: https://www.mdpi.com/1996-1944/13/17/3895]

10.1.1 History of LOM

LOM has its roots in the 1980s, when researchers at the Massachusetts Institute of Technology (MIT) began exploring the use of laminated materials for rapid prototyping. The first LOM machines were developed in the early 1990s and were primarily used for producing paper-based models. Over the years, the technology has evolved to include a wider range of materials, including metals, ceramics, and composites.

10.1.2 How LOM Works

The LOM process involves several steps:

- **Layer Preparation**: A layer of material is prepared, such as cutting a sheet of paper or metal to the desired shape. The layer may be cut using a laser, waterjet, or other cutting tool.

- **Adhesive Application**: An adhesive is applied to the layer, if necessary. The adhesive may be a liquid, paste, or film, and is used to bond the layer to the previous layer.

- **Layer Bonding**: The layer is bonded to the previous layer, using the adhesive or other means. The bonding process may involve heat, pressure, or other methods to ensure a strong bond.

- **Cutting or Shaping**: The layer is cut or shaped to form the desired shape. This may involve using a laser, waterjet, or other cutting tool to remove excess material.

- **Repeat**: Steps 1-4 are repeated until the part is complete. The layers are built up incrementally, with each layer bonded to the previous layer to form the final part.

Figure 47 displays the working principle of Laminated Object Manufacturing.

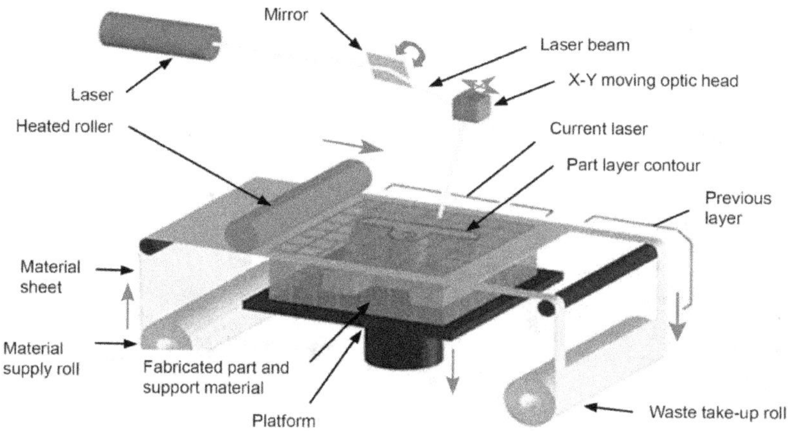

Figure 2: Principle of laminated object manufacturing

10.1.3 Types of LOM

There are several types of LOM, including:

- **Paper-Based LOM**: This type of LOM uses paper or paper-based materials, such as cardboard or paperboard, to create parts. Paper-based LOM is often used for rapid prototyping and model making.

- **Metal-Based LOM**: This type of LOM uses metal sheets or foils to create parts. Metal-based LOM is often used for producing functional components, such as engine parts or aerospace components.

- **Ceramic-Based LOM**: This type of LOM uses ceramic materials, such as clay or ceramic powder, to create parts. Ceramic-based LOM is often used for producing complex ceramic components, such as dental implants or aerospace components.

- **Composite-Based LOM**: This type of LOM uses composite materials, such as carbon fiber or fiberglass, to create parts. Composite-based LOM is often used for producing lightweight, high-strength components, such as aircraft parts or sports equipment.

Figure 48 displays the Products of Laminated Object Manufacturing.

Figure 3: Products of laminated object manufacturing

10.1.4 Advantages of LOM

LOM has several advantages, including:

- **High Accuracy**: LOM can produce parts with high accuracy and detail resolution.
- **Flexibility**: LOM can be used to produce a wide range of parts, from simple prototypes to complex production components.
- **Material Flexibility**: LOM can be used with a wide range of materials, including paper, metal, ceramic, and composite materials.
- **Low Cost**: LOM can be a cost-effective method for producing parts, especially for small batches or one-off prototypes.

10.1.5 Limitations of LOM

LOM also has some limitations, including:

- **Layer Thickness**: The layer thickness of LOM parts can be limited, which can affect the overall strength and durability of the part.

- **Material Properties**: The material properties of LOM parts can be affected by the bonding process, which can lead to reduced strength or durability.

- **Post-Processing**: LOM parts may require additional post-processing, such as machining or grinding, to achieve the desired surface finish.

10.1.6 Applications of LOM

LOM has a wide range of applications, including:

- **Rapid Prototyping**: LOM is often used for rapid prototyping, allowing designers and engineers to quickly test and iterate on their designs.

- **Production Components**: LOM can be used to produce functional components, such as engine parts or aerospace components.

- **Model Making**: LOM is often used for model making, allowing architects and designers to create detailed, accurate models of buildings and other structures.

- **Artistic Applications**: LOM can be used for artistic applications, such as creating complex sculptures or installations.

Laminated object manufacturing (LOM) is a versatile 3D printing technology that offers high accuracy, flexibility, and material flexibility. While it has some limitations, LOM is a powerful tool for producing a wide range of parts, from simple prototypes to complex production components. Whether one is a designer, engineer, or artist, LOM offers a unique and innovative way to create your ideas in 3D. Figure 49 illustrates the Laminated Object Manufacturing (LOM) process, a method used in additive manufacturing.

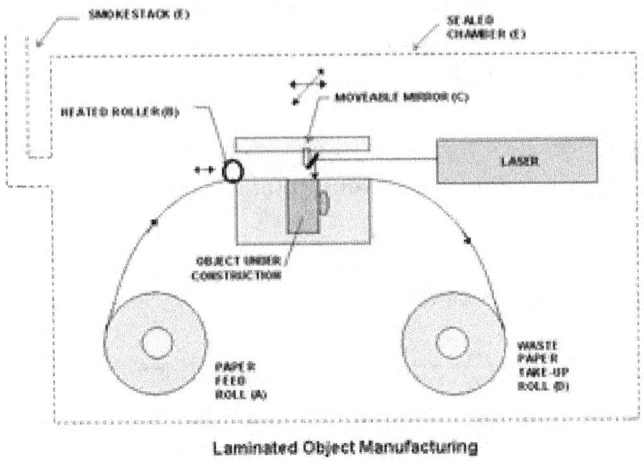

Figure 4: Laminated object manufacturing
[Source: https://www.livescience.com/40310-laminated-object-manufacturing.html]

10.2 Laminated Object Manufacturing (LOM)

Laminated object manufacturing (LOM) is a specific type of LOM technology that uses paper or cardboard sheets as the build material. The sheets are bonded together using an adhesive, and then cut or shaped to form the desired shape. LOM is a popular 3D printing technology due to its ease of use, low cost, and environmental benefits.

10.2.1 History of LOM

LOM has its roots in the 1980s, when researchers at the Massachusetts Institute of Technology (MIT) began exploring the use of laminated materials for rapid prototyping. The first LOM machines were developed in the early 1990s and were primarily used for producing paper-based models. Over the years, the technology has evolved to include a wider range of materials, including cardboard, foam, and even metal sheets.

10.2.2 How LOM Works

The LOM process involves several steps:

- **Paper Sheet Preparation**: A sheet of paper is prepared, such as cutting it to the desired shape. The sheet may be cut using a laser, waterjet, or other cutting tool.
- **Adhesive Application**: An adhesive is applied to the sheet. The adhesive may be a liquid, paste, or film, and is used to bond the sheet to the previous sheet.
- **Sheet Bonding**: The sheet is bonded to the previous sheet, using the adhesive. The bonding process may involve heat, pressure, or other methods to ensure a strong bond.
- **Cutting or Shaping**: The sheet is cut or shaped to form the desired shape. This may involve using a laser, waterjet, or other cutting tool to remove excess material.
- **Repeat**: Steps 1-4 are repeated until the part is complete. The sheets are built up incrementally, with each sheet bonded to the previous sheet to form the final part.

10.2.3 Advantages of LOM

LOM has several advantages, including:

- **Low Cost**: LOM is a low-cost 3D printing technology, making it accessible to a wide range of users.
- **Environmental Benefits**: LOM uses paper or cardboard sheets, which are biodegradable and recyclable, making it an environmentally friendly technology.
- **Ease of Use**: LOM is a relatively simple technology to use, making it accessible to users with limited technical expertise.

- **High Accuracy**: LOM can produce parts with high accuracy and detail resolution.

10.2.4 Limitations of LOM

LOM also has some limitations, including:

- **Material Properties**: The material properties of LOM parts can be limited, such as low strength and durability.
- **Layer Thickness**: The layer thickness of LOM parts can be limited, which can affect the overall strength and durability of the part.
- **Post-Processing**: LOM parts may require additional post-processing, such as machining or grinding, to achieve the desired surface finish.

10.2.5 Applications of LOM

LOM has a wide range of applications, including:

- **Rapid Prototyping**: LOM is often used for rapid prototyping, allowing designers and engineers to quickly test and iterate on their designs.
- **Model Making**: LOM is used for model making, allowing architects and designers to create detailed, accurate models of buildings and other structures.
- **Artistic Applications**: LOM can be used for artistic applications, such as creating complex sculptures or installations.
- **Educational Applications**: LOM is often used in educational settings, allowing students to learn about 3D printing and design principles.

10.2.6 Comparison of LOM and Other 3D Printing Technologies

LOM is often compared to other 3D printing technologies, such as fused deposition modeling (FDM) and stereolithography (SLA). While each technology has its own strengths and weaknesses, LOM is unique in its use of paper or cardboard sheets as the build material.

10.2.7 Future of LOM

The future of LOM is promising, with ongoing research and development aimed at improving the technology. Some areas of research include:

- **New Materials**: Researchers are exploring the use of new materials, such as bioplastics and recycled paper, to improve the sustainability and performance of LOM parts.
- **Improved Adhesives**: Researchers are developing new adhesives that can bond the sheets together more strongly and efficiently.
- **Increased Speed**: Researchers are working to increase the speed of the LOM process, making it more suitable for high-volume production.

Laminated object manufacturing (LOM) is a unique and innovative 3D printing technology that offers low cost, environmental benefits, and ease of use. While it has some limitations, LOM is a powerful tool for producing a wide range of parts, from simple prototypes to complex models. Whether you're a designer, engineer, or artist, LOM offers a new and exciting way to create your ideas in 3D. Figure 50 illustrates the Laminated Object Manufacturing (LOM) process, a method used in additive manufacturing. The diagram shows several key components: a laser, lenses, scanning mirrors, a crosshatch pattern on the part being manufactured, a heated roller, sheet material from a supply roll feeding into the system, and a take-up roll for excess material. The part being created is situated on a build platform. Arrows indicate the movement of materials and the direction of the laser's path. In this process, layers of adhesive-coated paper, plastic, or metal laminates are successively glued together and cut to shape with a laser or blade. The heated roller ensures the layers are firmly bonded, while the laser or blade cuts the desired shape. The waste material is collected by the take-up roll. This technique is known for its ability to produce relatively large parts quickly and cost-effectively, making it ideal for rapid prototyping.

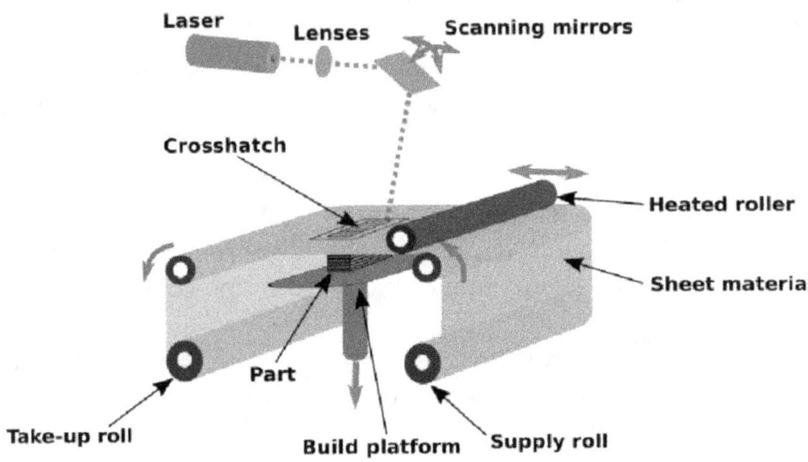

Figure 5: Laminated object manufacturing (LOM) [dx.doi.org/10.1007/978-3-030-10579-2_5]

10.3 Ultrasonic Additive Manufacturing (UAM)

Ultrasonic additive manufacturing (UAM) is a type of LOM technology that uses metal sheets as the build material. The sheets are bonded together using ultrasonic welding, and then cut or shaped to form the desired shape. UAM is a relatively new 3D printing technology that offers high accuracy, strong bonding, and low material waste.

10.3.1 History of UAM

UAM has its roots in the early 2000s, when researchers at the University of California, Los Angeles (UCLA) began exploring the use of ultrasonic welding for additive manufacturing. The first UAM machines were developed in the mid-2000s, and were primarily used for producing small, complex parts. Over the years, the technology has evolved to include a wider range of materials, including aluminum, copper, and titanium.

10.3.2 How UAM Works

The UAM process involves several steps:

- **Metal Sheet Preparation**: A sheet of metal is prepared, such as cutting it to the desired shape. The sheet may be cut using a laser, waterjet, or other cutting tool.

- **Ultrasonic Welding**: The sheet is bonded to the previous sheet using ultrasonic welding. The ultrasonic welding process involves using high-frequency ultrasonic vibrations to create a strong bond between the sheets.

- **Cutting or Shaping**: The sheet is cut or shaped to form the desired shape. This may involve using a laser, waterjet, or other cutting tool to remove excess material.

- **Repeat**: Steps 1-3 are repeated until the part is complete. The sheets are built up incrementally, with each sheet bonded to the previous sheet to form the final part.

10.3.3 Advantages of UAM

UAM has several advantages, including:

- **High Accuracy**: UAM can produce parts with high accuracy and detail resolution.

- **Strong Bonding**: The ultrasonic welding process creates a strong bond between the sheets, resulting in parts with high strength and durability.

- **Low Material Waste**: UAM produces minimal material waste, making it an environmentally friendly technology.
- **Flexibility**: UAM can be used with a wide range of metals, including aluminum, copper, and titanium.

10.3.4 Limitations of UAM

UAM also has some limitations, including:

- **Material Properties**: The material properties of UAM parts can be limited, such as low ductility and high porosity.
- **Layer Thickness**: The layer thickness of UAM parts can be limited, which can affect the overall strength and durability of the part.
- **Post-Processing**: UAM parts may require additional post-processing, such as machining or grinding, to achieve the desired surface finish.

10.3.5 Applications of UAM

UAM has a wide range of applications, including:

- **Aerospace**: UAM is often used in the aerospace industry to produce lightweight, high-strength components, such as engine parts and structural components.
- **Automotive**: UAM is used in the automotive industry to produce complex components, such as engine parts and transmission components.
- **Medical**: UAM is used in the medical industry to produce custom implants and surgical instruments.
- **Consumer Products**: UAM is used in the consumer products industry to produce complex components, such as phone cases and laptop parts.

10.3.6 Comparison of UAM and Other 3D Printing Technologies

UAM is often compared to other 3D printing technologies, such as selective laser sintering (SLS) and electron beam melting (EBM). While each technology has its own strengths and weaknesses, UAM is unique in its use of ultrasonic welding to bond the sheets together.

10.3.7 Future of UAM

The future of UAM is promising, with ongoing research and development aimed at improving the technology. Some areas of research include:

- **New Materials**: Researchers are exploring the use of new materials, such as advanced alloys and composites, to improve the performance and properties of UAM parts.

- **Improved Ultrasonic Welding**: Researchers are developing new ultrasonic welding techniques to improve the strength and reliability of the bonds between the sheets.

- **Increased Speed**: Researchers are working to increase the speed of the UAM process, making it more suitable for high-volume production.

Ultrasonic additive manufacturing (UAM) is a powerful 3D printing technology that offers high accuracy, strong bonding, and low material waste. While it has some limitations, UAM is a promising technology for producing a wide range of parts, from simple prototypes to complex production components. Whether you're a designer, engineer, or manufacturer, UAM offers a new and exciting way to create your ideas in 3D. Figure 51 illustrates the Ultrasonic Additive Manufacturing (UAM) process, a hybrid additive manufacturing technique. The diagram shows a rotating sonotrode applying normal force and ultrasonic oscillation to a foil that is to be welded onto a metal substrate. The foil and the substrate are supported by an anvil. The welding occurs at the surface contact points between the foil and the substrate, where the ultrasonic vibrations break up the oxide layer, allowing the metals to bond without melting. This process preserves the original properties of the materials and is useful for creating complex, multi-material parts.

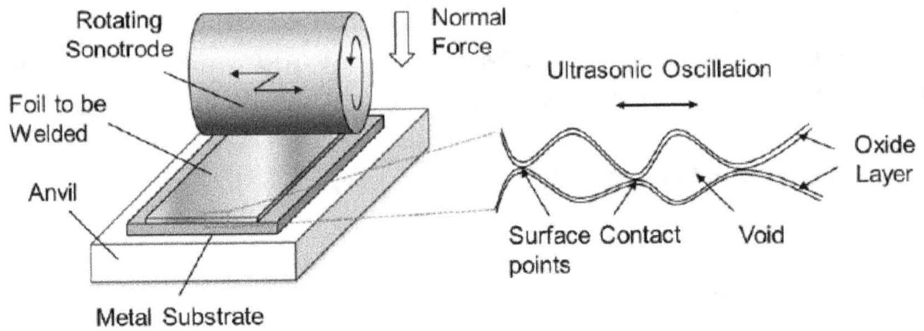

Figure 6: Ultrasonic additive manufacturing (UAM)
[Source: http://utw10945.utweb.utexas.edu/sites/default/files/2014-068-Li.pdf]

10.4 Applications and Limitations

Laminated object manufacturing (LOM) technologies, such as LOM and UAM, are commonly used for rapid prototyping, production of functional parts, and creation of complex geometries. These technologies offer high resolution and accuracy, making them suitable for producing fine details and intricate shapes.

10.4.1 Applications of LOM Technologies

LOM technologies have a wide range of applications, including:

- **Rapid Prototyping**: LOM technologies are often used for rapid prototyping, allowing designers and engineers to quickly test and iterate on their designs. The high accuracy and resolution of LOM technologies make them ideal for producing detailed prototypes.

- **Production of Functional Parts**: LOM technologies can be used to produce functional parts, such as mechanical components, electrical components, and medical devices.

The high accuracy and resolution of LOM technologies make them suitable for producing parts with complex geometries and fine details.

- **Creation of Complex Geometries:** LOM technologies are ideal for creating complex geometries, such as curved surfaces, intricate shapes, and hollow structures. The layer-by-layer build process of LOM technologies allows for the creation of complex geometries that would be difficult or impossible to produce using traditional manufacturing methods.

- **Artistic Applications:** LOM technologies are also used in artistic applications, such as creating sculptures, jewelry, and other decorative items. The high accuracy and resolution of LOM technologies make them ideal for producing intricate and detailed designs.

10.4.2 Limitations of LOM Technologies

While LOM technologies offer many advantages, they also have some limitations. For example:

- **Material Limitations:** The materials used in LOM technologies are typically limited to paper, cardboard, or metal sheets, which may have limited mechanical properties and may not be suitable for high-temperature or high-stress applications.

- **Build Volume Limitations:** The build volumes of LOM technologies are generally smaller than those of other 3D printing technologies, such as powder bed fusion or material jetting. This can limit the size of the parts that can be produced using LOM technologies.

- **Layer Thickness Limitations:** The layer thickness of LOM technologies can be limited, which can affect the overall strength and durability of the part.

- **Post-Processing Requirements**: LOM parts may require additional post-processing, such as machining or grinding, to achieve the desired surface finish.

- **Cost and Complexity**: LOM technologies can be more expensive and complex than other 3D printing technologies, which can limit their accessibility to some users.

10.4.3 Comparison of LOM Technologies with Other 3D Printing Technologies

LOM technologies are often compared to other 3D printing technologies, such as fused deposition modeling (FDM), stereolithography (SLA), and selective laser sintering (SLS). While each technology has its own strengths and weaknesses, LOM technologies are unique in their use of laminated materials and layer-by-layer build process.

10.4.4 Future of LOM Technologies

The future of LOM technologies is promising, with ongoing research and development aimed at improving the technology. Some areas of research include:

- **New Materials**: Researchers are exploring the use of new materials, such as advanced composites and nanomaterials, to improve the performance and properties of LOM parts.

- **Improved Build Speed**: Researchers are working to increase the build speed of LOM technologies, making them more suitable for high-volume production.

- **Increased Build Volume**: Researchers are developing new LOM technologies with larger build volumes, allowing for the production of larger parts and assemblies.

- **Improved Accuracy and Resolution: -** Researchers are working to improve the accuracy and resolution of LOM technologies, making them suitable for producing even more complex and detailed parts.

Laminated object manufacturing (LOM) technologies, such as LOM and UAM, offer high resolution and accuracy, making them suitable for producing fine details and intricate shapes. While they have some limitations, LOM technologies are a powerful tool for rapid prototyping, production of functional parts, and creation of complex geometries. As the technology continues to evolve, we can expect to see even more innovative applications of LOM technologies in a wide range of industries.

10.5 Questions

10.5.1 Long Response Questions

1. What are the key differences between laminated object manufacturing (LOM) and ultrasonic additive manufacturing (UAM)?
2. What are some common applications of laminated object manufacturing technologies?
3. What are the limitations of laminated object manufacturing technologies?
4. How does the resolution and accuracy of laminated object manufacturing technologies compare to other 3D printing technologies?
5. What are some potential future developments in laminated object manufacturing technologies?

10.5.2 Multiple Choice Questions

1) What material is primarily used in Laminated Object Manufacturing (LOM)?

a) Metal sheets
b) Paper or cardboard sheets
c) Plastic filament
d) Ceramic powder

Solution: - b) Paper or cardboard sheets

2) **In Laminated Object Manufacturing (LOM), what is used to bond the layers together?**
   a) Ultrasonic welding
   b) Adhesive
   c) Heat treatment
   d) Laser sintering
   Solution: - b) Adhesive
3) **Which type of LOM uses metal sheets or foils as the build material?**
   a) Paper-Based LOM
   b) Metal-Based LOM
   c) Ceramic-Based LOM
   d) Composite-Based LOM
   Solution: - b) Metal-Based LOM
4) **What is a common application of LOM technology?**
   a) Creating phone cases
   b) Producing dental implants
   c) Rapid prototyping
   d) Printing 3D food
   Solution: - c) Rapid prototyping
5) **Which of the following is NOT a type of LOM?**
   a) Paper-Based LOM
   b) Metal-Based LOM
   c) Composite-Based LOM
   d) Liquid-Based LOM
   Solution: - d) Liquid-Based LOM
6) **What is a key advantage of Laminated Object Manufacturing (LOM)?**
   a) High material cost
   b) Low accuracy
   c) High flexibility
   d) Complex post-processing
   Solution: - c) High flexibility
7) **Which process is used in Ultrasonic Additive Manufacturing (UAM) to bond metal sheets?**
   a) Laser cutting
   b) Ultrasonic welding
   c) Stereolithography
   d) Fused deposition modeling

Solution: - b) Ultrasonic welding

8) **Which material is commonly used in Ultrasonic Additive Manufacturing (UAM)?**
   a) Plastic filament
   b) Paper sheets
   c) Metal sheets
   d) Ceramic powder
   Solution: - c) Metal sheets

9) **What is one of the primary advantages of UAM?**
   a) High material waste
   b) Low bonding strength
   c) High accuracy
   d) Limited material flexibility
   Solution: - c) High accuracy

10) **Which technology is known for using high-frequency ultrasonic vibrations to bond materials?**
   a) Laminated Object Manufacturing
   b) Ultrasonic Additive Manufacturing
   c) Stereolithography
   d) Fused Deposition Modeling
   Solution: - b) Ultrasonic Additive Manufacturing

11) **Which step in the LOM process involves bonding the layers together?**
   a) Layer preparation
   b) Adhesive application
   c) Layer bonding
   d) Cutting or shaping
   Solution: - c) Layer bonding

12) **In which decade did LOM technology begin to evolve with the development of the first machines?**
   a) 1970s
   b) 1980s
   c) 1990s
   d) 2000s
   Solution: - c) 1990s

13) **What is one limitation of LOM technology related to material properties?**
   a) High material cost
   b) Limited material flexibility

c) Reduced strength and durability
d) Excessive material waste
Solution: - c) Reduced strength and durability

14) What type of LOM uses ceramic materials like clay or ceramic powder?
a) Paper-Based LOM
b) Metal-Based LOM
c) Ceramic-Based LOM
d) Composite-Based LOM
Solution: - c) Ceramic-Based LOM

15) What type of cutting tool is often used in the LOM process?
a) Ultrasonic cutter
b) Laser cutter
c) Waterjet cutter
d) Plasma cutter
Solution: - b) Laser cutter

16) What is a common application of UAM in the aerospace industry?
a) Producing engine parts
b) Creating plastic prototypes
c) Printing decorative items
d) Making paper models
Solution: - a) Producing engine parts

17) Which step in UAM involves the use of ultrasonic vibrations to bond sheets?
a) Metal sheet preparation
b) Ultrasonic welding
c) Cutting or shaping
d) Layer bonding
Solution: - b) Ultrasonic welding

18) What is a common material used in UAM that allows for strong bonding?
a) Titanium
b) Plastic
c) Paper
d) Ceramic
Solution: - a) Titanium

19) What is one limitation of UAM related to material properties?

a) High material cost
b) Low ductility and high porosity
c) High material waste
d) Limited speed

Solution: - b) Low ductility and high porosity

20) What technology is unique for using metal sheets and ultrasonic welding in its process?

a) Laminated Object Manufacturing
b) Electron Beam Melting
c) Ultrasonic Additive Manufacturing
d) Selective Laser Sintering

Solution: - c) Ultrasonic Additive Manufacturing

21) What is a significant environmental benefit of LOM technology?

a) High energy consumption
b) Use of non-recyclable materials
c) Use of biodegradable and recyclable materials
d) High material waste

Solution: - c) Use of biodegradable and recyclable materials

22) Which aspect of UAM is improved by developing new ultrasonic welding techniques?

a) Material cost
b) Bond strength and reliability
c) Layer thickness
d) Build volume

Solution: - b) Bond strength and reliability

23) What is one research area aimed at enhancing the future of LOM technology?

a) Reducing build speed
b) Increasing material waste
c) Exploring new materials like bioplastics
d) Limiting material types

Solution: - c) Exploring new materials like bioplastics

24) What challenge does the layer thickness in LOM pose for part production?

a) Increased material waste

b) Reduced strength and durability
c) Increased post-processing time
d) Limited accuracy

Solution: - b) Reduced strength and durability

25) How does UAM preserve the original properties of metals during the manufacturing process?

a) By melting the metals
b) By using high-temperature processes
c) By applying ultrasonic vibrations without melting
d) By chemically treating the metals

Solution: - c) By applying ultrasonic vibrations without melting

26) Which LOM technology development focuses on increasing the build volume?

a) Research on improved adhesives
b) Development of new build materials
c) Development of larger LOM machines
d) Improvements in build speed

Solution: - c) Development of larger LOM machines

27) What is a critical limitation of UAM that affects the final product quality?

a) High build speed
b) Limited material flexibility
c) High porosity and low ductility
d) Excessive build volume

Solution: - c) High porosity and low ductility

28) What aspect of LOM is often enhanced by using new adhesives?

a) Material properties
b) Layer bonding strength and efficiency
c) Build speed
d) Accuracy of cutting tools

Solution: - b) Layer bonding strength and efficiency

29) Which industry benefits from UAM's ability to produce high-strength components with minimal material waste?

a) Food industry
b) Textile industry
c) Aerospace industry
d) Furniture industry

Solution: - c) Aerospace industry

30) **What is one of the primary goals in the future research of UAM technology?**
   a) Increasing material waste
   b) Decreasing the material types used
   c) Improving ultrasonic welding techniques
   d) Reducing the material strength
   Solution: - c) Improving ultrasonic welding techniques

31) **Which of the following is NOT a type of Laminated Object Manufacturing (LOM)?**
   a) Paper-Based LOM
   b) Metal-Based LOM
   c) Composite-Based LOM
   d) Liquid Resin-Based LOM
   Solution: - d) Liquid Resin-Based LOM

32) **What is a primary advantage of Laminated Object Manufacturing (LOM)?**
   a) High material flexibility
   b) High resolution and accuracy
   c) Low cost
   d) No need for post-processing
   Solution: - c) Low cost

33) **What does the "UAM" in Ultrasonic Additive Manufacturing stand for?**
   a) Unique Additive Manufacturing
   b) Ultrasonic Alignment Manufacturing
   c) Ultrasonic Additive Manufacturing
   d) Ultra-Advanced Manufacturing
   Solution: - c) Ultrasonic Additive Manufacturing

34) **Which material is commonly used in Ultrasonic Additive Manufacturing (UAM)?**
   a) Paper
   b) Plastic
   c) Metal
   d) Ceramic
   Solution: - c) Metal

35) **What is a key feature of the Ultrasonic Additive Manufacturing (UAM) process?**

a) Use of laser sintering
b) Ultrasonic welding for bonding
c) Inkjet deposition
d) Material extrusion

Solution: - b) Ultrasonic welding for bonding

36) Which step is NOT part of the Laminated Object Manufacturing (LOM) process?

a) Layer Preparation
b) Ultrasonic Welding
c) Cutting or Shaping
d) Adhesive Application

Solution: - b) Ultrasonic Welding

37) What is a primary limitation of Laminated Object Manufacturing (LOM)?

a) High material cost
b) Limited layer thickness
c) Complexity in material bonding
d) Low accuracy in detailed parts

Solution: - b) Limited layer thickness

38) Which application is NOT commonly associated with Laminated Object Manufacturing (LOM)?

a) Aerospace components
b) Rapid prototyping
c) Model making
d) Artistic sculptures

Solution: - a) Aerospace components

39) What is a notable advantage of Ultrasonic Additive Manufacturing (UAM)?

a) High material waste
b) Limited metal types
c) Strong bonding between sheets
d) Large build volumes

Solution: - c) Strong bonding between sheets

40) What future improvement is being researched for Ultrasonic Additive Manufacturing (UAM)?

a) New types of adhesives
b) Improved ultrasonic welding techniques
c) Increased build volume of LOM
d) Development of liquid resin technologies

Solution: - b) Improved ultrasonic welding techniques

Part 3

Materials and Design

Chapter 11

Hybrid and Multi-Material 3D Printing

11.1 Introduction to Hybrid and Multi-Material 3D Printing

Hybrid and multi-material 3D printing technologies are a class of additive manufacturing (AM) processes that combine different materials or AM processes to create a single part. These technologies enable the creation of complex geometries and functional parts with unique properties, such as varying material properties, textures, or colors.

11.1.1 Definition of Hybrid and Multi-Material 3D Printing

Hybrid and multi-material 3D printing can be defined as the combination of two or more different materials or AM processes to create a single part. This can include the combination of different materials, such as metals, polymers, and ceramics, or the combination of different AM processes, such as fused deposition modeling (FDM) and stereolithography (SLA).

11.1.2 Advantages of Hybrid and Multi-Material 3D Printing

Hybrid and multi-material 3D printing technologies offer several advantages, including:

- **Increased Design Flexibility**: The ability to combine different materials and AM processes enables the creation of complex geometries and functional parts with unique properties. This allows designers and engineers to push the boundaries of what is possible with traditional manufacturing methods.

- **Improved Part Performance**: The combination of different materials can result in parts with improved mechanical, thermal, or electrical properties. For example, a part made from a combination of metal and polymer materials may have improved strength and durability, while also being lighter and more flexible.

- **Reduced Material Waste**: Hybrid and multi-material 3D printing can reduce material waste by using multiple materials with different properties. This can be particularly beneficial in industries where material waste is a significant concern, such as aerospace and automotive.

- **Enhanced Aesthetics**: Hybrid and multi-material 3D printing can also be used to create parts with unique textures, colors, and finishes. This can be particularly beneficial in industries where aesthetics is important, such as consumer products and architecture.

- **Improved Sustainability**: Hybrid and multi-material 3D printing can also be used to create parts with improved sustainability. For example, a part made from a combination of recycled materials and bioplastics may have a reduced environmental impact.

11.1.3 Types of Hybrid and Multi-Material 3D Printing

There are several types of hybrid and multi-material 3D printing, including:

- **Material Jetting**: This process involves the deposition of multiple materials through a jetting process. Material jetting is commonly used in applications such as jewelry making and dental implants.

- **Fused Deposition Modeling (FDM)**: This process involves the deposition of multiple materials through a heated extrusion process. FDM is commonly used in applications such as aerospace and automotive.

- **Stereolithography (SLA)**: This process involves the deposition of multiple materials through a laser-based process. SLA is commonly used in applications such as prototyping and model making.

- **Directed Energy Deposition (DED)**: This process involves the deposition of multiple materials through a directed energy source, such as a laser or electron beam. DED is commonly used in applications such as aerospace and energy.

11.1.4 Applications of Hybrid and Multi-Material 3D Printing

Hybrid and multi-material 3D printing has a wide range of applications, including:

- **Aerospace**: Hybrid and multi-material 3D printing is used in the aerospace industry to create lightweight, high-performance parts with unique properties.

- **Automotive**: Hybrid and multi-material 3D printing is used in the automotive industry to create parts with improved mechanical and thermal properties.

- **Consumer Products**: Hybrid and multi-material 3D printing is used in the consumer products industry to create parts with unique textures, colors, and finishes.

- **Medical**: Hybrid and multi-material 3D printing is used in the medical industry to create implants, prosthetics, and other medical devices with unique properties.

- **Energy**: Hybrid and multi-material 3D printing is used in the energy industry to create parts with improved thermal and electrical properties.

11.1.5 Challenges and Limitations of Hybrid and Multi-Material 3D Printing

While hybrid and multi-material 3D printing offers many advantages, it also has some challenges and limitations, including:

- **Material Compatibility**: One of the biggest challenges of hybrid and multi-material 3D printing is ensuring that the different materials are compatible with each other.

- **Process Complexity**: Hybrid and multi-material 3D printing can be a complex process, requiring specialized equipment and expertise.

- **Cost**: Hybrid and multi-material 3D printing can be more expensive than traditional manufacturing methods, particularly for small batches or one-off parts.

- **Scalability**: Hybrid and multi-material 3D printing can be difficult to scale up for large production runs.

Figure 52 illustrates various aspects of Hybrid and Multi-Material 3D Printing. The collage includes photographs, diagrams, and illustrations showcasing the equipment used, the processes involved, and examples of products created using these advanced techniques. Hybrid and multi-material 3D printing allows for the combination of different materials within a single print, enabling the creation of objects with customized properties and functionalities. This versatility is particularly useful in fields such as aerospace, medical devices, and consumer products, where complex geometries and material properties are often required.

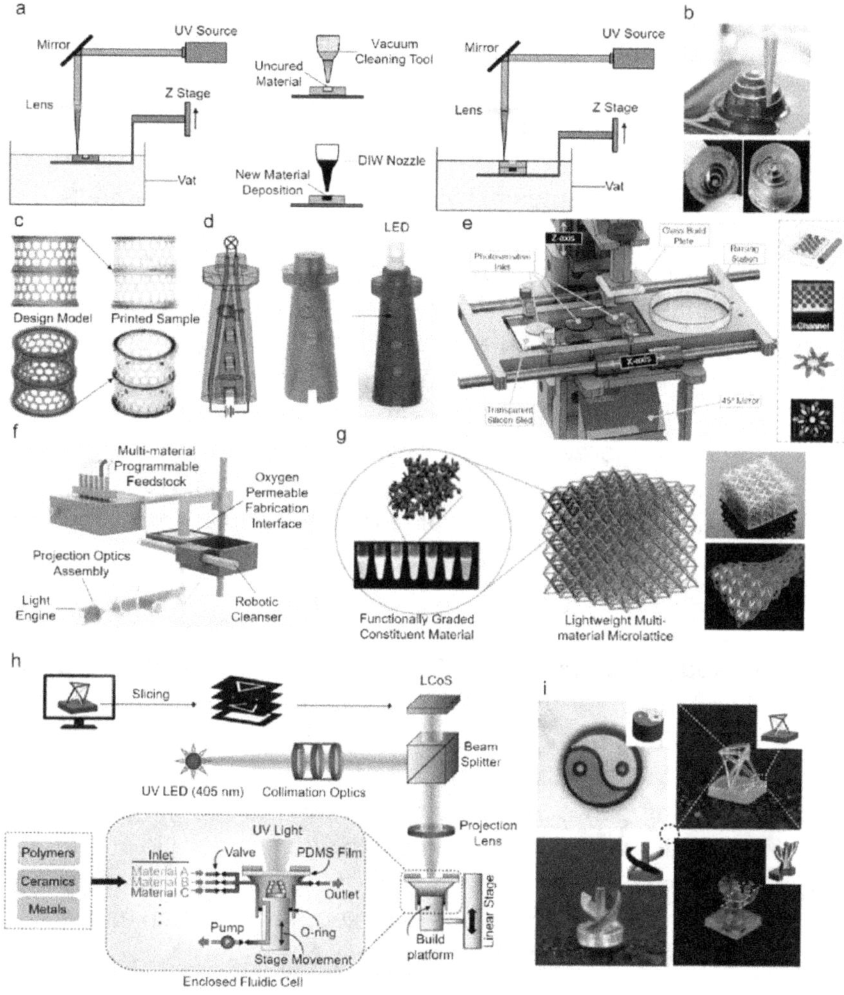

Figure 1: Hybrid and multi-material 3D printing
[Source: https://www.mdpi.com/2073-4360/14/21/4635]

11.2 Hybrid AM Processes

Hybrid AM processes combine different AM processes, such as powder bed fusion and directed energy deposition, to create a single part. These processes enable the creation of complex geometries and functional parts with unique properties. By combining different AM processes, hybrid AM processes can overcome the limitations of individual processes and offer a wider range of material properties, geometries, and functionalities.

11.2.1 Examples of Hybrid AM Processes

There are several examples of hybrid AM processes, including:

- **Powder Bed Fusion + Directed Energy Deposition**: Combining powder bed fusion with directed energy deposition enables the creation of parts with complex geometries and varying material properties. Powder bed fusion can be used to create the bulk of the part, while directed energy deposition can be used to add features or details with specific material properties. This hybrid process is particularly useful for creating parts with complex geometries, such as turbine blades or aerospace components.

- **Fused Deposition Modeling + Stereolithography**: Combining fused deposition modeling with stereolithography enables the creation of parts with complex geometries and high-resolution details. Fused deposition modeling can be used to create the bulk of the part, while stereolithography can be used to add fine details or features with high accuracy. This hybrid process is particularly useful for creating parts with complex geometries, such as medical devices or consumer products.

- **Selective Laser Sintering + Binder Jetting**: Combining selective laser sintering with binder jetting enables the creation of parts with complex geometries and varying material properties. Selective laser sintering can be used to create the bulk of the part, while binder jetting can be used to add features or details with specific material properties. This hybrid process is particularly useful for creating parts with complex geometries, such as automotive components or industrial equipment.

- **Laminated Object Manufacturing + Fused Deposition Modeling:** Combining laminated object manufacturing with fused deposition modeling enables the creation of parts with complex geometries and varying material properties. Laminated object manufacturing can be used to create the bulk of the part, while fused deposition modeling can be used to add features or details with specific material properties. This hybrid process is particularly useful for creating parts with complex geometries, such as architectural models or prototypes.

11.2.2 Advantages of Hybrid AM Processes

Hybrid AM processes offer several advantages, including:

- **Increased Design Flexibility**: Hybrid AM processes enable the creation of complex geometries and functional parts with unique properties, allowing designers and engineers to push the boundaries of what is possible with traditional manufacturing methods.

- **Improved Part Performance**: Hybrid AM processes can create parts with improved mechanical, thermal, or electrical properties, making them suitable for a wide range of applications.

- **Reduced Material Waste:** Hybrid AM processes can reduce material waste by using multiple materials with different properties, making them more sustainable and environmentally friendly.

- **Enhanced Aesthetics**: Hybrid AM processes can create parts with unique textures, colors, and finishes, making them suitable for applications where aesthetics are important.

11.2.3 Challenges and Limitations of Hybrid AM Processes

While hybrid AM processes offer many advantages, they also have some challenges and limitations, including:

- **Process Complexity**: Hybrid AM processes can be complex and require specialized equipment and expertise, making them more difficult to implement and operate.
- **Material Compatibility**: Hybrid AM processes require the use of multiple materials with different properties, which can be challenging to combine and process.
- **Cost**: Hybrid AM processes can be more expensive than traditional manufacturing methods, particularly for small batches or one-off parts.
- **Scalability**: Hybrid AM processes can be difficult to scale up for large production runs, making them more suitable for small-batch or prototype production.

11.2.4 Future of Hybrid AM Processes

The future of hybrid AM processes is promising, with ongoing research and development aimed at improving the technology. Some areas of research include:

- **New Material Combinations**: Researchers are exploring new material combinations and properties to create hybrid AM processes with improved performance and functionality.
- **Improved Process Control**: Researchers are working to improve process control and monitoring to ensure consistent and reliable results.
- **Increased Scalability**: Researchers are developing new hybrid AM processes that can be scaled up for large production runs, making them more suitable for industrial applications.
- **Cost Reduction**: Researchers are working to reduce the cost of hybrid AM processes, making them more accessible to a wider range of industries and applications.

Hybrid AM processes offer a powerful tool for creating complex geometries and functional parts with unique properties. By combining different AM processes, hybrid AM processes can overcome the limitations of individual processes and offer a wider range of material properties, geometries, and functionalities. While there are some challenges and limitations to this technology, the future of hybrid AM processes is promising, with ongoing research and development aimed at improving the technology. Figure 53 illustrates the Hybrid Additive Manufacturing (AM) Process, showcasing the integration of various technologies to enhance manufacturing capabilities. The flowchart is divided into three main columns:

- **3D Printing Technology**: This column represents the core 3D printing methods used in hybrid AM processes.

- **Material Extrusion Technology:** This column lists several post-processing techniques such as Abrasive Finishing, Chemical Finishing, CNC Machining, Hot Air Jet Polishing, Roller Burnishing, Media Blasting & Peening, Painting & Coating, and Resin Infiltration. These techniques are used to refine and enhance the printed parts.

- **Metal Additive Processes**: This column is further divided into two sub-columns:

- **Deposited Energy Manufacturing Processes**: Includes CNC Machining, Rolling & Analogous Mechanical Forming, and Shot Peening.

- **Powder Bed Fusion Technology**: Includes Thermal Assisted CNC Machining, Chemical Assisted CNC Machining, and Hot Isostatic Pressing.

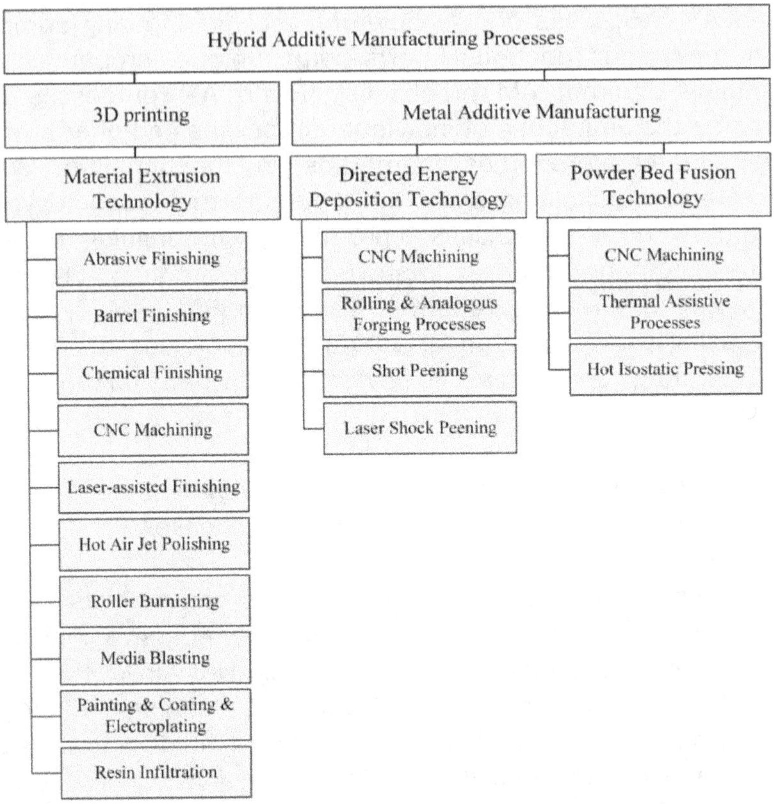

Figure 2: Hybrid AM process
[Source: https://link.springer.com/article/10.1007/s00170-021-06688-1]

11.3 Multi-Material AM Processes

Multi-material AM processes use multiple materials with different properties to create a single part. These processes enable the creation of parts with varying material properties, textures, or colors. By using multiple materials, designers and engineers can create parts with unique properties that cannot be achieved with a single material.

11.3.1 Examples of Multi-Material AM Processes

There are several examples of multi-material AM processes, including:

- **Material Jetting**: Material jetting uses multiple materials with different properties, such as color or texture, to create a single part. This process involves depositing droplets of material onto a build platform, where they are fused together to form a solid part. Material jetting is commonly used in applications such as jewelry making, dental implants, and consumer products.

- **Fused Deposition Modeling with Multiple Materials**: Fused deposition modeling can use multiple materials with different properties, such as color or texture, to create a single part. This process involves extruding multiple materials through a heated nozzle, which are then deposited onto a build platform. Fused deposition modeling with multiple materials is commonly used in applications such as aerospace, automotive, and industrial equipment.

- **Stereolithography with Multiple Materials**: Stereolithography can use multiple materials with different properties, such as color or texture, to create a single part. This process involves using a laser to cure liquid resin, which is then built-up layer by layer to form a solid part. Stereolithography with multiple materials is commonly used in applications such as prototyping, model making, and medical devices.

- **Binder Jetting with Multiple Materials**: Binder jetting can use multiple materials with different properties, such as color or texture, to create a single part. This process involves depositing a liquid binder onto a bed of powder, which is then built-up layer by layer to form a solid part. Binder jetting with multiple materials is commonly used in applications such as architectural models, prototypes, and industrial equipment.

11.3.2 Advantages of Multi-Material AM Processes

Multi-material AM processes offer several advantages, including:

- **Increased Design Flexibility**: Multi-material AM processes enable the creation of parts with varying material properties, textures, or colors, allowing designers and engineers to push the boundaries of what is possible with traditional manufacturing methods.

- **Improved Part Performance**: Multi-material AM processes can create parts with improved mechanical, thermal, or electrical properties, making them suitable for a wide range of applications.

- **Reduced Material Waste**: Multi-material AM processes can reduce material waste by using multiple materials with different properties, making them more sustainable and environmentally friendly.

- **Enhanced Aesthetics**: Multi-material AM processes can create parts with unique textures, colors, and finishes, making them suitable for applications where aesthetics are important.

11.3.3 Challenges and Limitations of Multi-Material AM Processes

While multi-material AM processes offer many advantages, they also have some challenges and limitations, including:

- **Material Compatibility**: Multi-material AM processes require the use of multiple materials with different properties, which can be challenging to combine and process.

- **Process Complexity**: Multi-material AM processes can be complex and require specialized equipment and expertise, making them more difficult to implement and operate.

- **Cost**: Multi-material AM processes can be more expensive than traditional manufacturing methods, particularly for small batches or one-off parts.

- **Scalability**: Multi-material AM processes can be difficult to scale up for large production runs, making them more suitable for small-batch or prototype production.

11.3.4 Applications of Multi-Material AM Processes

Multi-material AM processes have a wide range of applications, including:

- **Aerospace**: Multi-material AM processes are used in the aerospace industry to create lightweight, high-performance parts with unique properties.

- **Automotive**: Multi-material AM processes are used in the automotive industry to create parts with improved mechanical and thermal properties.

- **Consumer Products**: Multi-material AM processes are used in the consumer products industry to create parts with unique textures, colors, and finishes.

- **Medical**: Multi-material AM processes are used in the medical industry to create implants, prosthetics, and other medical devices with unique properties.

- **Industrial Equipment**: Multi-material AM processes are used in the industrial equipment industry to create parts with improved mechanical and thermal properties.

11.3.5 Future of Multi-Material AM Processes

The future of multi-material AM processes is promising, with ongoing research and development aimed at improving the technology. Some areas of research include:

- **New Material Combinations**: Researchers are exploring new material combinations and properties to create multi-material AM processes with improved performance and functionality.

- **Improved Process Control**: Researchers are working to improve process control and monitoring to ensure consistent and reliable results.
- **Increased Scalability**: Researchers are developing new multi-material AM processes that can be scaled up for large production runs, making them more suitable for industrial applications.
- **Cost Reduction**: Researchers are working to reduce the cost of multi-material AM processes, making them more accessible to a wider range of industries and applications.

Multi-material AM processes offer a powerful tool for creating complex geometries and functional parts with unique properties. By using multiple materials, designers and engineers can create parts with varying material properties, textures, or colors, making them suitable for a wide range of applications. While there are some challenges and limitations to this technology, the future of multi-material AM processes is promising, with ongoing research and development aimed at improving the technology. Figure 54 illustrates various Multi-Material Additive Manufacturing (AM) processes. The diagram showcases six different techniques, each labeled from a) to f):

a) Vat Photopolymerization: Utilizes a vat of liquid photopolymer resin that is selectively cured by light-activated polymerization. b) Material Jetting: Deposits droplets of build material selectively. c) Binder Jetting: Deposits a liquid bonding agent to join powder materials. d) Material Extrusion (FDM): Dispenses thermoplastic polymers through a nozzle or orifice. e) Material Extrusion: Extrudes a continuous filament of thermoplastic material from a nozzle or orifice. f) Powder Bed Fusion: Uses thermal energy to selectively fuse regions of a powder bed. g) Directed Energy Deposition: Employs focused thermal energy to fuse materials by melting them as they are deposited. h) Sheet Lamination: Bonds sheets of material to form an object.

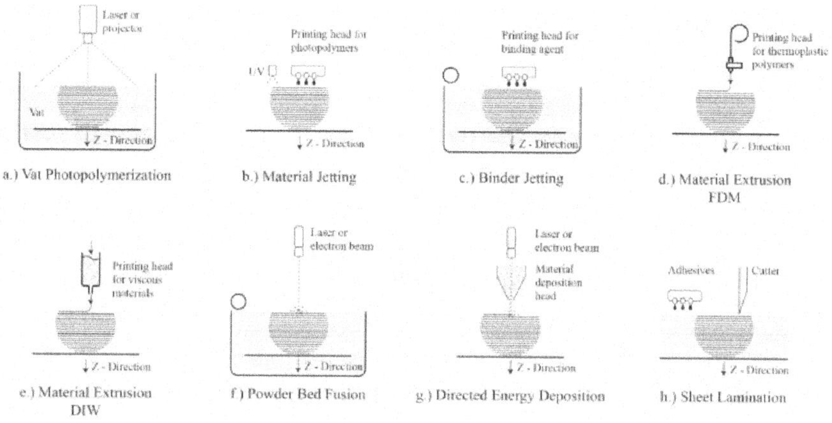

Figure 3: Multi-material AM process
[Source:
https://www.sciencedirect.com/science/article/pii/S2352710221014613]

11.4 Applications and Limitations

Hybrid and multi-material 3D printing technologies have several applications across various industries, including aerospace, biomedical, consumer products, and more. These technologies offer a range of benefits, including the ability to create complex geometries, customize products, and improve performance.

Aerospace Applications

Hybrid and multi-material 3D printing can be used to create lightweight, high-performance parts for aerospace applications. The ability to combine different materials with unique properties enables the creation of parts that meet the demanding requirements of the aerospace industry. For example, hybrid 3D printing can be used to create aircraft components with improved strength-to-weight ratios, reducing fuel consumption and emissions. Additionally, multi-material 3D printing can be used to create parts with integrated functionality, such as sensors or antennas, reducing the need for assembly and increasing overall system performance.

Biomedical Applications

Hybrid and multi-material 3D printing can be used to create customized implants and prosthetics with unique properties. The ability to combine different materials enables the creation of implants that mimic the natural properties of bone or tissue, improving biocompatibility and reducing the risk of rejection. For example, hybrid 3D printing can be used to create customized hip implants with a strong, porous structure that promotes bone growth and reduces the risk of implant failure. Additionally, multi-material 3D printing can be used to create prosthetics with integrated sensors or actuators, enabling more precise control and improved functionality.

Consumer Products Applications

Hybrid and multi-material 3D printing can be used to create customized consumer products with unique properties, such as color or texture. The ability to combine different materials enables the creation of products with improved performance, durability, and aesthetics. For example, hybrid 3D printing can be used to create customized phone cases with integrated functionality, such as wireless charging or shock absorption. Additionally, multi-material 3D printing can be used to create customized shoes with improved comfort, support, and style.

Other Applications

Hybrid and multi-material 3D printing technologies have a wide range of applications beyond aerospace, biomedical, and consumer products. These technologies can be used in industries such as:

- **Automotive**: Hybrid and multi-material 3D printing can be used to create customized car parts with improved performance, durability, and aesthetics.

- **Industrial equipment**: Hybrid and multi-material 3D printing can be used to create customized industrial equipment with improved performance, durability, and functionality.

- **Sports equipment**: Hybrid and multi-material 3D printing can be used to create customized sports equipment with improved performance, durability, and aesthetics.

11.4.1 Limitations of Hybrid and Multi-Material 3D Printing

While hybrid and multi-material 3D printing technologies offer a range of benefits, they also have some limitations. These limitations include:

- **Material compatibility**: The combination of different materials can result in material incompatibility issues, such as delamination or cracking. This can lead to reduced part performance, durability, and lifespan.

- **Process complexity**: Hybrid and multi-material 3D printing processes can be complex and require specialized equipment and expertise. This can increase production costs, reduce efficiency, and limit accessibility.

- **Cost**: Hybrid and multi-material 3D printing technologies can be more expensive than traditional manufacturing methods, particularly for small batches or one-off parts.

- **Scalability**: Hybrid and multi-material 3D printing technologies can be difficult to scale up for large production runs, making them more suitable for small-batch or prototype production.

11.4.2 Addressing the Limitations

To address the limitations of hybrid and multi-material 3D printing technologies, researchers and manufacturers are working to develop new materials, processes, and equipment. Some areas of research include:

- **Material development**: Researchers are developing new materials with improved compatibility, strength, and durability.

- **Process optimization**: Researchers are working to optimize hybrid and multi-material 3D printing processes to reduce complexity, improve efficiency, and increase scalability.

- **Equipment development**: Manufacturers are developing new equipment and machinery that can handle hybrid and multi-material 3D printing processes with improved precision, speed, and reliability.

Hybrid and multi-material 3D printing technologies have a wide range of applications across various industries, including aerospace, biomedical, consumer products, and more. While these technologies offer a range of benefits, they also have some limitations, including material compatibility, process complexity, cost, and scalability. To address these limitations, researchers and manufacturers are working to develop new materials, processes, and equipment. As the technology continues to evolve, we can expect to see even more innovative applications and solutions that take advantage of the unique benefits of hybrid and multi-material 3D printing.

11.5 Questions

11.5.1 Long Response Questions

1. What are the key differences between hybrid and multi-material 3D printing technologies?
2. What are some common applications of hybrid and multi-material 3D printing technologies?
3. What are the limitations of hybrid and multi-material 3D printing technologies?
4. How does the use of multiple materials affect the mechanical properties of a 3D printed part?
5. What are some potential future developments in hybrid and multi-material 3D printing technologies?

11.5.2 Multiple Choice Questions

1) What is hybrid additive manufacturing?

a) Combining different subtractive processes
b) Combining different formative processes
c) Combining traditional manufacturing processes with additive manufacturing
d) Using only additive manufacturing
Solution: - c) Combining traditional manufacturing processes with additive manufacturing

2) **What does multi-material AM involve?**
a) Using a single material
b) Using multiple materials in a single AM process
c) Only using metals
d) Only using plastics
Solution: - b) Using multiple materials in a single AM process

3) **Which industry commonly uses multi-material AM for creating customized implants?**
a) Aerospace
b) Automotive
c) Medical
d) Consumer products
Solution: - c) Medical

4) **What is a key challenge of multi-material 3D printing?**
a) Low cost
b) High material compatibility
c) Process simplicity
d) Material incompatibility
Solution: - d) Material incompatibility

5) **What is one benefit of hybrid AM?**
a) Only uses metals
b) Only uses plastics
c) Integration of multiple materials and processes in a single cycle
d) Requires no expertise
Solution: - c) Integration of multiple materials and processes in a single cycle

6) **Which of the following is a limitation of hybrid and multi-material 3D printing technologies?**
a) High efficiency

b) Material incompatibility
c) Low production costs
d) Easy scalability
Solution: - b) Material incompatibility

7) **In which industry is hybrid AM used to create lightweight parts to reduce fuel consumption and emissions?**
a) Medical
b) Consumer products
c) Aerospace
d) Automotive
Solution: - c) Aerospace

8) **What is one way researchers are addressing the limitations of hybrid and multi-material 3D printing?**
a) Reducing material compatibility
b) Increasing process complexity
c) Developing new materials
d) Decreasing process control
Solution: - c) Developing new materials

9) **What advantage does hybrid AM offer for producing parts for industrial equipment?**
a) Low cost
b) Improved mechanical and thermal properties
c) Simplified manufacturing process
d) Reduced precision
Solution: - b) Improved mechanical and thermal properties

10) **Which of the following best describes multi-material AM in the consumer products industry?**
a) Limited design options
b) Standardized products
c) Customized products with unique properties
d) Inconsistent performance
Solution: - c) Customized products with unique properties

11) **What is a potential future development in hybrid and multi-material 3D printing technologies?**
a) Increased process complexity
b) Reduced scalability
c) Cost reduction

d) Decreased material options

Solution: - c) Cost reduction

12) **How does the use of multiple materials affect the mechanical properties of a 3D printed part?**
    a) Reduces flexibility
    b) Eliminates customization
    c) Enables varying properties within a single part
    d) Increases process simplicity

    Solution: - c) Enables varying properties within a single part

13) **Which of the following is not a challenge associated with hybrid and multi-material 3D printing?**
    a) Material compatibility
    b) Process complexity
    c) High cost
    d) Limited material options

    Solution: - d) Limited material options

14) **What role does process optimization play in hybrid and multi-material 3D printing?**
    a) Reducing material options
    b) Increasing complexity
    c) Improving efficiency and scalability
    d) Reducing precision

    Solution: - c) Improving efficiency and scalability

15) **What is an example of a biomedical application of multi-material 3D printing?**
    a) Creating car parts
    b) Producing customized hip implants
    c) Manufacturing industrial equipment
    d) Developing sports equipment

    Solution: - b) Producing customized hip implants

16) **What is the primary advantage of integrating traditional manufacturing processes with additive manufacturing in hybrid AM?**
    a) Simplified design process
    b) Reduced need for specialized equipment
    c) Improved mechanical, thermal, and aesthetic properties
    d) Lower production costs

Solution: - c) Improved mechanical, thermal, and aesthetic properties

17) **In what way does hybrid AM contribute to the automotive industry?**
    a) By reducing the strength of car parts
    b) By simplifying the manufacturing process
    c) By creating parts with improved mechanical and thermal properties
    d) By limiting the design options
    Solution: - c) By creating parts with improved mechanical and thermal properties

18) **Which technology enables the creation of parts with integrated functionality, such as sensors or antennas, in aerospace applications?**
    a) Traditional subtractive manufacturing
    b) Formative processes
    c) Multi-material 3D printing
    d) Single-material 3D printing
    Solution: - c) Multi-material 3D printing

19) **What is one key consideration when developing new materials for hybrid and multi-material 3D printing?**
    a) Reducing durability
    b) Improving material compatibility
    c) Increasing process complexity
    d) Reducing mechanical properties
    Solution: - b) Improving material compatibility

20) **How can process control and monitoring be improved in multi-material AM?**
    a) By decreasing the number of materials used
    b) By simplifying the process
    c) By enhancing process control technologies
    d) By reducing precision
    Solution: - c) By enhancing process control technologies

21) **Which industry benefits from multi-material AM by creating customized phone cases with integrated functionality?**
    a) Aerospace
    b) Medical

c) Consumer products
d) Industrial equipment

Solution: - c) Consumer products

22) What challenge does process complexity in hybrid and multi-material 3D printing create?

a) Increased ease of operation
b) Reduced need for expertise
c) Higher production costs
d) Improved scalability

Solution: - c) Higher production costs

23) Which of the following is a future direction for hybrid and multi-material 3D printing?

a) Reducing material combinations
b) Improving cost-effectiveness
c) Limiting industrial applications
d) Decreasing process control

Solution: - b) Improving cost-effectiveness

24) How does hybrid AM improve the manufacturing of aerospace components?

a) By reducing strength-to-weight ratios
b) By increasing fuel consumption
c) By integrating multiple materials and processes
d) By limiting design options

Solution: - c) By integrating multiple materials and processes

25) What is a significant limitation of using multiple materials in 3D printing?

a) Increased ease of material selection
b) Simplified manufacturing process
c) Potential material incompatibility
d) Lower production costs

Solution: - c) Potential material incompatibility

26) How do multi-material AM processes impact product customization in consumer products?

a) They reduce customization options
b) They increase standardized production
c) They enable unique textures, colors, and finishes
d) They limit design freedom

Solution: - c) They enable unique textures, colors, and finishes

27) **What is a common application of hybrid AM in the medical field?**
   a) Creating aircraft components
   b) Manufacturing industrial equipment
   c) Producing customized implants
   d) Developing automotive parts
   Solution: - c) Producing customized implants

28) **Which of the following technologies is essential for addressing material incompatibility in hybrid and multi-material 3D printing?**
   a) Reducing process control
   b) Limiting material options
   c) Developing new materials
   d) Simplifying manufacturing processes
   Solution: - c) Developing new materials

29) **What is a key benefit of using multi-material AM for creating sports equipment?**
   a) Reducing product performance
   b) Standardizing production
   c) Enhancing durability and aesthetics
   d) Simplifying the design process
   Solution: - c) Enhancing durability and aesthetics

30) **Which of the following industries can benefit from improved scalability in hybrid and multi-material 3D printing?**
   a) Small-scale prototype production
   b) Large production runs
   c) One-off parts manufacturing
   d) Custom implant production
   Solution: - b) Large production runs

31) **How does hybrid AM enhance the performance of industrial equipment?**
   a) By reducing mechanical properties
   b) By simplifying manufacturing processes
   c) By improving mechanical and thermal properties
   d) By limiting design freedom

Solution: - c) By improving mechanical and thermal properties

32) What challenge does cost present in hybrid and multi-material 3D printing?

a) Reduces material compatibility issues
b) Increases accessibility to a wider range of industries
c) Can be more expensive than traditional methods
d) Simplifies the production process

Solution: - c) Can be more expensive than traditional methods

33) What is a key area of research for improving hybrid and multi-material 3D printing technologies?

a) Reducing mechanical properties
b) Increasing process complexity
c) Enhancing material compatibility
d) Limiting material combinations

Solution: - c) Enhancing material compatibility

34) In what way does process complexity affect hybrid and multi-material 3D printing?

a) Increases ease of use
b) Simplifies the design process
c) Presents challenges for scalability and cost-efficiency
d) Reduces expertise required

Solution: - c) Presents challenges for scalability and cost-efficiency

35) How can hybrid and multi-material 3D printing be applied in aerospace to improve part performance?

a) By reducing material combinations
b) By integrating multiple materials and processes
c) By simplifying manufacturing processes
d) By limiting design options

Solution: - b) By integrating multiple materials and processes

36) What is a potential benefit of hybrid AM in the medical field?

a) Standardized products
b) Reduced customization
c) Creation of customized implants with enhanced properties

d) Simplified design process

Solution: - c) Creation of customized implants with enhanced properties

37) How do researchers address the limitations of hybrid and multi-material 3D printing technologies?

a) By reducing material compatibility
b) By developing new materials and process optimizations
c) By increasing process complexity
d) By limiting material options

Solution: - b) By developing new materials and process optimizations

38) What is an example of a hybrid AM application in the automotive industry?

a) Standardized car parts
b) Customized hip implants
c) Lightweight parts with improved mechanical properties
d) Industrial equipment manufacturing

Solution: - c) Lightweight parts with improved mechanical properties

39) How does multi-material AM enhance product development in consumer products?

a) By reducing design freedom
b) By enabling unique textures, colors, and finishes
c) By limiting customization options
d) By increasing standardized production

Solution: - b) By enabling unique textures, colors, and finishes

40) What future direction is anticipated for hybrid and multi-material 3D printing technologies?

a) Reduced material options
b) Increased process complexity
c) Improved cost-effectiveness and scalability
d) Limited industrial applications

Solution: - c) Improved cost-effectiveness and scalability

Chapter 12

3D Printing Materials: Polymers and Composites

12.1 Introduction to 3D Printing Materials

3D printing materials play a crucial role in the additive manufacturing process, as they determine the final properties and performance of the printed part. The selection of the right material is critical to ensure that the part meets the required specifications and functional requirements. With the rapid growth of the 3D printing industry, the range of available materials has expanded significantly, offering designers and engineers a wide range of options to choose from.

In this chapter, we will focus on polymer and composite materials, which are commonly used in 3D printing. We will discuss their properties, advantages, and limitations, as well as their applications in various industries. Understanding the characteristics of these materials is essential to optimize the 3D printing process and produce high-quality parts that meet the desired performance and functional requirements.

12.1.1 Importance of Material Selection

Material selection is a critical step in the 3D printing process. The wrong material can lead to a part that fails to meet the required specifications, resulting in reduced performance, durability, and functionality. On the other hand, the right material can enable the creation of complex geometries, improve part performance, and reduce production costs.

The selection of the right material depends on several factors, including the application, functional requirements, and desired properties. For example, a part that requires high strength and stiffness may require a material with high tensile strength and modulus, such as a carbon fiber-reinforced polymer. On the other hand, a part that requires flexibility and impact resistance may require a material with high elongation and toughness, such as a rubber-like polymer.

12.1.2 Polymer Materials

Polymer materials are the most widely used materials in 3D printing. They offer a range of benefits, including ease of processing, low cost, and versatility. Polymers can be classified into several categories, including:

- **Thermoplastics**: These materials can be melted and reformed multiple times without undergoing any significant chemical change. Examples of thermoplastics include ABS, PLA, and Nylon.

- **Thermosets**: These materials undergo a chemical reaction during the curing process and cannot be melted and reformed. Examples of thermosets include epoxy and polyurethane.

- **Elastomers**: These materials exhibit elastic properties and can be stretched and deformed without breaking. Examples of elastomers include silicone and rubber-like polymers.

12.1.3 Composite Materials

Composite materials are made by combining two or more materials with different properties. They offer a range of benefits, including improved strength, stiffness, and durability. Composites can be classified into several categories, including:

- **Fiber-reinforced polymers**: These materials consist of fibers, such as carbon or glass, embedded in a polymer matrix. Examples of fiber-reinforced polymers include carbon fiber-reinforced PLA and glass fiber-reinforced Nylon.

- **Particle-reinforced polymers**: These materials consist of particles, such as silica or alumina, embedded in a polymer matrix. Examples of particle-reinforced polymers include silica-reinforced PLA and alumina-reinforced Nylon.

12.1.4 Properties of Polymer and Composite Materials

Polymer and composite materials exhibit a range of properties that make them suitable for various applications. Some of the key properties include:

- **Mechanical properties**: These include strength, stiffness, toughness, and elongation.

- **Thermal properties**: These include melting point, glass transition temperature, and thermal conductivity.

- **Electrical properties**: These include conductivity, resistivity, and dielectric strength.

- **Optical properties**: These include transparency, reflectivity, and refractive index.

12.1.5 Applications of Polymer and Composite Materials

Polymer and composite materials have a wide range of applications in various industries, including:

- **Aerospace**: Polymer and composite materials are used in aircraft and spacecraft components, such as wings, fuselages, and satellite components.

- **Automotive**: Polymer and composite materials are used in car parts, such as bumpers, dashboards, and engine components.

- **Medical**: Polymer and composite materials are used in medical devices, such as implants, prosthetics, and surgical instruments.

- **Consumer products**: Polymer and composite materials are used in consumer products, such as toys, appliances, and sporting goods.

In conclusion, polymer and composite materials play a critical role in the 3D printing process. Understanding their properties, advantages, and limitations is essential to optimize the 3D printing process and produce high-quality parts that meet the desired performance and functional requirements. By selecting the right material, designers and engineers can create complex geometries, improve part performance, and reduce production costs. In the next section, we will discuss the properties and applications of specific polymer and composite materials in more detail. Figure 55 showcases a variety of objects made from different 3D printing materials, each labeled with the type of material used. Here's a brief overview of the materials displayed:

- **PET**: A clear cup, demonstrating the transparency and durability of PET (Polyethylene Terephthalate).

- **PLA**: Colorful flowers and a blue dinosaur skeleton, highlighting the versatility and ease of use of PLA (Polylactic Acid).

- **WOOD + PLA**: A brown tree-like structure, showing how wood fibers can be combined with PLA for a wood-like appearance.

- **WOOD**: A brown table, emphasizing the natural look and feel of wood-based 3D printing materials.

- **FLEXI**: A black bracelet with red detailing, illustrating the flexibility and elasticity of flexible filaments.

- **LUMINOUS**: An ornamented ball, showcasing materials that glow in the dark.

- **THERMOCHROME**: A flower with color gradient petals, demonstrating materials that change color with temperature

- **ABS**: A red and yellow gift box, highlighting the strength and durability of ABS (Acrylonitrile Butadiene Styrene).

- **HIPS + PLA**: A blue box with red details and white snowflake patterns, showing the combination of HIPS (High Impact Polystyrene) and PLA for complex structures.

These materials offer a wide range of properties and applications, making 3D printing a versatile tool for various projects and industries.

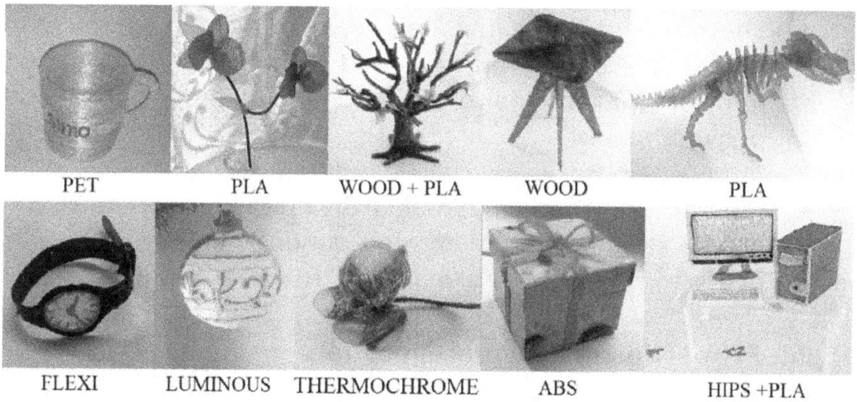

Figure 1: 3D printing materials
[Source: https://medium.com/@theyashrane/types-of-3d-printing-materials-used-in-3d-printing-6cbc55157c86]

12.2 Polymer Materials

Polymer materials are the most widely used 3D printing materials, accounting for over 70% of all 3D printing materials. They are available in various forms, including filaments, powders, and resins. Polymer materials offer a range of benefits, including ease of processing, low cost, and versatility. They can be used to create a wide range of products, from consumer goods to industrial components.

12.2.1 Types of Polymer Materials

There are several types of polymer materials used in 3D printing, each with its own unique properties and applications. Some of the most common types of polymer materials include:

- **Thermoplastics**: These materials can be melted and reformed multiple times without undergoing any significant chemical change. Examples of thermoplastics include ABS, PLA, and Nylon.

- **Thermosets**: These materials undergo a chemical reaction during the curing process and cannot be melted and reformed. Examples of thermosets include epoxy and polyurethane.

- **Elastomers**: These materials exhibit elastic properties and can be stretched and deformed without breaking. Examples of elastomers include silicone and rubber-like polymers.

12.2.2 Properties of Polymer Materials

Polymer materials exhibit a range of properties that make them suitable for various applications. Some of the key properties include:

- **Mechanical properties**: These include strength, stiffness, toughness, and elongation.

- **Thermal properties**: These include melting point, glass transition temperature, and thermal conductivity.

- **Chemical properties**: These include resistance to chemicals, solvents, and UV light.

- **Optical properties**: These include transparency, reflectivity, and refractive index.

12.2.3 Examples of Polymer Materials

Some examples of polymer materials used in 3D printing include:

- **Acrylonitrile Butadiene Styrene (ABS)**: A strong and impact-resistant material, commonly used in consumer products and automotive applications. ABS is a popular choice for 3D printing due to its ease of processing, low cost, and high strength-to-weight ratio.
- **Polylactic Acid (PLA)**: A biodegradable and renewable material, commonly used in packaging, medical devices, and consumer products. PLA is a popular choice for 3D printing due to its low cost, ease of processing, and environmental benefits.

- **Nylon**: A strong and flexible material, commonly used in industrial and consumer applications. Nylon is a popular choice for 3D printing due to its high strength, toughness, and resistance to abrasion and chemicals.

- **Polycarbonate (PC)**: A strong and impact-resistant material, commonly used in industrial and consumer applications. PC is a popular choice for 3D printing due to its high strength, stiffness, and resistance to heat and chemicals.

- **Polyethylene Terephthalate (PET)**: A strong and flexible material, commonly used in packaging, medical devices, and consumer products. PET is a popular choice for 3D printing due to its low cost, ease of processing, and high strength-to-weight ratio.

12.2.4 Applications of Polymer Materials

Polymer materials have a wide range of applications in various industries, including:

- **Consumer products**: Polymer materials are used in consumer products, such as toys, appliances, and sporting goods.

- **Automotive**: Polymer materials are used in car parts, such as bumpers, dashboards, and engine components.
- **Medical**: Polymer materials are used in medical devices, such as implants, prosthetics, and surgical instruments.
- **Industrial**: Polymer materials are used in industrial applications, such as machinery components, pipes, and fittings.
- **Aerospace**: Polymer materials are used in aircraft and spacecraft components, such as wings, fuselages, and satellite components.

12.2.5 Advantages and Limitations of Polymer Materials

Polymer materials offer several advantages, including:

- **Ease of processing**: Polymer materials can be easily melted and formed into complex shapes.
- **Low cost**: Polymer materials are generally less expensive than metal or ceramic materials.
- **Versatility**: Polymer materials can be used in a wide range of applications, from consumer products to industrial components. However, polymer materials also have some limitations, including:
- **Limited strength and stiffness**: Polymer materials may not have the same strength and stiffness as metal or ceramic materials.
- **Sensitivity to temperature and humidity**: Polymer materials may be sensitive to temperature and humidity, which can affect their properties and performance.

- **Limited durability**: Polymer materials may not have the same durability as metal or ceramic materials, and may be prone to degradation over time.

In conclusion, polymer materials are a popular choice for 3D printing due to their ease of processing, low cost, and versatility. They offer a range of benefits, including high strength, toughness, and resistance to chemicals and UV light. However, they also have some limitations, including limited strength and stiffness, sensitivity to temperature and humidity, and limited durability. By understanding the properties and limitations of polymer materials, designers and engineers can make informed decisions about the best material for their 3D printing project. In the next section, we will discuss composite materials and their applications in 3D printing. Figure 56 showcases a variety of objects made from different polymer materials. The collection includes rings, tubes, sheets, and molded parts in various shapes, sizes, and colors such as black, white, brown, red, and yellow. Some objects are transparent while others are opaque, highlighting the versatility of polymers in manufacturing. These materials are used in a wide range of applications due to their diverse properties, including flexibility, durability, and resistance to chemicals and heat.

Figure 2: Polymer materials
[Source: https://www.azom.com/article.aspx?ArticleID=21265]

12.3 Composite Materials

Composite materials are made by combining two or more materials, such as polymers and fibers, to create a material with unique properties. They offer improved mechanical, thermal, and electrical properties compared to traditional polymer materials. Composite materials are widely used in various industries, including aerospace, automotive, sports equipment, industrial, and construction.

12.3.1 Types of Composite Materials

There are several types of composite materials, including:

- **Fiber-reinforced polymers**: These materials consist of fibers, such as carbon, glass, or Kevlar, embedded in a polymer matrix. The fibers provide strength and stiffness, while the polymer matrix binds the fibers together and provides toughness.

- **Particle-reinforced polymers**: These materials consist of particles, such as silica or alumina, embedded in a polymer matrix. The particles provide improved mechanical and thermal properties.

- **Hybrid composites**: These materials combine two or more types of fibers or particles to create a material with unique properties.

12.3.2 Properties of Composite Materials

Composite materials exhibit a range of properties that make them suitable for various applications. Some of the key properties include:

- **Mechanical properties**: Composite materials offer improved strength, stiffness, toughness, and fatigue resistance compared to traditional polymer materials.

- **Thermal properties**: Composite materials offer improved thermal conductivity, thermal stability, and flame resistance.

- **Electrical properties**: Composite materials offer improved electrical conductivity, dielectric strength, and electromagnetic interference shielding.

- **Optical properties**: Composite materials offer improved transparency, reflectivity, and refractive index.

12.3.3 Examples of Composite Materials

Some examples of composite materials include:

- **Carbon Fiber-Reinforced Polymers (CFRP)**: A lightweight and high-strength material, commonly used in aerospace, automotive, and sports equipment applications. CFRP is a popular choice for 3D printing due to its high strength-to-weight ratio, corrosion resistance, and fatigue resistance.

- **Glass Fiber-Reinforced Polymers (GFRP)**: A strong and durable material, commonly used in industrial, automotive, and construction applications. GFRP is a popular choice for 3D printing due to its high strength, stiffness, and resistance to corrosion and fatigue.

- **Kevlar-Reinforced Polymers**: A high-strength and high-temperature material, commonly used in aerospace, automotive, and industrial applications. Kevlar-reinforced polymers are a popular choice for 3D printing due to their high strength, stiffness, and resistance to heat and flames.

- **Basalt Fiber-Reinforced Polymers**: A sustainable and eco-friendly material, commonly used in industrial, automotive, and construction applications. Basalt fiber-reinforced polymers are a popular choice for 3D printing due to their high strength, stiffness, and resistance to corrosion and fatigue.

12.3.4 Applications of Composite Materials

Composite materials have a wide range of applications in various industries, including:

- **Aerospace**: Composite materials are used in aircraft and spacecraft components, such as wings, fuselages, and satellite components.
- **Automotive**: Composite materials are used in car parts, such as bumpers, dashboards, and engine components.
- **Sports equipment**: Composite materials are used in sports equipment, such as bicycle frames, golf clubs, and tennis rackets.
- **Industrial**: Composite materials are used in industrial applications, such as machinery components, pipes, and fittings.
- **Construction**: Composite materials are used in construction applications, such as building facades, roofing, and insulation.

12.3.5 Advantages and Limitations of Composite Materials

Composite materials offer several advantages, including:

- **Improved mechanical properties**: Composite materials offer improved strength, stiffness, and toughness compared to traditional polymer materials.
- **Lightweight**: Composite materials are often lighter than traditional materials, making them ideal for applications where weight is a critical factor.
- **Corrosion resistance**: Composite materials offer improved corrosion resistance compared to traditional materials.

However, composite materials also have some limitations, including:

- **High cost**: Composite materials can be expensive to produce and process.

- **Complexity**: Composite materials can be complex to design and manufacture, requiring specialized equipment and expertise.

- **Interlaminar shear**: Composite materials can be prone to interlaminar shear, which can lead to delamination and failure.

In conclusion, composite materials are a popular choice for 3D printing due to their improved mechanical, thermal, and electrical properties. They offer a range of benefits, including high strength, stiffness, and resistance to corrosion and fatigue. However, they also have some limitations, including high cost, complexity, and interlaminar shear. By understanding the properties and limitations of composite materials, designers and engineers can make informed decisions about the best material for their 3D printing project. In the next section, we will discuss the design considerations for 3D printing with composite materials. Figure 57 illustrates the composition of composite materials through three diagrams:

- **Fiber/Filament Reinforcement**: Depicts a single long fiber or filament.

- **Matrix**: Shows a plain rectangular block representing the matrix material.

- **Composite**: Combines the fibers or filaments embedded within the matrix block to form a composite material.

This visual representation explains how composite materials are created by combining two different materials to leverage the properties of both components for enhanced performance.

Composites are widely used in various industries, including aerospace, automotive, and construction, due to their strength, lightweight, and durability.

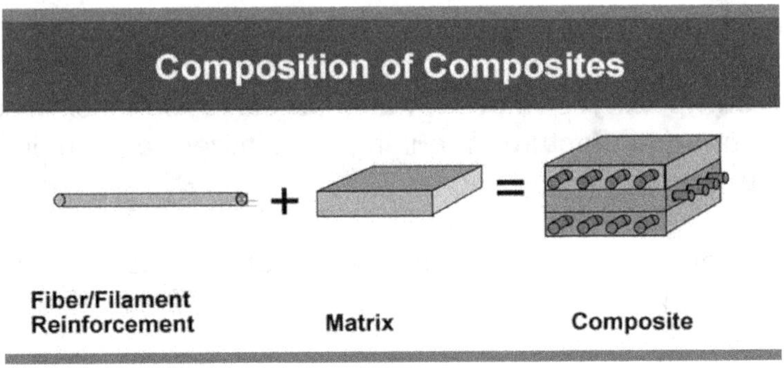

Figure 3: Composite materials
[Source: https://www.linkedin.com/pulse/importance-applications-composite-material-ahmad-husnain-76umf/]

12.4 Material Properties and Selection

When selecting a 3D printing material, it is essential to consider the material properties, such as mechanical strength, thermal resistance, and electrical conductivity. The material selection process involves evaluating the material properties against the required specifications and functional requirements of the part. By understanding the material properties and their impact on the performance of the part, designers and engineers can make informed decisions about the best material for their 3D printing project.

Mechanical Strength

Mechanical strength is the ability of the material to withstand mechanical loads and stresses. It is a critical factor in the design and selection of 3D printing materials. The mechanical strength of a material depends on several factors, including the material type, the printing process, and the printing parameters.

Some of the key mechanical properties to consider include:

- **Tensile strength**: The maximum stress that a material can withstand before breaking. Tensile strength is an important consideration for parts that will be subjected to pulling or stretching forces.

- **Flexural strength**: The maximum stress that a material can withstand before bending or breaking. Flexural strength is an important consideration for parts that will be subjected to bending or flexural forces.

- **Compressive strength**: The maximum stress that a material can withstand before compressing or collapsing. Compressive strength is an important consideration for parts that will be subjected to compressive forces.

- **Impact strength**: The ability of a material to withstand high-speed impacts. Impact strength is an important consideration for parts that will be subjected to impact forces, such as those used in aerospace or automotive applications.

- **Fatigue strength**: The ability of a material to withstand repeated loading and unloading. Fatigue strength is an important consideration for parts that will be subjected to cyclic loading, such as those used in mechanical components.

Thermal Resistance

Thermal resistance is the ability of the material to withstand high temperatures and thermal stresses. It is a critical factor in the design and selection of 3D printing materials, especially for applications that involve high temperatures or rapid temperature changes.

Some of the key thermal properties to consider include:

- **Melting point**: The temperature at which a material transitions from a solid to a liquid. Melting point is an important consideration for parts that will be subjected to high temperatures.

- **Glass transition temperature**: The temperature at which a material transitions from a brittle state to a flexible state. Glass transition temperature is an important consideration for parts that will be subjected to temperature changes.

- **Thermal conductivity**: The ability of a material to conduct heat. Thermal conductivity is an important consideration for parts that will be subjected to heat transfer, such as those used in electronics or thermal management applications.

- **Thermal expansion coefficient:** The rate at which a material expands or contracts with temperature changes. Thermal expansion coefficient is an important consideration for parts that will be subjected to temperature changes.

Electrical Conductivity

Electrical conductivity is the ability of a material to conduct electricity. It is a critical factor in the design and selection of 3D printing materials, especially for applications that involve electrical circuits or components.

Some of the key electrical properties to consider include:

- **Electrical conductivity**: The ability of a material to conduct electricity. Electrical conductivity is an important consideration for parts that will be used in electrical circuits or components.

- **Dielectric constant**: The ability of a material to store electrical energy. Dielectric constant is an important consideration for parts that will be used in capacitors or other electrical components.

- **Volume resistivity**: The resistance of a material to electrical current flow. Volume resistivity is an important consideration for parts that will be used in electrical circuits or components.

- **Surface resistivity:** The resistance of a material to electrical current flow on its surface. Surface resistivity is an important consideration for parts that will be used in electrical circuits or components.

Other Material Properties

In addition to mechanical strength, thermal resistance, and electrical conductivity, there are several other material properties to consider when selecting a 3D printing material. These include:

- **Optical properties:** The ability of a material to transmit or reflect light. Optical properties are an important consideration for parts that will be used in optical applications, such as lenses or mirrors.

- **Chemical resistance:** The ability of a material to withstand chemical corrosion or degradation. Chemical resistance is an important consideration for parts that will be exposed to chemicals or corrosive environments.

- **Biocompatibility:** The ability of a material to be used in medical or biological applications without causing harm to living tissues. Biocompatibility is an important consideration for parts that will be used in medical devices or implants.

- **Aesthetics:** The appearance and feel of a material. Aesthetics are an important consideration for parts that will be used in consumer products or other applications where appearance is important.

12.4.1 Material Selection Process

The material selection process involves evaluating the material properties against the required specifications and functional requirements of the part. It is a critical step in the 3D printing process, as the material selection can impact the performance, durability, and cost of the part.

The material selection process typically involves the following steps:

- **Define the functional requirements**: Identify the functional requirements of the part, including its mechanical, thermal, electrical, and other properties.
- **Identify potential materials**: Identify potential materials that meet the functional requirements of the part.
- **Evaluate material properties**: Evaluate the material properties of each potential material, including its mechanical strength, thermal resistance, electrical conductivity, and other properties.
- **Compare materials**: Compare the material properties of each potential material to determine which material best meets the functional requirements of the part.
- **Select the material**: Select the material that best meets the functional requirements of the part.

12.4.2 Material Testing

Material testing is an essential step in the material selection process. It involves testing the material properties under various conditions to ensure that they meet the required specifications and functional requirements of the part.

Some of the common material testing methods include:

- **Tensile testing**: A test that measures the tensile strength, elongation, and modulus of elasticity of a material.
- **Flexural testing**: A test that measures the flexural strength and modulus of elasticity of a material.

- **Impact testing**: A test that measures the impact strength of a material.
- **Thermal testing**: A test that measures the thermal conductivity, glass transition temperature, and melting point of a material.
- **Electrical testing**: A test that measures the electrical conductivity, dielectric constant, and volume resistivity of a material.

In conclusion, the material properties and selection are critical factors in the 3D printing process. By understanding the material properties and their impact on the performance of the part, designers and engineers can make informed decisions about the best material for their 3D printing project. By following the material selection process and conducting material testing, designers and engineers can ensure that their parts meet the required specifications and functional requirements.

12.5 Questions

12.5.1 Long Response Questions

1. What are the key differences between polymer and composite materials?
2. What are some common applications of polymer materials in 3D printing?
3. What are some common applications of composite materials in 3D printing?
4. How do material properties affect the selection of a 3D printing material?
5. What are some potential future developments in 3D printing materials?

12.5.2 Multiple Choice Questions

1) **Which 3D printing material is characterized by its easy printability and biodegradability?**

a) PLA
b) ABS
c) PETG
d) Nylon
Solution: - a) PLA

2) **What material is commonly used as a support structure in FDM 3D printing due to its solubility in limonene?**
a) PLA
b) ABS
c) PVA
d) TPU
Solution: - c) PVA

3) **Which filament type is known for its flexibility and is often used for making phone cases and toys?**
a) PLA
b) ABS
c) PETG
d) TPU
Solution: - d) TPU

4) **Which 3D printing material is often used for high-temperature applications due to its thermal stability?**
a) PLA
b) ABS
c) PETG
d) PEEK
Solution: - d) PEEK

5) **Which material is generally preferred for creating strong and durable 3D prints for mechanical parts?**
a) PLA
b) ABS
c) PETG
d) Nylon
Solution: - d) Nylon

6) **What does the acronym ABS stand for in 3D printing materials?**
a) Acrylonitrile Butadiene Styrene
b) Acrylonitrile Butylene Styrene
c) Acrylonitrile Butadiene Styrene Resin
d) Acrylonitrile Butadiene Synthetic

Solution: - a) Acrylonitrile Butadiene Styrene

7) **Which material is most commonly used in 3D printing due to its ease of use and low cost?**
   a) PLA
   b) ABS
   c) PETG
   d) Resin
   Solution: - a) PLA

8) **Which polymer is known for its chemical resistance and is commonly used in industrial applications?**
   a) PLA
   b) ABS
   c) PETG
   d) Nylon
   Solution: - c) PETG

9) **Which 3D printing material is typically used for high-resolution and detailed prints?**
   a) PLA
   b) ABS
   c) PETG
   d) Resin
   Solution: - d) Resin

10) **Which material is often used in 3D printing due to its low melting temperature and ease of use?**
   a) PLA
   b) ABS
   c) PETG
   d) Nylon
   Solution: - a) PLA

11) **Which of the following materials is often used to achieve a wood-like finish in 3D printed objects?**
   a) PLA
   b) ABS
   c) Wood filament
   d) PETG
   Solution: - c) Wood filament

12) **Which material is typically used for its high strength-to-weight ratio and flexibility?**
   a) PLA

b) ABS
c) Nylon
d) TPU
Solution: - c) Nylon

13) Which material is commonly used for creating prototypes due to its good balance of strength and ease of printing?
a) PLA
b) ABS
c) PETG
d) Resin
Solution: - b) ABS

14) What property makes PLA an ideal choice for beginners in 3D printing?
a) High strength
b) Ease of printing
c) High temperature resistance
d) Chemical resistance
Solution: - b) Ease of printing

15) Which polymer is used in 3D printing for its high impact resistance and is commonly employed in automotive parts?
a) PLA
b) ABS
c) PETG
d) Nylon
Solution: - b) ABS

16) Which filament is known for its resistance to UV light and outdoor durability?
a) PLA
b) ABS
c) PETG
d) Nylon
Solution: - c) PETG

17) What is a key advantage of using carbon fiber composites in 3D printing?
a) Increased flexibility
b) Enhanced strength and rigidity
c) Reduced weight
d) Low cost

Solution: - b) Enhanced strength and rigidity

18) **Which material is often used in high-performance applications requiring low friction and wear resistance?**
   a) PLA
   b) ABS
   c) Nylon
   d) TPU
   Solution: - c) Nylon

19) **Which of the following materials is known for its high thermal resistance and is often used in aerospace applications?**
   a) PLA
   b) ABS
   c) PETG
   d) PEEK
   Solution: - d) PEEK

20) **Which polymer is used in 3D printing for creating flexible parts and is known for its high elasticity?**
   a) PLA
   b) ABS
   c) TPU
   d) PETG
   Solution: - c) TPU

21) **What is a primary disadvantage of using ABS in 3D printing, especially in an indoor environment?**
   a) High melting temperature
   b) Emission of unpleasant fumes
   c) Low strength
   d) Poor layer adhesion
   Solution: - b) Emission of unpleasant fumes

22) **Which 3D printing material is ideal for applications requiring high mechanical strength and rigidity, such as in engineering prototypes?**
   a) PLA
   b) ABS
   c) PETG
   d) Carbon fiber composites
   Solution: - d) Carbon fiber composites

23) **What is a common use for high-temperature polymers like PEEK in 3D printing?**
   a) Consumer product prototypes
   b) Medical implants
   c) Decorative items
   d) Low-strength parts
   Solution: - b) Medical implants

24) **Which material is characterized by its high resistance to chemicals and is used in environments with harsh conditions?**
   a) PLA
   b) ABS
   c) PETG
   d) Nylon
   Solution: - c) PETG

25) **In what scenario would you choose a composite filament over a standard polymer filament?**
   a) When cost is the primary concern
   b) For enhanced mechanical properties
   c) When printing speed is critical
   d) For aesthetic finishes
   Solution: - b) For enhanced mechanical properties

26) **Which polymer's high melting temperature makes it suitable for printing parts exposed to high heat?**
   a) PLA
   b) ABS
   c) PETG
   d) PEEK
   Solution: - d) PEEK

27) **What is the primary benefit of using polycarbonate (PC) filament in 3D printing?**
   a) Low cost
   b) High transparency
   c) High impact resistance
   d) Easy printability
   Solution: - c) High impact resistance

28) **Which 3D printing material is often used to produce highly detailed and intricate parts due to its low viscosity?**
   a) PLA

b) ABS
c) Resin
d) Nylon

Solution: - c) Resin

29) What is a significant challenge associated with using Nylon filament in 3D printing?

a) High cost
b) High moisture absorption
c) Low strength
d) High printing temperature

Solution: - b) High moisture absorption

30) Which material is frequently chosen for creating functional parts due to its superior mechanical properties and durability?

a) PLA
b) ABS
c) PETG
d) Carbon fiber composites

Solution: - d) Carbon fiber composites

31) What feature of a material like HIPS makes it useful as a dissolvable support material in dual-extrusion 3D printing?

a) High melting point
b) Low cost
c) Solubility in limonene
d) High rigidity

Solution: - c) Solubility in limonene

32) Which material is preferred for applications requiring a combination of strength, flexibility, and resistance to impact?

a) PLA
b) ABS
c) PETG
d) Nylon

Solution: - d) Nylon

33) What is a common application for 3D printing with metal composites?

a) Decorative items
b) Functional prototypes

c) Low-strength parts
d) Temporary molds
Solution: - b) Functional prototypes

34) Which filament type is known for its ability to create transparent parts with high clarity?
a) PLA
b) ABS
c) PETG
d) Nylon
Solution: - c) PETG

35) In which of the following applications is PEEK filament most likely used?
a) Household items
b) Educational models
c) Aerospace components
d) Art pieces
Solution: - c) Aerospace components

36) Which material's properties make it suitable for use in high-strength, high-impact applications?
a) PLA
b) ABS
c) PETG
d) Carbon fiber composites
Solution: - d) Carbon fiber composites

37) What advantage does using a filament with metal additives offer in 3D printing?
a) Increased flexibility
b) Enhanced electrical conductivity
c) Lower print temperature
d) Improved aesthetic appearance
Solution: - b) Enhanced electrical conductivity

38) What is a notable feature of using composite filaments like those with carbon fiber?
a) Increased flexibility
b) Enhanced strength and stiffness
c) Improved color variety
d) Reduced printing time
Solution: - b) Enhanced strength and stiffness

39) **Which polymer is specifically known for its high resistance to chemicals and is used in specialized industrial applications?**
   a) PLA
   b) ABS
   c) PETG
   d) PEEK
   Solution: - d) PEEK

40) **What is a primary challenge when working with Nylon filament in a 3D printing environment?**
   a) Difficulty in achieving high precision
   b) High moisture absorption requiring careful storage
   c) Low strength compared to other filaments
   d) Poor adhesion to the print bed
   Solution: - b) High moisture absorption requiring careful storage

Chapter 13

3D Printing Materials: Metals and Ceramics

13.1 Introduction to Metal and Ceramic 3D Printing Materials

Metal and ceramic 3D printing materials are increasingly being used in various industries, including aerospace, automotive, healthcare, and consumer products. These materials offer unique properties, such as high strength, corrosion resistance, and thermal conductivity, making them ideal for applications where traditional materials fall short.

13.1.1 Properties of Metal 3D Printing Materials

Metal 3D printing materials are known for their high strength, durability, and resistance to corrosion and wear. Some of the key properties of metal 3D printing materials include:

- **High strength-to-weight ratio**: Metal 3D printing materials offer a high strength-to-weight ratio, making them ideal for applications where weight reduction is critical, such as in aerospace and automotive industries.

- **Corrosion resistance**: Metal 3D printing materials can be designed to resist corrosion, making them suitable for applications where exposure to harsh environments is a concern.

- **Thermal conductivity**: Metal 3D printing materials have high thermal conductivity, making them ideal for applications where heat transfer is critical, such as in heat exchangers and thermal management systems.

- **Electrical conductivity**: Metal 3D printing materials can be designed to have high electrical conductivity, making them suitable for applications where electrical conductivity is critical, such as in electrical circuits and components.

13.1.2 Properties of Ceramic 3D Printing Materials

Ceramic 3D printing materials are known for their high hardness, corrosion resistance, and thermal insulation properties. Some of the key properties of ceramic 3D printing materials include:

- **High hardness:** Ceramic 3D printing materials are extremely hard, making them resistant to wear and abrasion.
- **Corrosion resistance**: Ceramic 3D printing materials are highly resistant to corrosion, making them suitable for applications where exposure to harsh environments is a concern.
- **Thermal insulation**: Ceramic 3D printing materials have low thermal conductivity, making them ideal for applications where thermal insulation is critical, such as in heat shields and thermal protection systems.
- **Biocompatibility**: Ceramic 3D printing materials can be designed to be biocompatible, making them suitable for applications in healthcare and medical devices.

13.1.3 Advantages of Metal and Ceramic 3D Printing Materials

Metal and ceramic 3D printing materials offer several advantages over traditional materials, including:

- **Complex geometries**: Metal and ceramic 3D printing materials can be used to create complex geometries and structures that cannot be produced with traditional manufacturing methods.
- **Customization**: Metal and ceramic 3D printing materials can be customized to meet specific application

requirements, such as corrosion resistance or thermal conductivity.

- **Reduced material waste**: Metal and ceramic 3D printing materials can be produced with minimal material waste, making them a more sustainable option.

- **Improved performance**: Metal and ceramic 3D printing materials can offer improved performance compared to traditional materials, such as higher strength-to-weight ratios and improved thermal conductivity.

13.1.4 Limitations of Metal and Ceramic 3D Printing Materials

While metal and ceramic 3D printing materials offer several advantages, they also have some limitations, including:

- **High cost**: Metal and ceramic 3D printing materials can be expensive, making them less accessible to some industries and applications.

- **Limited availability**: Metal and ceramic 3D printing materials may not be widely available, making it difficult to source them.

- **Post-processing requirements**: Metal and ceramic 3D printing materials may require post-processing, such as machining or grinding, to achieve the desired surface finish and properties.

- **Material properties variability**: The properties of metal and ceramic 3D printing materials can vary depending on the printing process and parameters, making it critical to control the printing process to achieve consistent results.

13.1.5 Applications of Metal and Ceramic 3D Printing Materials

Metal and ceramic 3D printing materials are being used in a wide range of industries, including:

- **Aerospace**: Metal 3D printing materials are being used in aerospace applications, such as engine components and satellite parts, due to their high strength-to-weight ratio and resistance to corrosion.

- **Automotive**: Metal 3D printing materials are being used in automotive applications, such as engine components and exhaust systems, due to their high strength-to-weight ratio and resistance to corrosion.

- **Healthcare**: Ceramic 3D printing materials are being used in healthcare applications, such as implants and surgical instruments, due to their biocompatibility and resistance to corrosion.

- **Consumer products**: Metal and ceramic 3D printing materials are being used in consumer products, such as jewelry and cookware, due to their unique properties and aesthetic appeal.

In the next section, we will explore the different types of metal and ceramic 3D printing materials, including their properties, advantages, and limitations. Figure 58 showcases a variety of objects made from metal and ceramic materials using 3D printing technology. The collection includes:

- **Vases with intricate designs**: Highlighting the precision and detail achievable with 3D printing.

- **Geometrically complex structures**: Demonstrating the ability to create shapes that would be difficult or impossible with traditional manufacturing methods.

- **Twisted sculptures**: Resembling fabric or paper, showcasing the flexibility and creativity possible with these materials.

- **Lattice-like patterns**: Illustrating the strength and lightweight properties of 3D printed metal and ceramic objects.

- These examples emphasize the versatility and advanced capabilities of 3D printing in producing detailed and varied items from different materials.

Figure 1: Metal and ceramic 3D printing materials
[Source: https://www.3dnatives.com/en/3d-printing-material110420174/]

13.2 Metal Materials

Metal 3D printing materials are used in a wide range of applications, including aerospace, automotive, and healthcare. They offer high strength, corrosion resistance, and thermal conductivity, making them ideal for applications where traditional materials fall short.

13.2.1 Properties of Metal 3D Printing Materials

Metal 3D printing materials are known for their high strength, durability, and resistance to corrosion and wear. Some of the key properties of metal 3D printing materials include:

- **High strength-to-weight ratio**: Metal 3D printing materials offer a high strength-to-weight ratio, making them ideal for applications where weight reduction is critical, such as in aerospace and automotive industries.

- **Corrosion resistance**: Metal 3D printing materials can be designed to resist corrosion, making them suitable for applications where exposure to harsh environments is a concern.

- **Thermal conductivity**: Metal 3D printing materials have high thermal conductivity, making them ideal for applications where heat transfer is critical, such as in heat exchangers and thermal management systems.

- **Electrical conductivity**: Metal 3D printing materials can be designed to have high electrical conductivity, making them suitable for applications where electrical conductivity is critical, such as in electrical circuits and components.

13.2.2 Examples of Metal Materials

Examples of metal materials include:

- **Titanium**: A strong and lightweight material, commonly used in aerospace and medical applications. Titanium is known for its high strength-to-weight ratio, corrosion resistance, and biocompatibility, making it an ideal material for applications such as aircraft components, medical implants, and surgical instruments.

- **Aluminum**: A lightweight and corrosion-resistant material, commonly used in aerospace and automotive applications.

Aluminum is known for its high strength-to-weight ratio, corrosion resistance, and thermal conductivity, making it an ideal material for applications such as aircraft components, engine parts, and heat exchangers.

- **Steel**: A strong and durable material, commonly used in industrial and consumer applications. Steel is known for its high strength, corrosion resistance, and thermal conductivity, making it an ideal material for applications such as machinery components, tools, and cookware.

- **Copper**: A highly conductive material, commonly used in electrical and thermal applications. Copper is known for its high electrical conductivity, thermal conductivity, and corrosion resistance, making it an ideal material for applications such as electrical circuits, heat exchangers, and thermal management systems.

- **Inconel**: A high-temperature resistant material, commonly used in aerospace and industrial applications. Inconel is known for its high strength, corrosion resistance, and thermal conductivity, making it an ideal material for applications such as turbine blades, heat exchangers, and thermal management systems.

- **Stainless Steel**: A corrosion-resistant material, commonly used in food processing, medical, and industrial applications. Stainless steel is known for its high corrosion resistance, durability, and ease of cleaning, making it an ideal material for applications such as cookware, medical instruments, and industrial equipment.

- **Bronze**: A strong and corrosion-resistant material, commonly used in bearings, bushings, and other wear-resistant applications. Bronze is known for its high strength, corrosion resistance, and wear resistance, making it an ideal material for applications such as engine components, gears, and other mechanical components.

- **Brass**: A corrosion-resistant material, commonly used in decorative fixtures, musical instruments, and other applications where aesthetics are important. Brass is known for its high corrosion resistance, durability, and attractive appearance, making it an ideal material for applications such as door handles, faucets, and other decorative fixtures.

13.2.3 Advantages of Metal 3D Printing Materials

Metal 3D printing materials offer several advantages over traditional materials, including:

- **Complex geometries**: Metal 3D printing materials can be used to create complex geometries and structures that cannot be produced with traditional manufacturing methods.

- **Customization**: Metal 3D printing materials can be customized to meet specific application requirements, such as corrosion resistance or thermal conductivity.

- **Reduced material waste**: Metal 3D printing materials can be produced with minimal material waste, making them a more sustainable option.

- **Improved performance**: Metal 3D printing materials can offer improved performance compared to traditional materials, such as higher strength-to-weight ratios and improved thermal conductivity.

13.2.4 Limitations of Metal 3D Printing Materials

While metal 3D printing materials offer several advantages, they also have some limitations, including:

- **High cost**: Metal 3D printing materials can be expensive, making them less accessible to some industries and applications.

- **Limited availability**: Metal 3D printing materials may not be widely available, making it difficult to source them.

- **Post-processing requirements**: Metal 3D printing materials may require post-processing, such as machining or grinding, to achieve the desired surface finish and properties.

- **Material properties variability**: The properties of metal 3D printing materials can vary depending on the printing process and parameters, making it critical to control the printing process to achieve consistent results.

13.2.5 Applications of Metal 3D Printing Materials

Metal 3D printing materials are being used in a wide range of industries, including:

- **Aerospace**: Metal 3D printing materials are being used in aerospace applications, such as engine components and satellite parts, due to their high strength-to-weight ratio and resistance to corrosion.

- **Automotive**: Metal 3D printing materials are being used in automotive applications, such as engine components and exhaust systems, due to their high strength-to-weight ratio and resistance to corrosion.

- **Healthcare**: Metal 3D printing materials are being used in healthcare applications, such as implants and surgical instruments, due to their biocompatibility and resistance to corrosion.

- **Industrial**: Metal 3D printing materials are being used in industrial applications, such as machinery components and tools, due to their high strength and durability.

- **Consumer Goods**: Metal 3D printing materials are being used in consumer goods applications, such as jewelry and home decor, due to their attractive appearance and durability.

- **Electrical and Electronics**: Metal 3D printing materials are being used in electrical and electronics applications, such as electrical connectors and heat sinks, due to their high electrical conductivity and thermal conductivity.

Metal 3D printing materials offer a wide range of benefits, including high strength, corrosion resistance, and thermal conductivity. They can be used to create complex geometries and structures that cannot be produced with traditional manufacturing methods, and can be customized to meet specific application requirements. While metal 3D printing materials have some limitations, such as high cost and limited availability, they are being used in a wide range of industries, from aerospace and automotive to healthcare and consumer goods. As the technology continues to evolve, it is likely that metal 3D printing materials will become even more accessible and widely used. Figure 59 showcases two intricate metal objects, likely produced using advanced manufacturing techniques such as 3D printing:

- **Left Object**: A spherical item with a geometric pattern cut into its surface and several flanges. This design suggests it might be a part of machinery or a piping system.

- **Right Object**: An irregularly shaped, skeletal frame-like structure with a complex web of supports and struts. This could be a component designed for lightweight strength in engineering applications.

These examples highlight the capabilities of modern manufacturing to create complex shapes and structures that may not be possible with traditional methods.

Figure 2: Metal materials
[Source: https://vexmatech.com/titanium-material.html]

13.3 Ceramic Materials

Ceramic 3D printing materials are used in a wide range of applications, including aerospace, healthcare, and consumer products. They offer high thermal resistance, corrosion resistance, and biocompatibility, making them ideal for applications where traditional materials fall short.

13.3.1 Properties of Ceramic 3D Printing Materials

Ceramic 3D printing materials are known for their unique properties, including:

- **High thermal resistance**: Ceramic 3D printing materials can withstand high temperatures, making them ideal for applications where thermal resistance is critical, such as in aerospace and industrial applications.

- **Corrosion resistance**: Ceramic 3D printing materials are resistant to corrosion, making them suitable for applications where exposure to harsh environments is a concern.

- **Biocompatibility**: Ceramic 3D printing materials are biocompatible, making them ideal for applications where they come into contact with living tissue, such as in medical implants and devices.

- **Hardness and wear resistance**: Ceramic 3D printing materials are hard and wear-resistant, making them suitable for applications where wear and tear is a concern, such as in industrial and consumer applications.

13.3.2 Examples of Ceramic Materials

Examples of ceramic materials include:

- **Zirconia**: A strong and biocompatible material, commonly used in dental and medical applications. Zirconia is known for its high strength, toughness, and biocompatibility, making it an ideal material for applications such as dental implants, surgical instruments, and medical devices.

- **Alumina**: A hard and corrosion-resistant material, commonly used in industrial and consumer applications. Alumina is known for its high hardness, corrosion resistance, and thermal conductivity, making it an ideal material for applications such as industrial machinery, cutting tools, and heat exchangers.

- **Silicon Carbide**: A hard and thermal-resistant material, commonly used in aerospace and industrial applications. Silicon carbide is known for its high hardness, thermal resistance, and corrosion resistance, making it an ideal material for applications such as rocket nozzles, heat shields, and industrial machinery.

- **Titanium Dioxide**: A biocompatible and corrosion-resistant material, commonly used in medical and consumer applications. Titanium dioxide is known for its high biocompatibility, corrosion resistance, and UV resistance, making it an ideal material for applications such as medical implants, surgical instruments, and cosmetics.

- **Ceramic Matrix Composites**: A composite material made from ceramic fibers and a ceramic matrix, commonly used in aerospace and industrial applications. Ceramic matrix composites are known for their high strength, toughness, and thermal resistance, making them an ideal material for applications such as aircraft components, rocket nozzles, and heat shields.

13.3.3 Advantages of Ceramic 3D Printing Materials

Ceramic 3D printing materials offer several advantages over traditional materials, including:

- **Complex geometries**: Ceramic 3D printing materials can be used to create complex geometries and structures that cannot be produced with traditional manufacturing methods.

- **Customization**: Ceramic 3D printing materials can be customized to meet specific application requirements, such as thermal resistance or biocompatibility.

- **Reduced material waste**: Ceramic 3D printing materials can be produced with minimal material waste, making them a more sustainable option.

- **Improved performance**: Ceramic 3D printing materials can offer improved performance compared to traditional materials, such as higher thermal resistance and corrosion resistance.

13.3.4 Limitations of Ceramic 3D Printing Materials

While ceramic 3D printing materials offer several advantages, they also have some limitations, including:

- **High cost**: Ceramic 3D printing materials can be expensive, making them less accessible to some industries and applications.
- **Limited availability**: Ceramic 3D printing materials may not be widely available, making it difficult to source them.
- **Post-processing requirements**: Ceramic 3D printing materials may require post-processing, such as machining or grinding, to achieve the desired surface finish and properties.
- **Material properties variability**: The properties of ceramic 3D printing materials can vary depending on the printing process and parameters, making it critical to control the printing process to achieve consistent results.

13.3.5 Applications of Ceramic 3D Printing Materials

Ceramic 3D printing materials are being used in a wide range of industries, including:

- **Aerospace**: Ceramic 3D printing materials are being used in aerospace applications, such as rocket nozzles and heat shields, due to their high thermal resistance and corrosion resistance.
- **Healthcare**: Ceramic 3D printing materials are being used in healthcare applications, such as medical implants and surgical instruments, due to their biocompatibility and corrosion resistance.
- **Industrial**: Ceramic 3D printing materials are being used in industrial applications, such as industrial machinery and cutting tools, due to their high hardness and wear resistance.
- **Consumer Products**: Ceramic 3D printing materials are being used in consumer products, such as cosmetics and jewelry, due to their biocompatibility and aesthetic appeal.

Ceramic 3D printing materials offer a wide range of benefits, including high thermal resistance, corrosion resistance, and biocompatibility. They can be used to create complex geometries and structures that cannot be produced with traditional manufacturing methods and can be customized to meet specific application requirements. While ceramic 3D printing materials have some limitations, such as high cost and limited availability, they are being used in a wide range of industries, from aerospace and healthcare to industrial and consumer products. As the technology continues to evolve, it is likely that ceramic 3D printing materials will become even more accessible and widely used. Figure 60 showcases a variety of ceramic materials and objects, each with different shapes and textures. The collection includes items with geometric patterns, such as a honeycomb structure, spherical objects with dimpled surfaces, and others with porous or lattice-like designs. This assortment highlights the versatility and range of designs possible with ceramic materials, which are known for their durability, heat resistance, and insulating properties. Ceramics are used in a wide range of applications, from traditional pottery and tiles to advanced engineering components in aerospace and electronics. Their ability to withstand high temperatures and resist chemical erosion makes them ideal for many industrial uses.

Figure 3: Ceramic materials
[Source: https://all3dp.com/2/vat-photopolymerization-dlp-advanced-ceramics/]

13.4 Material Properties and Selection

When selecting a metal or ceramic 3D printing material, it is essential to consider the material properties, such as mechanical strength, thermal resistance, and corrosion resistance. The material selection process involves evaluating the material properties against the required specifications and functional requirements of the part.

13.4.1 Material Properties

Material properties are the characteristics of a material that determine its behavior under various conditions. Some key material properties to consider when selecting a metal or ceramic 3D printing material include:

- **Mechanical strength**: The ability of the material to withstand mechanical loads and stresses. Mechanical strength is a critical property in many applications, such as aerospace, automotive, and industrial equipment.

- **Thermal resistance**: The ability of the material to withstand high temperatures and thermal stresses. Thermal resistance is essential in applications where the material will be exposed to high temperatures, such as in aerospace, power generation, and chemical processing.

- **Corrosion resistance**: The ability of the material to resist corrosion and degradation. Corrosion resistance is critical in applications where the material will be exposed to harsh environments, such as in marine, chemical, and oil and gas industries.

- **Electrical conductivity**: The ability of the material to conduct electricity. Electrical conductivity is important in applications such as electronic, electrical engineering, and electromagnetism.

- **Magnetic properties**: The ability of the material to be magnetized and attract other magnets. Magnetic properties are important in applications such as electrical engineering, electromagnetism, and medical devices.

- **Optical properties**: The ability of the material to transmit, absorb, or reflect light. Optical properties are important in applications such as optics, photonics, and biomedical devices.

- **Biocompatibility**: The ability of the material to be compatible with living tissue. Biocompatibility is critical in medical applications, such as implants, surgical instruments, and medical devices.

- **Density**: The mass per unit volume of the material. Density is important in applications where weight is a critical factor, such as in aerospace and automotive industries.

13.4.2 Material Selection Process

The material selection process involves evaluating the material properties against the required specifications and functional requirements of the part. The following steps can be followed to select the appropriate metal or ceramic 3D printing material:

- **Define the application requirements**: Identify the functional requirements of the part, including the operating conditions, loads, and stresses.

- **Determine the material properties**: Evaluate the material properties required for the application, including mechanical strength, thermal resistance, corrosion resistance, and other properties.

- **Research material options**: Research different metal and ceramic 3D printing materials that meet the required material properties.

- **Evaluate material properties**: Evaluate the material properties of each option, including their strengths, weaknesses, and limitations.

- **Select the material**: Select the material that best meets the required material properties and functional requirements of the part.

- **Verify the material selection**: Verify the material selection through testing and simulation to ensure that it meets the required specifications and functional requirements.

13.4.3 Material Selection Tools

Several material selection tools are available to aid in the material selection process, including:

- **Material property databases**: Databases that provide information on the material properties of different metals and ceramics.

- **Material selection software**: Software that allows users to input the required material properties and functional requirements to select the appropriate material.

- **Material selection charts**: Charts that provide a visual representation of the material properties and allow users to select the appropriate material based on their requirements.

Material properties and selection are critical considerations in metal and ceramic 3D printing. By understanding the material properties and following a structured material selection process, designers and engineers can select the appropriate material for their application, ensuring that the part meets the required specifications and functional requirements. With the increasing availability of metal and ceramic 3D printing materials, the material selection process is becoming increasingly complex, and the use of material selection tools and databases is becoming more important. Figure 61 illustrates the relationship between material properties and selection in a cyclical process. The diagram features three interlocking gears labeled Function, Process, and Material:

- **Function**: Dictates the choice of material and shape process.
- **Process**: Interacts with shape.
- **Material**: Initial lab research.
- Below the gears, a curved arrow labeled **Starting Materials** on one end and **Product** on the other represents the stages of development:
- **Advanced Lab Research** (blue sphere)
- **Test, Improve Production** (green sphere)
- **Initial Property Considerations** (yellow sphere) - Includes environmental impact, product function, choice of material, and degradation.
- **Final Product Testing** (red sphere)

This cyclical process emphasizes how materials are chosen based on their function, which then dictates their processing. The process influences material choices, leading to further development and testing in a continuous loop that results in product creation.

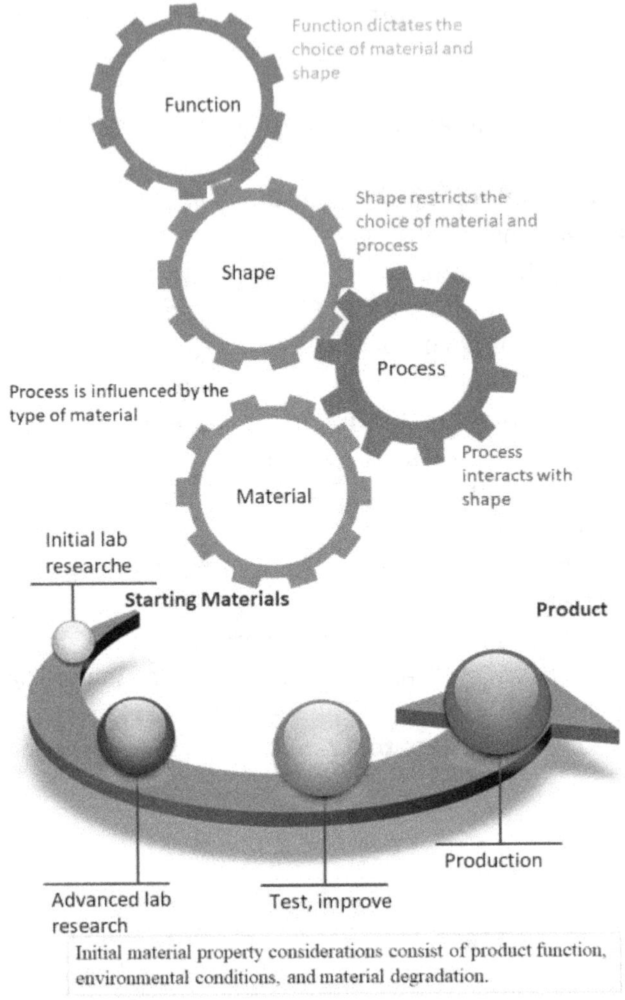

Figure 4: Material Properties and Selection
[Source: https://www.sciencedirect.com/topics/materials-science/material-selection]

13.5 Questions

13.5.1 Long Response Questions

1. What are the key differences between metal and ceramic 3D printing materials?

2. What are some common applications of metal 3D printing materials?
3. What are some common applications of ceramic 3D printing materials?
4. How do material properties affect the selection of a metal or ceramic 3D printing material?
5. What are some potential future developments in metal and ceramic 3D printing materials?

13.5.2 Multiple Choice Questions

1) **Which of the following is a common metal used in 3D printing?**
   a) Polylactic Acid (PLA)
   b) Titanium
   c) Nylon
   d) Polyethylene
   Solution: - b) Titanium
2) **Which metal is known for its high corrosion resistance and is commonly used in medical implants?**
   a) Aluminum
   b) Cobalt-chrome
   c) Iron
   d) Bronze
   Solution: - b) Cobalt-chrome
3) **What is a primary advantage of using metals in 3D printing compared to plastics?**
   a) Lower density
   b) Higher strength
   c) Lower cost
   d) Easier to recycle
   Solution: - b) Higher strength
4) **Which ceramic material is frequently used in 3D printing for its high heat resistance?**
   a) Alumina
   b) Polycarbonate
   c) Nylon
   d) PLA
   Solution: - a) Alumina

5) **What is the typical use of metal 3D printing in the aerospace industry?**
   a) Packaging
   b) Decorative items
   c) Structural components
   d) Furniture
   Solution: - c) Structural components
6) **Which of the following metals is often used for creating jewelry through 3D printing?**
   a) Steel
   b) Silver
   c) Cast iron
   d) Brass
   Solution: - b) Silver
7) **What property makes ceramics ideal for high-temperature applications in 3D printing?**
   a) High thermal conductivity
   b) Low thermal expansion
   c) High electrical conductivity
   d) High malleability
   Solution: - b) Low thermal expansion
8) **Which metal is primarily used in additive manufacturing for creating strong and lightweight parts?**
   a) Lead
   b) Gold
   c) Titanium
   d) Tin
   Solution: - c) Titanium
9) **What is a common post-processing technique for metal 3D printed parts to enhance their surface finish?**
   a) Painting
   b) Sandblasting
   c) Polishing
   d) Coating with plastic
   Solution: - c) Polishing
10) **In ceramic 3D printing, what is often used as a binder material?**
   a) Wax
   b) Polymer

c) Metal powder
d) Sand
Solution: - b) Polymer

11) What is a common challenge in 3D printing with metals?
a) Limited material availability
b) Difficulty in achieving fine details
c) High cost of equipment
d) Low resolution
Solution: - c) High cost of equipment

12) Which of the following metals can be used in laser sintering processes?
a) Copper
b) Aluminum
c) Stainless steel
d) Lead
Solution: - c) Stainless steel

13) Which ceramic material is often used for dental implants?
a) Zirconia
b) Alumina
c) Silicon carbide
d) Silica
Solution: - a) Zirconia

14) What is one of the main benefits of using ceramics in 3D printing for engineering applications?
a) High ductility
b) High thermal and chemical stability
c) Low cost
d) High flexibility
Solution: - b) High thermal and chemical stability

15) Which metal is known for its high strength-to-weight ratio and is frequently used in aerospace components?
a) Tungsten
b) Aluminum
c) Steel
d) Nickel
Solution: - b) Aluminum

16) In the context of 3D printing, what does SLM stand for?
a) Selective Laser Melting

b) Standard Laser Modelling
c) Sintered Laser Material
d) Scanned Laser Method

Solution: - a) Selective Laser Melting

17) Which of the following metals is commonly used in high-performance engine components?

a) Bronze
b) Magnesium
c) Nickel-based alloys
d) Zinc

Solution: - c) Nickel-based alloys

18) What is the primary reason for using metal powders in 3D printing processes?

a) To increase the weight of the final product
b) To enhance the flexibility of the material
c) To facilitate the melting and solidification process
d) To reduce the printing cost

Solution: - c) To facilitate the melting and solidification process

19) Which of the following is a key consideration when selecting ceramics for 3D printing?

a) Electrical conductivity
b) Thermal conductivity
c) Impact resistance
d) Flexibility

Solution: - b) Thermal conductivity

20) Which technique is used for creating metal parts by melting metal powder layer by layer?

a) Stereolithography
b) Fused Deposition Modeling
c) Direct Metal Laser Sintering
d) Binder Jetting

Solution: - c) Direct Metal Laser Sintering

21) Which of the following metals is particularly challenging to 3D print due to its high thermal conductivity and reactivity?

a) Tungsten
b) Gold
c) Platinum

d) Iridium

Solution: - a) Tungsten

22) What effect does high thermal expansion have on metal 3D printed parts during cooling?

a) Improved structural integrity
b) Increased risk of warping
c) Enhanced surface finish
d) Reduced mechanical properties

Solution: - b) Increased risk of warping

23) In the context of 3D printing with ceramics, what is the main purpose of using a support structure during printing?

a) To improve the thermal properties of the ceramic
b) To hold the ceramic material in place and prevent deformation
c) To enhance the surface finish of the ceramic
d) To increase the print speed

Solution: - b) To hold the ceramic material in place and prevent deformation

24) Which metal is known for its superalloy properties and is often used in extreme environments such as jet engines?

a) Iron
b) Cobalt
c) Nickel
d) Silver

Solution: - c) Nickel

25) What is the primary role of a binder in the binder jetting process for ceramics?

a) To increase the melting point of the ceramic
b) To consolidate ceramic powders into a solid form
c) To improve the color of the ceramic part
d) To enhance the mechanical properties of the ceramic

Solution: - b) To consolidate ceramic powders into a solid form

26) In metal 3D printing, what is the effect of using a laser with a higher power setting?

a) Reduced penetration depth
b) Increased powder consumption
c) Increased melting rate

d) Reduced thermal efficiency

Solution: - c) Increased melting rate

27) Which ceramic material is preferred for high-precision 3D printing applications due to its low thermal expansion and high strength?

a) Zirconium dioxide
b) Silicon nitride
c) Aluminum oxide
d) Barium titanate

Solution: - b) Silicon nitride

28) Which technique in metal 3D printing involves the use of an electron beam to melt and fuse metal powders?

a) Electron Beam Melting (EBM)
b) Selective Laser Sintering (SLS)
c) Fused Deposition Modeling (FDM)
d) Direct Ink Writing (DIW)

Solution: - a) Electron Beam Melting (EBM)

29) In ceramic 3D printing, what is the purpose of post-printing sintering?

a) To improve the color of the ceramic
b) To remove any remaining binder material
c) To increase the porosity of the ceramic
d) To decrease the mechanical strength of the ceramic

Solution: - b) To remove any remaining binder material

30) Which metal alloy is commonly used in aerospace applications due to its combination of high strength, low weight, and resistance to high temperatures?

a) Aluminum-lithium alloy
b) Steel-cobalt alloy
c) Nickel-cobalt alloy
d) Copper-silver alloy

Solution: - a) Aluminum-lithium alloy

31) What is the primary challenge associated with printing high-temperature ceramics using traditional methods?

a) Difficulty in achieving high resolution
b) Limited availability of ceramic powders
c) High energy requirements for sintering
d) Difficulty in removing binder materials

Solution: - c) High energy requirements for sintering

32) **Which metal property is particularly enhanced in applications involving 3D printed parts exposed to corrosive environments?**
  a) Ductility
  b) Hardness
  c) Corrosion resistance
  d) Thermal conductivity
  Solution: - c) Corrosion resistance

33) **Which process involves adding layers of metal powder and selectively melting them with a laser or electron beam?**
  a) Powder Bed Fusion
  b) Material Extrusion
  c) Binder Jetting
  d) Direct Energy Deposition
  Solution: - a) Powder Bed Fusion

34) **Which of the following factors does NOT directly influence the mechanical properties of 3D printed metal parts?**
  a) Layer thickness
  b) Print speed
  c) Ambient temperature
  d) Post-processing methods
  Solution: - c) Ambient temperature

35) **Which material is commonly used in the binder material for ceramic 3D printing processes to provide initial shape and strength?**
  a) Plastic resin
  b) Metal alloy
  c) Ceramic powder
  d) Organic binder
  Solution: - d) Organic binder

36) **What is the primary advantage of using high-temperature ceramics in aerospace applications?**
  a) Low cost
  b) High mechanical flexibility
  c) High thermal stability and resistance to extreme temperatures

d) High electrical conductivity

Solution: - c) High thermal stability and resistance to extreme temperatures

37) Which metal is used for its excellent combination of high strength, low weight, and resistance to high temperatures in jet engines and gas turbines?

a) Nickel-based superalloys

b) Cast iron

c) Lead

d) Magnesium alloy

Solution: - a) Nickel-based superalloys

38) In ceramic 3D printing, what is the role of the green body before sintering?

a) To improve the mechanical strength

b) To provide structural support during the initial stages of processing

c) To reduce the thermal expansion of the ceramic

d) To enhance the aesthetic appearance

Solution: - b) To provide structural support during the initial stages of processing

39) Which of the following techniques is used for creating high-resolution metal parts with intricate designs?

a) Stereolithography (SLA)

b) Selective Laser Melting (SLM)

c) Fused Deposition Modeling (FDM)

d) Digital Light Processing (DLP)

Solution: - b) Selective Laser Melting (SLM)

40) Which ceramic material is most suitable for applications requiring high hardness and wear resistance?

a) Silicon carbide

b) Alumina

c) Zirconia

d) Boron nitride

Solution: - a) Silicon carbide

Chapter 14

Post-Processing and Finishing Techniques

14.1 Introduction to Post-Processing and Finishing

Post-processing and finishing are critical steps in the 3D printing process, as they can significantly impact the final quality and functionality of the printed part. Post-processing involves removing support materials, treating the surface, and machining or assembling the part to achieve the desired shape and functionality.

Support Material Removal

Support materials are used in 3D printing to provide additional support to the part during the printing process. These materials are typically removed after the printing process is complete, and the part is removed from the build platform. Support material removal can be a time-consuming and labor-intensive process, and it is essential to use the appropriate tools and techniques to ensure that the support materials are removed without damaging the part.

There are several support material removal techniques used in 3D printing, including:

- **Manual removal:** Manual removal involves using hand tools, such as pliers, knives, or sandpaper, to remove the support materials. This technique is suitable for small parts or parts with simple geometries.

- **Water jet cutting:** Water jet cutting uses a high-pressure stream of water to remove the support materials. This technique is suitable for parts with complex geometries or parts that require precise removal of the support materials.

- **Chemical removal:** Chemical removal involves using chemicals to dissolve the support materials. This technique is suitable for parts with intricate geometries or parts that require the removal of support materials in hard-to-reach areas.

Surface Treatment and Finishing

Surface treatment and finishing are critical steps in the 3D printing process, as they can significantly impact the final quality and functionality of the printed part. Surface treatment and finishing techniques are used to improve the surface finish, texture, and appearance of the part.

There are several surface treatment and finishing techniques used in 3D printing, including:

- **Sanding**: Sanding involves using sandpaper or other abrasive materials to smooth the surface of the part. Sanding can be done manually or using automated equipment.

- **Polishing**: Polishing involves using a polishing compound or a polishing wheel to create a smooth and shiny surface. Polishing can be done manually or using automated equipment.

- **Painting**: Painting involves applying a coating of paint or other materials to the surface of the part. Painting can be done manually or using automated equipment.

- **Plating**: Plating involves applying a thin layer of metal or other materials to the surface of the part. Plating can be done using electroplating or electroless plating techniques.

- **Coating**: Coating involves applying a thin layer of materials, such as powder coatings or liquid coatings, to the surface of the part. Coating can be done using spraying or dipping techniques.

Machining and Assembly

Machining and assembly are critical steps in the 3D printing process, as they can significantly impact the final shape and functionality of the printed part. Machining and assembly techniques are used to achieve the desired shape and functionality of the part, such as drilling holes, tapping threads, or assembling multiple parts.

There are several machining and assembly techniques used in 3D printing, including:

- **Drilling**: Drilling involves using a drill bit to create holes in the part. Drilling can be done manually or using automated equipment.

- **Tapping**: Tapping involves using a tap to create threads in the holes. Tapping can be done manually or using automated equipment.

- **Milling**: Milling involves using a milling cutter to remove material from the part. Milling can be done manually or using automated equipment.

- **Assembly**: Assembly involves joining multiple parts together to create a complete assembly. Assembly can be done using various techniques, such as welding, soldering, or using fasteners.

Post-processing and finishing are critical steps in the 3D printing process, as they can significantly impact the final quality and functionality of the printed part. Support material removal, surface treatment and finishing, and machining and assembly are the three main categories of post-processing and finishing techniques used in 3D printing. By understanding these techniques and selecting the appropriate tools and techniques, designers and engineers can achieve the desired shape and functionality of the printed part, ensuring that it meets the required specifications and functional requirements. With the increasing availability of 3D printing materials and technologies, the post-processing and finishing process is becoming increasingly complex, and the use of automated equipment and software tools is becoming more important. Figure 62 illustrates the various stages involved in post-processing and finishing for 3D printed objects. These steps are crucial for ensuring the final product meets the required specifications and quality standards.

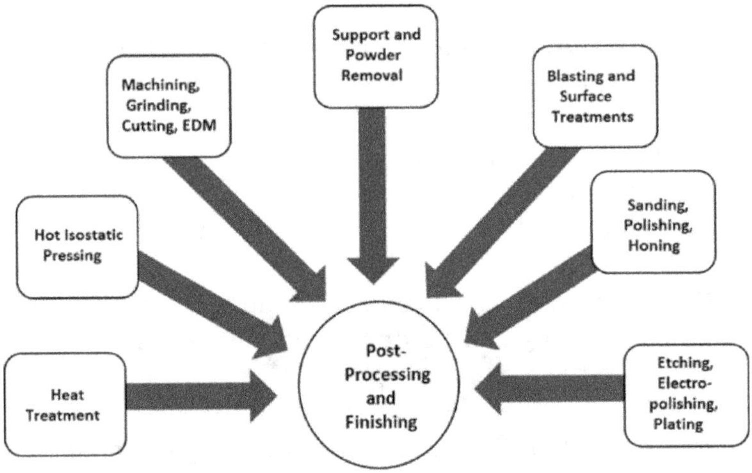

Figure 1: Post-processing and finishing
[Source:
https://ebrary.net/158228/engineering/post_processing_finishing]

14.2 Support Material Removal

Support materials are used to hold the part in place during the 3D printing process. These materials provide additional support to the part, allowing it to maintain its shape and structure during the printing process. However, these materials must be removed after printing to reveal the final part. Support material removal is a critical step in the 3D printing process, as it can significantly impact the final quality and functionality of the printed part.

Mechanical Removal

Mechanical removal involves using tools or machines to physically remove the support material. This technique is commonly used for support materials that are easily removable, such as those used in Fused Deposition Modeling (FDM) and Stereolithography (SLA) printing.

There are several mechanical removal techniques used in 3D printing, including:

- **Manual removal**: Manual removal involves using hand tools, such as pliers, knives, or sandpaper, to remove the support material. This technique is suitable for small parts or parts with simple geometries.

- **Automated removal**: Automated removal involves using machines, such as CNC machines or robotic arms, to remove the support material. This technique is suitable for large parts or parts with complex geometries.

- **Vibration-assisted removal**: Vibration-assisted removal involves using vibration to loosen the support material, making it easier to remove. This technique is suitable for parts with intricate geometries or parts that require precise removal of the support material.

Chemical Dissolution

Chemical dissolution involves using solvents or chemicals to dissolve the support material. This technique is commonly used for support materials that are difficult to remove mechanically, such as those used in Selective Laser Sintering (SLS) and Selective Laser Melting (SLM) printing.

There are several chemical dissolution techniques used in 3D printing, including:

- **Solvent-based Dissolution:** - Solvent-based dissolution involves using a solvent to dissolve the support material. This technique is suitable for parts with simple geometries and small sizes.

- **Enzyme-based Dissolution:** - Enzyme-based dissolution involves using enzymes to break down the support material. This technique is suitable for parts with complex geometries or parts that require precise removal of the support material.

- **Acid-based Dissolution:** - Acid-based dissolution involves using acids to dissolve the support material. This technique is suitable for parts with intricate geometries or parts that require aggressive removal of the support material.

Thermal Removal

Thermal removal involves using heat or thermal energy to melt or vaporize the support material. This technique is commonly used for support materials that are sensitive to heat, such as those used in FDM and SLA printing.

There are several thermal removal techniques used in 3D printing, including:

- **Heat gun removal**: Heat gun removal involves using a heat gun to apply heat to the support material, causing it to melt or vaporize. This technique is suitable for parts with simple geometries and small sizes.

- **Oven removal**: Oven removal involves using an oven to apply heat to the support material, causing it to melt or vaporize. This technique is suitable for parts with complex geometries or parts that require precise removal of the support material.

- **Laser removal**: Laser removal involves using a laser to apply heat to the support material, causing it to melt or vaporize. This technique is suitable for parts with intricate geometries or parts that require aggressive removal of the support material.

Challenges and Limitations

Support material removal can be a challenging and time-consuming process, especially for parts with complex geometries or intricate details. Some of the challenges and limitations of support material removal include:

- **Material compatibility**: The support material must be compatible with the part material to ensure that it can be removed without damaging the part.

- **Removal efficiency**: The removal technique must be efficient to minimize the time and cost of the removal process.

- **Part damage**: The removal technique must be gentle to avoid damaging the part, especially for parts with delicate or intricate features.

Support material removal is a critical step in the 3D printing process, as it can significantly impact the final quality and functionality of the printed part. Mechanical removal, chemical dissolution, and thermal removal are the three main categories of support material removal techniques used in 3D printing. By understanding these techniques and selecting the appropriate tools and techniques, designers and engineers can achieve the desired shape and functionality of the printed part, ensuring that it meets the required specifications and functional requirements. With the increasing availability of 3D printing materials and technologies, the support material removal process is becoming increasingly complex, and the use of automated equipment and software tools is becoming more important. Figure 63 illustrates various Support Material Removal Processes used in 3D printing. The flowchart categorizes these processes into three main types:

Machining:

- **Circular**: Includes Turning and Drilling.
- **Other**: Includes Milling, Shaping, Planing, Broaching, and Gear Generating.
- **Abrasive**:
- **Bonded**: Includes Grinding and Honing.
- **Loose**: Includes Lapping, Polishing, and Buffing.
- **Non-Traditional**: This category is not further subdivided in the image.

These processes are essential for removing support structures and refining the final product to achieve the desired finish and functionality.

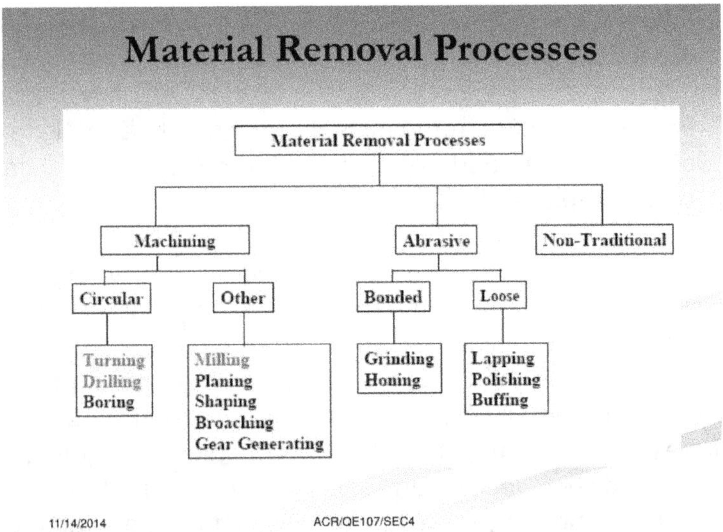

Figure 2: Support material removal processes
[Source: https://www.slideserve.com/karen-mcneil/qe-107-workshop-technology-section-4-metal-cutting]

14.3 Surface Treatment and Finishing

Surface treatment and finishing techniques are used to enhance the appearance and functionality of the printed part. These techniques are essential to improve the surface quality, durability, and overall performance of the part. In this section, we will explore the various surface treatment and finishing techniques used in 3D printing, including sanding and polishing, painting and coating, and anodizing and plating.

Sanding and Polishing

Sanding and polishing are surface treatment techniques used to smooth and refine the surface of the printed part. These techniques involve using abrasive materials, such as sandpaper or polishing compounds, to remove imperfections and create a high-gloss finish.

There are several sanding and polishing techniques used in 3D printing, including:

- **Manual sanding**: Manual sanding involves using hand tools, such as sandpaper or sanding blocks, to smooth the surface of the part. This technique is suitable for small parts or parts with simple geometries.

- **Automated sanding**: Automated sanding involves using machines, such as CNC sanders or robotic arms, to smooth the surface of the part. This technique is suitable for large parts or parts with complex geometries.

- **Polishing compounds**: Polishing compounds, such as wax or ceramic-based compounds, are used to create a high-gloss finish on the surface of the part. These compounds are applied using a polishing wheel or cloth.

The benefits of sanding and polishing include:

- **Improved surface finish**: Sanding and polishing can improve the surface finish of the part, making it smoother and more aesthetically pleasing.

- **Increased durability**: Sanding and polishing can increase the durability of the part by removing imperfections and creating a more uniform surface.

- **Enhanced functionality**: Sanding and polishing can enhance the functionality of the part by improving its surface properties, such as friction or wear resistance.

Painting and Coating

Painting and coating are surface treatment techniques used to protect the surface of the printed part and enhance its appearance. These techniques involve applying a layer of paint or coating to the surface of the part, which can provide additional benefits, such as corrosion resistance or UV protection.

There are several painting and coating techniques used in 3D printing, including:

- **Spray painting**: Spray painting involves using a spray gun to apply a layer of paint to the surface of the part. This technique is suitable for parts with simple geometries and small sizes.

- **Dip coating**: Dip coating involves submerging the part in a tank of coating material, which is then cured in an oven. This technique is suitable for parts with complex geometries or parts that require a uniform coating.

- **Electroplating**: Electroplating involves using an electric current to deposit a layer of material, such as chrome or nickel, onto the surface of the part. This technique is suitable for parts that require a high-gloss finish or additional corrosion resistance.

The benefits of painting and coating include:

- **Corrosion resistance**: Painting and coating can provide corrosion resistance to the part, protecting it from environmental factors such as moisture or chemicals.

- **UV protection**: Painting and coating can provide UV protection to the part, protecting it from the damaging effects of ultraviolet light.

- **Aesthetics**: Painting and coating can enhance the appearance of the part, making it more visually appealing.

Anodizing and Plating

Anodizing and plating are surface treatment techniques used to enhance the surface properties of the printed part. These techniques involve applying a thin layer of material, such as aluminum oxide or titanium nitride, to the surface of the part, which can provide additional benefits, such as increased hardness or corrosion resistance.

There are several anodizing and plating techniques used in 3D printing, including:

- **Anodizing**: Anodizing involves using an electrochemical process to create a thin layer of aluminum oxide on the surface of the part. This technique is suitable for parts made from aluminum or its alloys.

- **Electroplating**: Electroplating involves using an electric current to deposit a layer of material, such as chrome or nickel, onto the surface of the part. This technique is suitable for parts that require a high-gloss finish or additional corrosion resistance.

- **Physical vapor deposition (PVD)**: PVD involves using a vacuum chamber to deposit a thin layer of material, such as titanium nitride or chromium nitride, onto the surface of the part. This technique is suitable for parts that require a high-gloss finish or additional hardness.

The benefits of anodizing and plating include:

- **Increased hardness**: Anodizing and plating can increase the hardness of the part, making it more resistant to wear and tear.
- **Corrosion resistance**: Anodizing and plating can provide corrosion resistance to the part, protecting it from environmental factors such as moisture or chemicals.
- **Enhanced functionality**: Anodizing and plating can enhance the functionality of the part by improving its surface properties, such as friction or wear resistance.

Challenges and Limitations

Surface treatment and finishing techniques can be challenging and time-consuming, especially for parts with complex geometries or intricate details. Some of the challenges and limitations of surface treatment and finishing include:

- **Material compatibility**: The surface treatment or finishing technique must be compatible with the part material to ensure that it does not damage the part.
- **Surface preparation**: The surface of the part must be properly prepared before applying a surface treatment or finishing technique, which can be time-consuming and labor-intensive.
- **Cost and time**: Surface treatment and finishing techniques can be costly and time-consuming, especially for large parts or parts with complex geometries.

Surface treatment and finishing techniques are essential to enhance the appearance and functionality of the printed part. Sanding and polishing, painting and coating, and anodizing and plating are the three main categories of surface treatment and finishing techniques used in 3D printing. By understanding these techniques and selecting the appropriate tools and techniques, designers and engineers can achieve the desired surface quality and functionality of the printed part, ensuring that it meets the required specifications and functional requirements. With the increasing availability of 3D printing materials and technologies, the surface treatment and finishing process is becoming increasingly complex, and the use of automated equipment and software tools is becoming more important. Figure 64 categorizes different surface finishing methods into three main types:

Wet Process Methods

- **Electrical Methods**: Includes techniques like Electro-plating, Anodizing, and Electro-polishing/Passivating.
- **Non-Electrical Methods**: Encompasses Chemical processes such as Etching and Chemical Polishing.

Dry Process Methods

- **Mechanical Methods**: Includes Painting, Thermal Spraying, and Mechanical Plating.
- **In-Vacuum Coating Methods**: Features Vacuum Deposition, Sputtering, and Ion Plating.
- **Diffusion Coating Methods**: Includes processes like Carburizing, Nitriding, Hot Dipping, and Surface Hardening.

These methods are crucial for enhancing the appearance, durability, and performance of materials in various manufacturing applications.

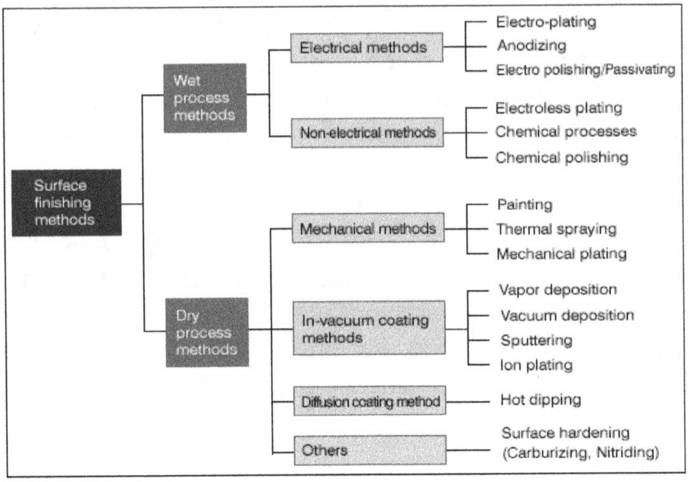

Figure 3: Surface treatment and finishing
[Source: https://www.misumi-techcentral.com/tt/en/surface/2009/07/0001-hydrogen-embrittlement.html]

14.4 Machining and Assembly

Machining and assembly techniques are used to modify the printed part to achieve the desired shape and functionality. These techniques are essential to refine the printed part and prepare it for its intended application. In this section, we will explore the various machining and assembly techniques used in 3D printing, including CNC machining, drilling and tapping, and assembly.

CNC Machining

CNC (Computer Numerical Control) machining is a machining technique used to precision-machine the printed part. This technique involves using computer-controlled machines to remove material from the part, creating a precise and accurate shape. CNC machining is commonly used in 3D printing to create complex geometries, precise holes, and smooth surfaces.

There are several CNC machining techniques used in 3D printing, including:

- **Milling**: Milling involves using a rotating cutting tool to remove material from the part. This technique is suitable for creating complex shapes and geometries.

- **Turning**: Turning involves using a rotating cutting tool to remove material from a rotating part. This technique is suitable for creating cylindrical or symmetrical shapes.

- **Grinding**: Grinding involves using an abrasive wheel to remove material from the part. This technique is suitable for creating high-precision surfaces and edges.

The benefits of CNC machining include:

- **High precision**: CNC machining can achieve high precision and accuracy, making it suitable for applications that require tight tolerances.

- **Flexibility**: CNC machining can be used to create complex shapes and geometries, making it suitable for a wide range of applications.

- **Efficiency**: CNC machining can be faster and more efficient than traditional machining techniques, making it suitable for high-volume production.

Drilling and Tapping

Drilling and tapping are machining techniques used to create holes and threads in the printed part. These techniques involve using drills and taps to remove material from the part, creating a precise and accurate hole or thread. Drilling and tapping are commonly used in 3D printing to create holes for fasteners, screws, or other components.

There are several drilling and tapping techniques used in 3D printing, including:

- **Drilling**: Drilling involves using a rotating cutting tool to create a hole in the part. This technique is suitable for creating holes of various sizes and shapes.

- **Tapping**: Tapping involves using a rotating cutting tool to create a thread in the part. This technique is suitable for creating threads of various sizes and shapes.

The benefits of drilling and tapping include:

- **Precision**: Drilling and tapping can achieve high precision and accuracy, making it suitable for applications that require tight tolerances.

- **Flexibility**: Drilling and tapping can be used to create holes and threads of various sizes and shapes, making it suitable for a wide range of applications.

- **Efficiency**: Drilling and tapping can be faster and more efficient than traditional machining techniques, making it suitable for high-volume production.

Assembly

Assembly is a technique used to combine multiple parts into a single unit. This technique involves using fasteners, adhesives, or other methods to join the parts together, creating a functional and operational assembly. Assembly is commonly used in 3D printing to create complex systems, mechanisms, or products.

There are several assembly techniques used in 3D printing, including:

- **Fastening**: Fastening involves using fasteners, such as screws or bolts, to join the parts together. This technique is suitable for creating strong and durable assemblies.

- **Adhesive bonding**: Adhesive bonding involves using adhesives, such as glues or epoxies, to join the parts together. This technique is suitable for creating strong and durable assemblies.

- **Welding**: Welding involves using heat or pressure to join the parts together. This technique is suitable for creating strong and durable assemblies.

The benefits of assembly include:

- **Functionality**: Assembly can create functional and operational assemblies, making it suitable for a wide range of applications.

- **Customization**: Assembly can be used to create customized assemblies, making it suitable for applications that require unique or specialized designs.

- **Efficiency**: Assembly can be faster and more efficient than traditional manufacturing techniques, making it suitable for high-volume production.

Design Considerations

When designing a part for machining and assembly, there are several considerations to keep in mind. These include:

- **Part geometry**: The part geometry must be designed to accommodate the machining or assembly technique, taking into account factors such as material removal, tool access, and part stability.

- **Material selection**: The material selection must be compatible with the machining or assembly technique, taking into account factors such as material strength, durability, and machinability.

- **Tolerance and fit**: The tolerance and fit of the parts must be designed to ensure proper assembly and functionality.

- **Surface finish**: The surface finish of the parts must be designed to ensure proper assembly and functionality.

Challenges and Limitations

Machining and assembly techniques can be challenging and time-consuming, especially for parts with complex geometries or intricate details. Some of the challenges and limitations of machining and assembly include:

- **Material compatibility**: The machining or assembly technique must be compatible with the part material to ensure that it does not damage the part.
- **Surface preparation**: The surface of the part must be properly prepared before machining or assembly, which can be time-consuming and labor-intensive.

- **Cost and time**: Machining and assembly techniques can be costly and time-consuming, especially for large parts or parts with complex geometries.

Machining and assembly techniques are essential to modify the printed part to achieve the desired shape and functionality. These techniques are commonly used in 3D printing to create complex geometries, precise holes, and smooth surfaces. By understanding these techniques and selecting the appropriate tools and techniques, designers and engineers can achieve the desired shape and functionality of the printed part, ensuring that it meets the required specifications and functional requirements. However, machining and assembly techniques can be challenging and time-consuming, especially for parts with complex geometries or intricate details. Therefore, it is essential to consider the design considerations and challenges and limitations of these techniques to ensure successful and efficient machining and assembly. Figure 65 illustrates various CNC (Computer Numerical Control) Milling Operations. The diagram includes different processes such as:

- **Pocketing**: Removing material from a defined area on the surface of a workpiece.

- **Slotting**: Cutting a slot or groove into the workpiece.
- **Face Milling**: Creating a flat surface on the workpiece.
- **Chamfering**: Cutting a beveled edge on the workpiece.
- **Boring**: Enlarging a hole that has already been drilled or cast.

Each process is depicted with an illustration showing the milling action on a workpiece, including labels for parts like the tool rotation, machined surface, and workpiece. This image is relevant for understanding the different machining operations that can be performed using CNC technology.

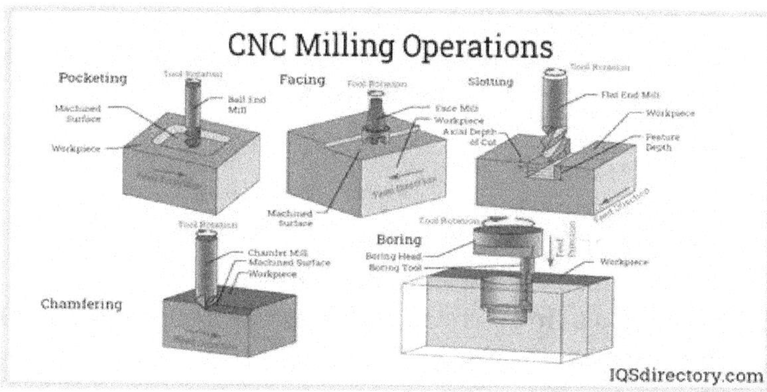

Figure 4: Machining and assembly
[Source: https://www.iqsdirectory.com/articles/cnc-machining.html]

14.5 Questions

14.5.1 Long Response Questions

1. What are the key steps involved in post-processing and finishing?
2. What are some common support material removal techniques?
3. What are some common surface treatment and finishing techniques?
4. What are some common machining and assembly techniques?

5. How do post-processing and finishing techniques impact the final quality and functionality of the printed part?

14.5.2 Multiple Choice Questions

1) **What is the main goal of post-processing in 3D printing?**
   a) To print the object
   b) To improve the object's appearance and functionality
   c) To design the object
   d) To create the object's CAD model
   Solution: - b) To improve the object's appearance and functionality
2) **Which of the following is a common post-processing technique for removing support structures?**
   a) Sanding
   b) Water curing
   c) Mechanical removal
   d) Painting
   Solution: - c) Mechanical removal
3) **What type of material is typically used for sandblasting in post-processing?**
   a) Sand
   b) Metal powder
   c) Glass beads
   d) Plastic pellets
   Solution: - c) Glass beads
4) **Which post-processing technique involves applying a coating to improve surface finish?**
   a) Heat treatment
   b) Painting
   c) Machining
   d) Polishing
   Solution: - b) Painting
5) **Which process is used to smooth the surface of a 3D-printed object using abrasive materials?**
   a) Polishing
   b) Smoothing
   c) Sanding
   d) Cutting

Solution: - c) Sanding

6) **In which technique are objects submerged in a liquid to enhance their finish?**
   a) Resin curing
   b) Chemical smoothing
   c) Thermal annealing
   d) Laser sintering
   Solution: - b) Chemical smoothing

7) **What is the purpose of chemical smoothing in 3D printing?**
   a) To reduce dimensional inaccuracies
   b) To make the object stronger
   c) To smooth the surface of the object
   d) To change the color of the object
   Solution: - c) To smooth the surface of the object

8) **Which of the following is a method used to improve the mechanical properties of a 3D-printed part?**
   a) Sanding
   b) Heat treatment
   c) Painting
   d) Electroplating
   Solution: - b) Heat treatment

9) **Which post-processing method uses ultraviolet (UV) light to cure the material?**
   a) Resin curing
   b) Sandblasting
   c) Heat treatment
   d) Chemical smoothing
   Solution: - a) Resin curing

10) **What does "deburring" refer to in the context of post-processing?**
   a) Removing excess material
   b) Adding color to the object
   c) Enhancing surface gloss
   d) Filling cracks and gaps
   Solution: - a) Removing excess material

11) **Which material is often used in chemical smoothing for FDM 3D-printed parts?**
   a) Acetone

b) Alcohol

c) Resin

d) Water

Solution: - a) Acetone

12) What is the primary advantage of using a vapor smoothing technique?

a) It increases the object's strength

b) It improves dimensional accuracy

c) It smooths and polishes the surface without manual labor

d) It changes the object's color

Solution: - c) It smooths and polishes the surface without manual labor

13) In what situation is thermal annealing particularly useful?

a) For increasing the print speed

b) For improving layer adhesion

c) For enhancing color accuracy

d) For reducing support material

Solution: - b) For improving layer adhesion

14) What is the primary purpose of applying a clear coat finish in post-processing?

a) To add color

b) To improve surface smoothness

c) To protect the object and enhance its appearance

d) To increase the object's strength

Solution: - c) To protect the object and enhance its appearance

15) Which method is commonly used to apply a metallic finish to a 3D-printed part?

a) Painting

b) Electroplating

c) Sandblasting

d) Chemical smoothing

Solution: - b) Electroplating

16) Which technique involves using a hot air gun to smooth the surface of a 3D-printed part?

a) Heat treatment

b) Resin curing

c) Thermal annealing

d) Chemical smoothing

Solution: - c) Thermal annealing

17) **Which of the following post-processing techniques is used to create a high-gloss finish?**
    a) Painting
    b) Sanding
    c) Polishing
    d) Heat treatment

    Solution: - c) Polishing

18) **What is the role of ultrasonic cleaning in post-processing?**
    a) To smooth the surface
    b) To remove fine particles and residues
    c) To apply a color coat
    d) To anneal the material

    Solution: - b) To remove fine particles and residues

19) **Which type of 3D-printed material is most likely to benefit from resin curing?**
    a) PLA
    b) ABS
    c) PETG
    d) Resin

    Solution: - d) Resin

20) **What is a common post-processing method for improving the strength of SLA (stereolithography) printed parts?**
    a) Painting
    b) Heat curing
    c) Sanding
    d) Chemical smoothing

    Solution: - b) Heat curing

21) **Which post-processing technique can be used to alter the mechanical properties and increase the density of a 3D-printed part?**
    a) Electroplating
    b) Heat treatment
    c) Chemical vapor deposition
    d) Sandblasting

    Solution: - b) Heat treatment

22) **What is the primary function of using a solubility-based support material in post-processing?**
   a) To improve surface finish
   b) To simplify the removal of support structures
   c) To enhance mechanical properties
   d) To add color to the part
   Solution: - b) To simplify the removal of support structures

23) **How does post-print thermal annealing affect the mechanical properties of FDM-printed PLA parts?**
   a) It decreases strength and increases brittleness
   b) It increases strength and reduces brittleness
   c) It increases flexibility and reduces strength
   d) It has no significant effect
   Solution: - b) It increases strength and reduces brittleness

24) **Which post-processing method can be used to bond 3D-printed parts with high precision?**
   a) Resin curing
   b) Laser welding
   c) Chemical smoothing
   d) Heat treating
   Solution: - b) Laser welding

25) **What is the primary purpose of using a cryogenic treatment in post-processing?**
   a) To increase the hardness of the material
   b) To smooth the surface of the part
   c) To enhance the object's color
   d) To remove support structures
   Solution: - a) To increase the hardness of the material

26) **Which post-processing technique is specifically used for improving the dimensional accuracy of parts printed with a high shrinkage material?**
   a) Heat treatment
   b) Chemical vapor smoothing
   c) Sanding
   d) Ultrasonic cleaning
   Solution: - a) Heat treatment

27) What does "electroforming" involve in the context of post-processing for 3D printing?

a) Applying a thin metal coating to the part
b) Removing support structures
c) Smoothing the surface using chemicals
d) Enhancing part strength using heat

Solution: - a) Applying a thin metal coating to the part

28) Which technique involves the use of a laser to selectively melt or fuse particles in post-processing?

a) Laser sintering
b) Chemical vapor deposition
c) Heat treatment
d) Electroplating

Solution: - a) Laser sintering

29) How does chemical vapor smoothing affect the dimensional stability of a 3D-printed part?

a) It increases dimensional accuracy
b) It decreases dimensional accuracy
c) It has no effect on dimensional stability
d) It improves flexibility

Solution: - b) It decreases dimensional accuracy

30) Which advanced post-processing technique involves embedding a 3D-printed part into a molten metal bath to create a dense and strong surface layer?

a) Metal coating
b) Metal casting
c) Metal plating
d) Metal injection molding

Solution: - a) Metal coating

31) What is the impact of UV light post-processing on SLA (stereolithography) printed parts?

a) It increases flexibility
b) It enhances color accuracy
c) It improves mechanical strength and durability
d) It reduces surface finish quality

Solution: - c) It improves mechanical strength and durability

32) In what scenario would a post-processing technique like "vapor smoothing" be most advantageous?

a) For parts requiring high strength
b) For parts needing a high level of surface smoothness and finish
c) For enhancing the color of the part
d) For improving dimensional accuracy

Solution: - b) For parts needing a high level of surface smoothness and finish

33) Which post-processing method can be used to enhance the thermal resistance of a 3D-printed part?

a) Thermal annealing
b) Chemical smoothing
c) Resin curing
d) Sanding

Solution: - a) Thermal annealing

34) What role does post-processing play in the development of biomedical implants using 3D printing?

a) It ensures aesthetic qualities
b) It improves biocompatibility and surface finish
c) It enhances color accuracy
d) It simplifies the printing process

Solution: - b) It improves biocompatibility and surface finish

35) Which post-processing technique is used to enhance the clarity and transparency of 3D-printed parts?

a) Chemical vapor smoothing
b) Painting
c) Electroplating
d) Sandblasting

Solution: - a) Chemical vapor smoothing

36) What is the main benefit of using a "polymer infiltration" technique in post-processing?

a) To improve color accuracy
b) To enhance part strength and reduce porosity
c) To smooth the surface
d) To remove support structures

Solution: - b) To enhance part strength and reduce porosity

37) Which post-processing method involves embedding a printed part in a polymer to improve its properties?

a) Polymer infiltration
b) Heat treatment
c) Chemical vapor deposition
d) Sanding

Solution: - a) Polymer infiltration

38) What does "post-print machining" typically involve?

a) Using hand tools to smooth surfaces
b) Employing CNC machines for precise adjustments and finishing
c) Applying coatings
d) Removing support structures

Solution: - b) Employing CNC machines for precise adjustments and finishing

39) Which of the following techniques is used to create a high-quality finish on metal 3D-printed parts?

a) Powder coating
b) Electroplating
c) Vapor smoothing
d) Painting

Solution: - b) Electroplating

40) What is the main advantage of using "automated sanding" in post-processing?

a) It allows for precise material removal with minimal labor
b) It improves the color of the part
c) It increases the part's flexibility
d) It reduces the part's weight

Solution: - a) It allows for precise material removal with minimal labor

Chapter 15

Design for Additive Manufacturing

15.1 Introduction to Design for Additive Manufacturing

Design for Additive Manufacturing (DFAM) is a critical aspect of 3D printing, as it enables designers to take full advantage of the unique capabilities of additive manufacturing. DFAM involves designing parts and products that are optimized for the additive manufacturing process, rather than simply adapting traditional designs for 3D printing. This approach allows designers to create complex geometries, reduce material usage, and improve product performance, making it an essential aspect of 3D printing.

Traditional design methods often focus on designing for manufacturing processes such as casting, machining, or injection molding. However, these methods may not be suitable for additive manufacturing, which offers a unique set of capabilities and constraints. DFAM takes into account the specific characteristics of additive manufacturing, such as layer-by-layer deposition, material properties, and build orientation, to create optimized designs that can only be produced using 3D printing.

The benefits of DFAM include:

- **Increased complexity**: DFAM enables designers to create complex geometries and shapes that would be difficult or impossible to produce using traditional manufacturing methods.

- **Material reduction**: DFAM can help reduce material usage by optimizing the design for minimal material consumption, reducing waste, and minimizing environmental impact.

- **Improved performance**: DFAM can improve product performance by optimizing the design for specific properties, such as strength, stiffness, or thermal conductivity.

- **Reduced assembly**: DFAM can reduce the need for assembly by designing parts that can be printed as a single unit, reducing production time and cost.

In this chapter, we will explore the principles and guidelines of DFAM, including topology optimization, lattice structures, multi-material and multi-functional designs, and design for assembly and disassembly.

Topology Optimization

Topology optimization is a design approach that uses algorithms and simulations to optimize the internal structure of a part or product. This approach can help reduce material usage, improve performance, and create complex geometries that would be difficult to produce using traditional manufacturing methods.

Topology optimization involves using software tools to analyze the design requirements and constraints, such as stress, strain, and material properties. The software then generates a optimized design that meets the requirements and constraints, often resulting in a complex and organic shape.

The benefits of topology optimization include:

- **Material reduction**: Topology optimization can help reduce material usage by optimizing the internal structure of the part or product.

- **Improved performance**: Topology optimization can improve product performance by optimizing the design for specific properties, such as strength, stiffness, or thermal conductivity.

- **Increased complexity**: Topology optimization can create complex geometries and shapes that would be difficult or impossible to produce using traditional manufacturing methods.

Lattice Structures

Lattice structures are a type of design that uses a network of interconnected nodes and struts to create a lightweight and strong structure. Lattice structures can be used to create complex geometries and shapes, and can be optimized for specific properties, such as strength, stiffness, or thermal conductivity.

The benefits of lattice structures include:

- **Light weighting**: Lattice structures can help reduce material usage and weight, making them ideal for applications where weight is a critical factor.

- **Improved performance**: Lattice structures can improve product performance by optimizing the design for specific properties, such as strength, stiffness, or thermal conductivity.

- **Increased complexity**: Lattice structures can create complex geometries and shapes that would be difficult or impossible to produce using traditional manufacturing methods.

Multi-Material and Multi-Functional Designs

Multi-material and multi-functional designs involve using different materials and functions to create a single part or product. This approach can help improve product performance, reduce material usage, and create complex geometries and shapes.

The benefits of multi-material and multi-functional designs include:

- **Improved performance**: Multi-material and multi-functional designs can improve product performance by optimizing the design for specific properties, such as strength, stiffness, or thermal conductivity.

- **Material reduction**: Multi-material and multi-functional designs can help reduce material usage by optimizing the design for minimal material consumption.

- **Increased complexity**: Multi-material and multi-functional designs can create complex geometries and shapes that would be difficult or impossible to produce using traditional manufacturing methods.

Design for Assembly and Disassembly

Design for assembly and disassembly involves designing parts and products that can be easily assembled and disassembled. This approach can help reduce production time and cost, improve product performance, and create complex geometries and shapes.

The benefits of design for assembly and disassembly include:

- **Reduced production time and cost**: Design for assembly and disassembly can help reduce production time and cost by simplifying the assembly process.

- **Improved performance**: Design for assembly and disassembly can improve product performance by optimizing the design for specific properties, such as strength, stiffness, or thermal conductivity.

- **Increased complexity**: Design for assembly and disassembly can create complex geometries and shapes that would be difficult or impossible to produce using traditional manufacturing methods.

In conclusion, DFAM is a critical aspect of 3D printing, as it enables designers to take full advantage of the unique capabilities of additive manufacturing. By following the principles and guidelines of DFAM, designers can create complex geometries, reduce material usage, and improve product performance, making it an essential aspect of 3D printing. However, DFAM can be challenging and time-consuming, especially for designers who are new to the field. Therefore, it is essential to consider the design considerations and challenges and limitations of DFAM to ensure successful and efficient design. Figure 66 outlines a three-step process for designing parts using additive manufacturing techniques:

Functional Specification

Create Design Space & Mechanical Interfaces: Define the design specifications.

Define Load Cases & Boundary Conditions: Establish the conditions under which the part will operate.

Define Targets & Constraints: Set the goals and limitations for the design.

Concepts Generation & Trade-off

Variants Creation: Develop different design options.

Structural Pre-Validation: Perform initial checks to ensure the designs meet structural requirements.

Design Concept Generation: Create detailed design concepts.

Explore Concept: Evaluate and refine the design concepts.

Detailed Modeling and Optimization

Detail Design (For Additive Manufacturing): Finalize the design for additive manufacturing.

Parametric Refinement: Fine-tune the design parameters.

Structural Validation: Confirm the design meets all structural requirements.

Design Review/Conceptual Process Validation: Review and validate the design process.

The flowchart also includes an iterative loop for "Trade-off & Conceptual Selection," indicating that the process involves continuous refinement based on trade-offs and concept selection.

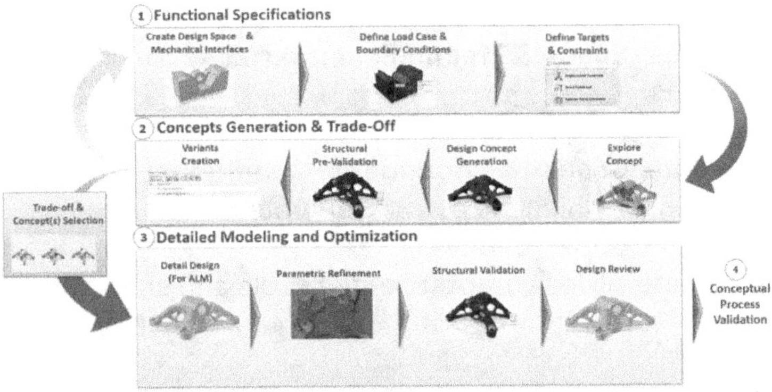

Figure 1: Design for additive manufacturing
[Source: https://www.technia.nl/blog/how-do-you-design-for-manufacturing-excellence-in-the-world-of-additive-manufacturing/]

15.2 Design Principles and Guidelines

DFAM involves a range of design principles and guidelines that are specific to additive manufacturing. These principles and guidelines are essential to ensure that designs are optimized for the additive manufacturing process, taking advantage of its unique capabilities and minimizing its limitations.

Design for Complexity

Additive manufacturing enables the creation of complex geometries and structures that would be difficult or impossible to produce using

traditional manufacturing methods. This is due to the layer-by-layer deposition process, which allows for the creation of intricate shapes and structures without the need for molds or tooling.
Designing for complexity involves considering the following principles:

- **Freeform geometry**: Additive manufacturing enables the creation of freeform geometries that are not limited by traditional manufacturing constraints.

- **Internal structures**: Additive manufacturing allows for the creation of internal structures, such as lattices or honeycombs, that can provide strength and functionality while minimizing material usage.

- **Curved surfaces**: Additive manufacturing enables the creation of curved surfaces that can be used to create complex shapes and structures.

Design for Minimal Material Usage

Additive manufacturing allows for the creation of lightweight structures that use minimal material while maintaining strength and functionality. This is achieved through the use of optimized designs that minimize material usage while maximizing performance.

Designing for minimal material usage involves considering the following principles:

- **Topology optimization**: Topology optimization involves using algorithms and simulations to optimize the internal structure of a part or product, minimizing material usage while maintaining strength and functionality.

- **Lattice structures**: Lattice structures involve using a network of interconnected nodes and struts to create a lightweight and strong structure.

- **Hollow structures**: Hollow structures involve creating parts or products with hollow interiors, reducing material usage while maintaining strength and functionality.

Design for Functional Integration

Additive manufacturing enables the integration of multiple functions and components into a single part or product. This is achieved through the use of complex geometries and structures that can perform multiple functions simultaneously.

Designing for functional integration involves considering the following principles:

- **Multi-functional design**: Multi-functional design involves designing parts or products that can perform multiple functions simultaneously, reducing the need for assembly and improving performance.

- **Component consolidation**: Component consolidation involves integrating multiple components into a single part or product, reducing assembly time and cost.

- **Functional grading**: Functional grading involves designing parts or products with varying material properties and structures to optimize performance and functionality.

Design for Assembly and Disassembly

Additive manufacturing enables the creation of parts and products that can be easily assembled and disassembled. This is achieved through the use of complex geometries and structures that can be designed for easy assembly and disassembly.

Designing for assembly and disassembly involves considering the following principles:

- **Modular design**: Modular design involves designing parts or products that can be easily assembled and disassembled, reducing production time and cost.
- **Snap-fit design**: Snap-fit design involves designing parts or products that can be easily assembled and disassembled using snap-fit connections.
- **Tool-free assembly**: Tool-free assembly involves designing parts or products that can be easily assembled and disassembled without the need for tools.

Design for Post-Processing

Additive manufacturing often requires post-processing techniques, such as machining or grinding, to achieve the desired surface finish and accuracy. Designing for post-processing involves considering the following principles:

- **Surface finish**: Surface finish involves designing parts or products with a surface finish that can be easily post-processed to achieve the desired accuracy and quality.
- **Material selection**: Material selection involves selecting materials that can be easily post-processed, reducing production time and cost.
- **Design for accessibility**: Design for accessibility involves designing parts or products that can be easily accessed for post-processing, reducing production time and cost.

In conclusion, DFAM involves a range of design principles and guidelines that are specific to additive manufacturing. By considering these principles and guidelines, designers can create complex geometries, minimize material usage, integrate multiple functions, and design for assembly and disassembly, post-processing, and other considerations. This enables the creation of optimized designs that take advantage of the unique capabilities of additive manufacturing, while minimizing its limitations. Figure 67 illustrates eight different 3D printed models, each representing a specific design principle or guideline for 3D printing:

- **Max Aspect Ratio**: A tall, thin tower demonstrating the maximum height-to-width ratio that can be achieved.
- **Max Bridging Distance**: A structure showing the longest distance that can be bridged without support.
- **Min Supported Walls**: A model with thin walls that require support during printing.
- **Min Unsupported Walls**: A model with thin walls that can stand without support.
- **Min Vertical Hole**: A model illustrating the smallest vertical hole that can be printed.
- **Max Horizontal Hole**: A model showing the largest horizontal hole that can be printed.
- **Max Overhang Angle**: A cantilevered structure demonstrating the maximum angle at which an overhang can be printed without support.
- **Moving Part Tolerance**: A model with moving parts, showing the necessary tolerance for parts to move freely.

These principles are crucial for designers and engineers to ensure structural integrity and printability when creating objects for 3D printing.

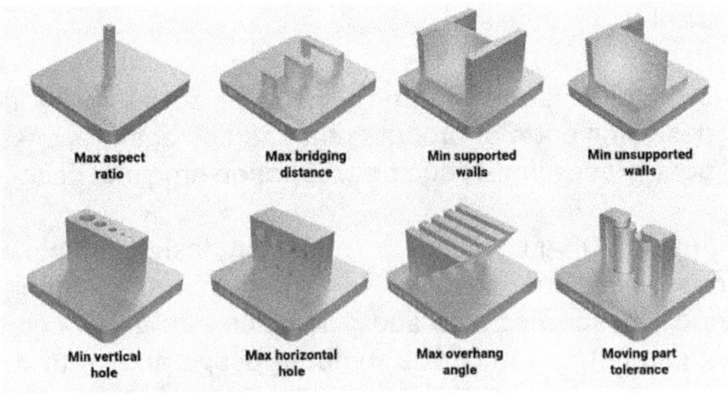

Figure 2: Design principles and guidelines
[Source: https://altair.com/newsroom/articles/four-key-principles-of-design-for-additive-manufacturing-dfam]

15.3 Topology Optimization and Lattice Structures

Topology optimization is a design approach that uses algorithms and simulation to optimize the internal structure of a part or product for maximum performance and minimal material usage. This approach is particularly useful for additive manufacturing, as it enables the creation of complex geometries and structures that would be difficult or impossible to produce using traditional manufacturing methods.

Topology optimization involves using software tools to analyze the design requirements and constraints, such as stress, strain, and material properties. The software then generates a optimized design that meets the requirements and constraints, often resulting in a complex and organic shape.

The benefits of topology optimization include:

- **Material reduction**: Topology optimization can help reduce material usage by optimizing the internal structure of the part or product.

- **Improved performance**: Topology optimization can improve product performance by optimizing the design for specific properties, such as strength, stiffness, or thermal conductivity.

- **Increased complexity**: Topology optimization can create complex geometries and shapes that would be difficult or impossible to produce using traditional manufacturing methods.

Lattice structures are a type of topology-optimized design that use repeating patterns of interconnected nodes and struts to create strong and lightweight structures. Lattice structures can be used to create a wide range of products, from aerospace components to medical implants.

The benefits of lattice structures include:

- **Lightweighting**: Lattice structures can help reduce material usage and weight, making them ideal for applications where weight is a critical factor.

- **Improved performance**: Lattice structures can improve product performance by optimizing the design for specific properties, such as strength, stiffness, or thermal conductivity.

- **Increased complexity**: Lattice structures can create complex geometries and shapes that would be difficult or impossible to produce using traditional manufacturing methods.

Types of Lattice Structures

There are several types of lattice structures, each with its own unique characteristics and benefits. Some of the most common types of lattice structures include:

- **Body-centered cubic (BCC) lattices**: BCC lattices are a type of lattice structure that uses a repeating pattern of interconnected nodes and struts to create a strong and lightweight structure.

- **Face-centered cubic (FCC) lattices**: FCC lattices are a type of lattice structure that uses a repeating pattern of interconnected nodes and struts to create a strong and lightweight structure.

- **Hexagonal lattices**: Hexagonal lattices are a type of lattice structure that uses a repeating pattern of interconnected nodes and struts to create a strong and lightweight structure.

- **Triangular lattices**: Triangular lattices are a type of lattice structure that uses a repeating pattern of interconnected nodes and struts to create a strong and lightweight structure.

Design Considerations for Lattice Structures

When designing lattice structures, there are several considerations that must be taken into account. These include:

- **Material selection**: The material selected for the lattice structure must be able to withstand the stresses and strains imposed by the design.

- **Node and strut design**: The design of the nodes and struts must be optimized for maximum strength and stiffness.

- **Lattice density**: The density of the lattice structure must be optimized for maximum performance and minimal material usage.

- **Manufacturing constraints**: The design of the lattice structure must take into account the manufacturing constraints of the additive manufacturing process.

Applications of Lattice Structures

Lattice structures have a wide range of applications, including:

- **Aerospace**: Lattice structures are used in aerospace applications to create lightweight and strong components, such as aircraft frames and engine components.

- **Medical**: Lattice structures are used in medical applications to create implants and devices, such as hip and knee replacements.

- **Automotive**: Lattice structures are used in automotive applications to create lightweight and strong components, such as engine mounts and suspension components.

- **Consumer products**: Lattice structures are used in consumer products to create lightweight and strong components, such as bicycle frames and sporting goods.

In conclusion, topology optimization and lattice structures are powerful design tools that can be used to create complex geometries and structures that would be difficult or impossible to produce using traditional manufacturing methods. By understanding the benefits and design considerations of lattice structures, designers can create optimized designs that take advantage of the unique capabilities of additive manufacturing. Figure 68 illustrates the process of topology optimization and the creation of lattice structures, divided into two main sections:

Topology Optimization Process

- **Initial Setup**: A gray cube with a downward load of 300 MPa and a fixed corner constraint.
- **Optimized Structure**: A mesh-like structure where material has been removed to reduce weight while maintaining structural integrity.
- **Smooth**: The final, smooth, organic form of the optimized structure.

Lattice Structure Modelling

- **Complex 3D Lattice Structure**: An intricate lattice structure within a bounding box measuring 20 mm x 20 mm x 22 mm.
- **Unit Lattice Structure**: Detailed views of the unit lattice structures that make up the larger lattice model, showing how these units are mirrored and repeated to create complex geometries.

These visualizations explain how engineers can optimize materials for strength and efficiency using computational methods, which is crucial in fields like aerospace, automotive design, and biomedical implants where weight reduction without compromising strength is essential.

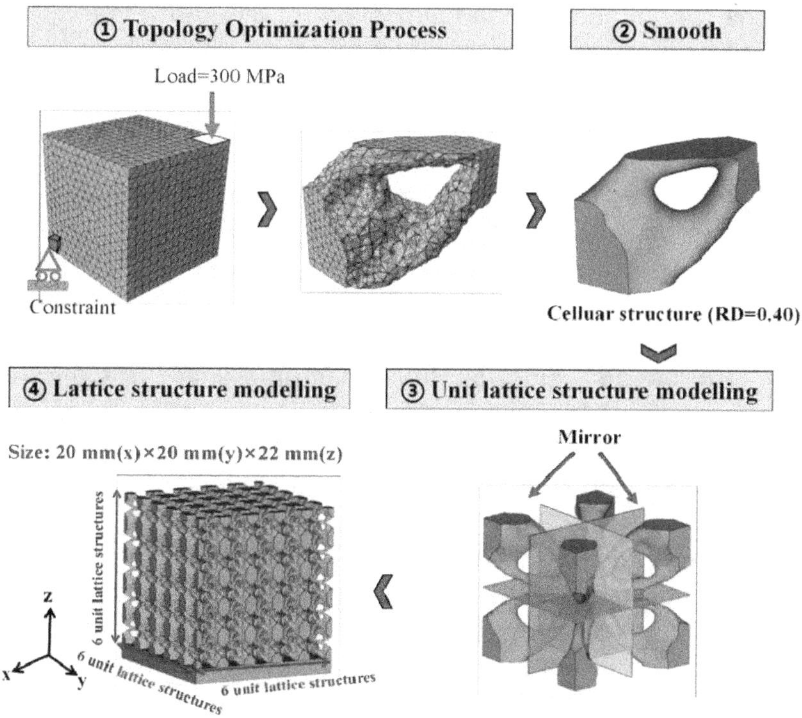

Figure 3: Topology optimization and lattice structures
[Source: https://www.mdpi.com/1996-1944/16/4/1700]

15.4 Multi-Material and Multi-Functional Designs

Additive manufacturing enables the creation of multi-material and multi-functional designs that combine different materials and functions into a single part or product. This can include the use of different materials for different components or functions, as well as the integration of sensors, actuators, and other functional elements.

Multi-Material Designs

Multi-material designs involve the use of different materials for different components or functions within a single part or product. This can be achieved through the use of multiple material deposition systems, which allow for the deposition of different materials in a single build process.

The benefits of multi-material designs include:

- **Improved performance**: Multi-material designs can improve product performance by allowing for the use of materials with specific properties in specific areas of the part or product.
- **Increased functionality**: Multi-material designs can increase functionality by allowing for the integration of different materials with different properties, such as conductivity, thermal resistance, or optical transparency.
- **Reduced material usage**: Multi-material designs can reduce material usage by allowing for the use of materials with specific properties in specific areas of the part or product, rather than using a single material throughout.

Multi-Functional Designs

Multi-functional designs involve the integration of sensors, actuators, and other functional elements into a single part or product. This can include the use of embedded sensors to monitor temperature, pressure, or other environmental factors, as well as the use of actuators to control movement or other functions.

The benefits of multi-functional designs include:

- **Improved performance**: Multi-functional designs can improve product performance by allowing for real-time monitoring and control of environmental factors.
- **Increased functionality**: Multi-functional designs can increase functionality by allowing for the integration of multiple functions into a single part or product.

- **Reduced complexity**: Multi-functional designs can reduce complexity by eliminating the need for multiple components and assemblies.

Design Considerations for Multi-Material and Multi-Functional Designs

When designing multi-material and multi-functional designs, there are several considerations that must be taken into account. These include:

- **Material compatibility**: The materials used in the design must be compatible with each other, in terms of their thermal, mechanical, and chemical properties.

- **Functional integration**: The functional elements, such as sensors and actuators, must be integrated into the design in a way that allows for optimal performance and functionality.

- **Manufacturing constraints**: The design must take into account the manufacturing constraints of the additive manufacturing process, including the limitations of material deposition and resolution.

- **Testing and validation**: The design must be tested and validated to ensure that it meets the required performance and functionality standards.

Applications of Multi-Material and Multi-Functional Designs

Multi-material and multi-functional designs have a wide range of applications, including:

- **Aerospace**: Multi-material and multi-functional designs are used in aerospace applications to create lightweight and high-performance components, such as aircraft frames and engine components.

- **Medical**: Multi-material and multi-functional designs are used in medical applications to create implants and devices, such as hip and knee replacements, that can monitor and respond to environmental factors.

- **Automotive**: Multi-material and multi-functional designs are used in automotive applications to create lightweight and high-performance components, such as engine mounts and suspension components.

- **Consumer products**: Multi-material and multi-functional designs are used in consumer products to create innovative and functional products, such as smart home devices and wearable technology.

Case Studies

Several companies have successfully implemented multi-material and multi-functional designs in their products, including:

- **GE Aviation**: GE Aviation has developed a multi-material jet engine component that combines ceramic and metal materials to improve performance and reduce weight.

- **Medtronic**: Medtronic has developed a multi-functional pacemaker that integrates sensors and actuators to monitor and respond to heart rhythms.

- **BMW**: BMW has developed a multi-material car door that combines carbon fiber and metal materials to improve performance and reduce weight.

In conclusion, multi-material and multi-functional designs are powerful tools that can be used to create innovative and functional products that take advantage of the unique capabilities of additive manufacturing. By understanding the benefits and design considerations of multi-material and multi-functional designs, designers can create optimized designs that meet the required performance and functionality standards. Figure 69 illustrates the concept of multi-material and multi-functional designs, specifically using a turbine jet engine blade as an example. The image is divided into two main sections:

Tailored Properties via Multi-Material and Microstructure

This section shows three colored microstructure images, indicating different material compositions or structures at a microscopic level. These tailored properties are achieved by manipulating materials to optimize specific functions.

Turbine Blade with Annotations

The right side features a photograph of a turbine blade with annotations pointing to specific features:

- **A**: High impact-resistant area.
- **B**: Fatigue-resistant area.
- **C**: Creep-resistant area.

Below this, a close-up image shows an "Inner lattice structure," which contributes to the blade's lightweight and conductivity properties.

This image highlights how advanced engineering techniques can create components with specific properties optimized for certain functions, making them more efficient and durable for high-performance applications like jet engines.

Figure: Example of a component with tailored functional properties (shown here: turbine jet engine blisk).

Figure 4: Multi-material and multi-functional designs
[Source: https://www.m2i.nl/workshop-enabling-multi-functional-design-through-multi-material-additive-manufacturing-engage-in-am/]

15.5 Design for Assembly and Disassembly

Design for assembly and disassembly involves designing parts and products that can be easily assembled and disassembled for maintenance, repair, and recycling. This can include the use of snap-fit joints, modular components, and other design features that enable easy assembly and disassembly.

Benefits of Design for Assembly and Disassembly

The benefits of designing for assembly and disassembly include:

- **Reduced assembly time and cost**: Designing for assembly and disassembly can reduce the time and cost associated with assembling and disassembling products.
- **Improved maintainability**: Designing for assembly and disassembly can make it easier to maintain and repair products, reducing downtime and increasing overall efficiency.
- **Increased recyclability**: Designing for assembly and disassembly can make it easier to recycle products, reducing waste and promoting sustainability.
- **Enhanced product customization**: Designing for assembly and disassembly can enable product customization, allowing customers to easily modify or upgrade products to meet their specific needs.

Design Strategies for Assembly and Disassembly

Several design strategies can be used to facilitate assembly and disassembly, including:

- **Modular design**: Modular design involves breaking down a product into smaller, independent components that can be easily assembled and disassembled.

- **Snap-fit joints**: Snap-fit joints are a type of joint that can be easily assembled and disassembled without the need for fasteners or adhesives.

- **Quick-release mechanisms**: Quick-release mechanisms are design features that enable easy disassembly, such as clips, latches, or other types of release mechanisms.

- **Standardized components**: Standardized components can be used to simplify assembly and disassembly, reducing the number of unique parts and making it easier to interchange components.

Design Considerations for Assembly and Disassembly

When designing for assembly and disassembly, several considerations must be taken into account, including:

- **Material selection**: The materials used in the design must be compatible with the assembly and disassembly process, and must be able to withstand the stresses and strains associated with repeated assembly and disassembly.

- **Tolerances and clearances**: The tolerances and clearances between components must be carefully controlled to ensure easy assembly and disassembly.

- **Component accessibility**: Components must be designed to be easily accessible for assembly and disassembly, reducing the need for specialized tools or expertise.

- **Disassembly sequence**: The disassembly sequence must be carefully planned to ensure that components can be easily removed and reassembled in the correct order.

Additive Manufacturing and Assembly

Additive manufacturing can be used to create complex geometries and structures that would be difficult or impossible to produce using traditional manufacturing methods. This can enable the creation of products that are optimized for assembly and disassembly, with features such as:

- **Internal structures**: Additive manufacturing can be used to create internal structures that provide additional strength and stability, while also facilitating assembly and disassembly.
- **Customized components**: Additive manufacturing can be used to create customized components that are optimized for specific applications, reducing the need for standardized components.
- **Integrated features**: Additive manufacturing can be used to integrate features such as snap-fit joints, quick-release mechanisms, and other design elements that facilitate assembly and disassembly.

Case Studies

Several companies have successfully implemented design for assembly and disassembly in their products, including:

- **IKEA**: IKEA has developed a range of products that are designed for easy assembly and disassembly, using modular components and snap-fit joints to simplify the assembly process.
- **Tesla**: Tesla has developed electric vehicles that are designed for easy maintenance and repair, using modular components and quick-release mechanisms to simplify the disassembly process.
- **Dell**: Dell has developed computer systems that are designed for easy upgrade and customization, using standardized components and snap-fit joints to simplify the assembly and disassembly process.

In conclusion, design for assembly and disassembly is an important consideration in the development of products that are optimized for maintenance, repair, and recycling. By using design strategies such as modular design, snap-fit joints, and quick-release mechanisms, designers can create products that are easier to assemble and disassemble, reducing costs and improving overall efficiency. Figure 70 illustrates various snap fit designs used for assembly and disassembly, categorized into four types:

Cantilever

Filled & Unfilled Materials: Shows two variations of cantilever snap fits, one for filled materials and one for unfilled materials.

Cylindrical

Depicts a cylindrical snap fit with a partial cutaway view to show internal features.

On a Block (Solid Wall)

Illustrates a snap fit being inserted into a solid wall.

On a Plate (Thin Wall)

Shows a snap fit being inserted into slots on a thin plate, with numbered steps indicating the order of assembly. These designs are essential for creating products that are easy to assemble and disassemble, which is crucial for maintenance, repair, and recycling.

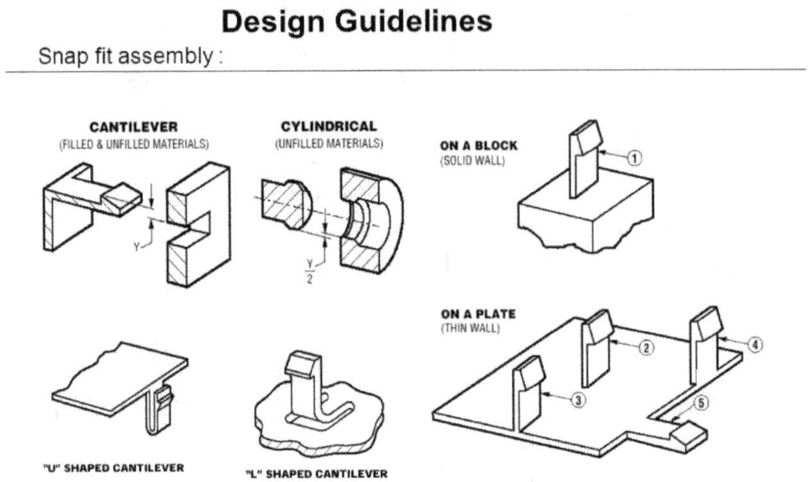

Figure 5: Design for assembly and disassembly
[Source: https://creativeidee.github.io/design-photo/basf-snap-fit-design-guide/]

15.6 Questions

15.6.1 Long Response Questions

1. What are the key principles and guidelines of design for additive manufacturing?
2. How does topology optimization enable the creation of optimized internal structures?
3. What are the benefits of using lattice structures in additive manufacturing?
4. How can multi-material and multi-functional designs be used to create complex products?
5. What are the benefits of designing for assembly and disassembly in additive manufacturing?

15.6.2 Multiple Choice Questions

1) **What does DFAM stand for?**
   a) Design for Automated Manufacturing
   b) Design for Additive Manufacturing

c) Design for Advanced Materials

d) Design for Assembly and Manufacturing

Solution: - b) Design for Additive Manufacturing

2) **What is one major benefit of using DFAM in 3D printing?**

a) Increased complexity

b) Reduced material usage

c) Improved performance

d) All of the above

Solution: - d) All of the above

3) **Which design approach uses algorithms to optimize the internal structure of a part?**

a) Functional Integration

b) Topology Optimization

c) Modular Design

d) Multi-material Design

Solution: - b) Topology Optimization

4) **What is a characteristic feature of lattice structures?**

a) They use a single material

b) They are designed for low weight and high strength

c) They are always solid

d) They require no support during printing

Solution: - b) They are designed for low weight and high strength

5) **Which principle involves reducing the need for assembly in DFAM?**

a) Design for Minimal Material Usage

b) Design for Assembly and Disassembly

c) Design for Functional Integration

d) Design for Post-Processing

Solution: - b) Design for Assembly and Disassembly

6) **What is a key advantage of using hollow structures in 3D printing?**

a) They are more expensive

b) They are heavier

c) They reduce material usage

d) They are less complex

Solution: - c) They reduce material usage

7) **Which type of lattice structure has a repeating pattern of interconnected nodes and struts in a cube?**

a) Face-centered cubic (FCC)
b) Hexagonal
c) Body-centered cubic (BCC)
d) Triangular

Solution: - c) Body-centered cubic (BCC)

8) **What is an example of a multi-functional design in DFAM?**

a) A single material part
b) A part with integrated multiple functions
c) A simple geometric shape
d) A non-optimized structure

Solution: - b) A part with integrated multiple functions

9) **Which DFAM principle involves using complex geometries to create parts?**

a) Design for Minimal Material Usage
b) Design for Complexity
c) Design for Functional Integration
d) Design for Post-Processing

Solution: - b) Design for Complexity

10) **In DFAM, what does the "snap-fit design" principle refer to?**

a) Designing parts for tool-free assembly
b) Designing parts for easy post-processing
c) Designing parts for quick fitting without tools
d) Designing parts with complex internal structures

Solution: - c) Designing parts for quick fitting without tools

11) **What is the main goal of topology optimization in additive manufacturing?**

a) To create aesthetic designs
b) To reduce material usage while maintaining performance
c) To simplify assembly processes
d) To increase part weight

Solution: - b) To reduce material usage while maintaining performance

12) **Which of the following is NOT a type of lattice structure mentioned in the chapter?**

a) Hexagonal
b) Triangular

c) Octagonal

d) Face-centered cubic (FCC)

Solution: - c) Octagonal

13) What does the term "design for post-processing" involve in DFAM?

a) Designing parts for easy assembly

b) Designing parts to reduce material usage

c) Designing parts that will require additional finishing or machining

d) Designing parts with complex internal structures

Solution: - c) Designing parts that will require additional finishing or machining

14) Which design principle helps in reducing production time and cost by simplifying the assembly process?

a) Design for Assembly and Disassembly

b) Design for Minimal Material Usage

c) Design for Complexity

d) Design for Functional Integration

Solution: - a) Design for Assembly and Disassembly

15) In DFAM, what does "functional grading" refer to?

a) Designing parts with varying material properties

b) Designing parts for post-processing

c) Designing parts with uniform material properties

d) Designing parts to be easily disassembled

Solution: - a) Designing parts with varying material properties

16) Which DFAM principle is focused on creating parts with minimal material while maintaining structural integrity?

a) Design for Complexity

b) Design for Minimal Material Usage

c) Design for Functional Integration

d) Design for Assembly and Disassembly

Solution: - b) Design for Minimal Material Usage

17) What is the primary benefit of using a modular design approach in additive manufacturing?

a) Increased weight

b) Simplified post-processing

c) Easier assembly and disassembly

d) Increased material usage

Solution: - c) Easier assembly and disassembly

18) Which of the following is a consideration when designing lattice structures?

a) Maximizing node size
b) Minimizing strut density
c) Material selection
d) Avoiding internal structures

Solution: - c) Material selection

19) What does the iterative loop in the DFAM process represent?

a) Continuous improvement based on trade-offs and concept selection
b) One-time optimization
c) Linear design steps without review
d) Sequential validation only

Solution: - a) Continuous improvement based on trade-offs and concept selection

20) Which lattice structure type uses a repeating pattern of interconnected nodes arranged in a hexagon?

a) Triangular lattice
b) Face-centered cubic (FCC) lattice
c) Hexagonal lattice
d) Body-centered cubic (BCC) lattice

Solution: - c) Hexagonal lattice

21) How does DFAM differ from traditional design methods in terms of complexity?

a) DFAM avoids complexity
b) DFAM requires less complex designs
c) DFAM enables more complex geometries and structures
d) Traditional design methods support complex designs better

Solution: - c) DFAM enables more complex geometries and structures

22) What is the role of parametric refinement in the DFAM process?

a) To finalize the design for additive manufacturing
b) To evaluate design concepts
c) To adjust and fine-tune design parameters for optimization

d) To perform initial structural validation

Solution: - c) To adjust and fine-tune design parameters for optimization

23) In topology optimization, what typically happens to the material distribution within the design space?

a) It remains uniform

b) It is removed or redistributed to minimize material while maintaining performance

c) It is increased to add more features

d) It is replaced with additional material

Solution: - b) It is removed or redistributed to minimize material while maintaining performance

24) Which design approach focuses on integrating multiple functions into a single part to reduce assembly time?

a) Design for Minimal Material Usage

b) Design for Functional Integration

c) Design for Complexity

d) Design for Post-Processing

Solution: - b) Design for Functional Integration

25) What is a common application of lattice structures in aerospace engineering?

a) Creating complex surface finishes

b) Designing lightweight and strong components

c) Reducing the number of parts in an assembly

d) Enhancing color properties of components

Solution: - b) Designing lightweight and strong components

26) Which type of lattice structure uses a pattern of nodes and struts in a repeating triangular arrangement?

a) Hexagonal lattice

b) Triangular lattice

c) Face-centered cubic (FCC) lattice

d) Body-centered cubic (BCC) lattice

Solution: - b) Triangular lattice

27) What challenge does the "Design for Post-Processing" principle address in 3D printing?

a) Simplifying assembly processes

b) Ensuring the part can be easily finished or machined after printing

c) Reducing material usage
d) Increasing part strength during printing
Solution: - b) Ensuring the part can be easily finished or machined after printing

28) How does functional grading contribute to the performance of a 3D printed part?

a) By maintaining uniform material properties throughout the part
b) By allowing for varying material properties to optimize performance
c) By simplifying the design process
d) By increasing the material cost
Solution: - b) By allowing for varying material properties to optimize performance

29) What is the primary purpose of using multi-material design in DFAM?

a) To create aesthetic variations
b) To integrate different functional properties into a single part
c) To reduce the cost of materials
d) To simplify post-processing
Solution: - b) To integrate different functional properties into a single part

30) What is a key consideration when designing parts with complex geometries in DFAM?

a) Ensuring ease of manual assembly
b) Simplifying internal structures
c) Accounting for support material and printing constraints
d) Reducing the part's weight
Solution: - c) Accounting for support material and printing constraints

31) What role does iterative testing play in the DFAM process?

a) It ensures that the design is finalized without changes.
b) It helps to refine and optimize the design through multiple iterations.
c) It focuses on initial concept creation only.
d) It finalizes material selection.

Solution: - b) It helps to refine and optimize the design through multiple iterations.

32) **What type of design is most suitable for complex geometries that traditional manufacturing methods cannot produce?**

a) Simple geometric shapes
b) Modular design
c) Additive manufacturing (AM) design
d) Traditional manufacturing design

Solution: - c) Additive manufacturing (AM) design

33) **Which design approach in DFAM is used to optimize part strength while minimizing material usage?**

a) Functional Integration
b) Topology Optimization
c) Multi-material Design
d) Design for Assembly

Solution: - b) Topology Optimization

34) **What does "design for manufacturability" focus on in the context of DFAM?**

a) Simplifying design for cost-effectiveness
b) Ensuring the part is easy to produce with additive manufacturing technologies
c) Enhancing post-processing techniques
d) Integrating multiple functions into one part

Solution: - b) Ensuring the part is easy to produce with additive manufacturing technologies

35) **In DFAM, what does "design for build orientation" refer to?**

a) Optimizing the orientation of the part during printing for better mechanical properties and reduced support material
b) Designing parts that do not require any supports
c) Focusing on the final appearance of the part
d) Creating parts that are easy to disassemble

Solution: - a) Optimizing the orientation of the part during printing for better mechanical properties and reduced support material

36) **Which principle in DFAM involves designing parts to minimize the need for additional finishing work?**

a) Design for Post-Processing
b) Design for Complexity
c) Design for Functional Integration
d) Design for Assembly

Solution: - a) Design for Post-Processing

37) How does "design for performance" in DFAM ensure a part meets its functional requirements?

a) By focusing on aesthetic features
b) By optimizing material properties and structural design for specific functions
c) By simplifying the assembly process
d) By reducing manufacturing costs

Solution: - b) By optimizing material properties and structural design for specific functions

38) Which DFAM strategy helps to reduce post-processing time and effort?

a) Multi-material Design
b) Topology Optimization
c) Design for Minimal Material Usage
d) Design for Build Orientation

Solution: - d) Design for Build Orientation

39) What is a critical factor to consider when designing parts for additive manufacturing in terms of build volume?

a) The overall cost of materials
b) The part's dimensions relative to the printer's build volume
c) The color accuracy of the material
d) The number of parts in the design

Solution: - b) The part's dimensions relative to the printer's build volume

40) Which design aspect ensures that a 3D-printed part is optimized for strength and efficiency while using the least amount of material?

a) Functional Integration
b) Topology Optimization
c) Multi-material Design
d) Modular Design

Solution: - b) Topology Optimization

Chapter 16

Design for Additive Manufacturing (continued)

16.1 Design for Material Properties and Anisotropy

Additive manufacturing enables the creation of parts and products with unique material properties and anisotropy. Material properties such as strength, stiffness, and toughness can be tailored to specific applications, while anisotropy can be used to create directional properties that enhance performance. This level of customization is not possible with traditional manufacturing methods, making additive manufacturing an attractive option for designers and engineers.

Designing for material properties and anisotropy involves understanding the relationships between material composition, processing conditions, and part geometry. This includes:

Material Selection

Selecting materials with specific properties and anisotropy for specific applications is a critical aspect of design for additive manufacturing. Different materials exhibit unique properties, such as strength, stiffness, toughness, and thermal conductivity, that can be leveraged to achieve specific performance goals. For example, titanium alloys are often used in aerospace applications due to their high strength-to-weight ratio, while copper alloys are used in heat exchangers due to their high thermal conductivity.

In addition to material properties, anisotropy must also be considered during material selection. Anisotropy refers to the direction-dependent properties of a material, which can be influenced by the manufacturing process. For example, in fused deposition modeling (FDM), the layer-by-layer deposition process can create anisotropic properties, such as higher strength in the x-y plane than in the z-axis. Understanding the anisotropic properties of a material is essential to designing parts that take advantage of these properties.

Process Optimization

Optimizing processing conditions is critical to achieving desired material properties and anisotropy. Processing conditions, such as temperature, pressure, and layer thickness, can significantly impact the final properties of a part. For example, in selective laser sintering (SLS), the laser power and scan speed can be adjusted to achieve specific material properties, such as density and surface roughness.

Process optimization involves understanding the relationships between processing conditions and material properties. This can be achieved through experimentation, simulation, and modeling. For example, finite element analysis (FEA) can be used to simulate the effects of different processing conditions on material properties, allowing designers to optimize the process before producing a physical part.

Geometry Optimization

Optimizing part geometry is essential to taking advantage of material properties and anisotropy. Geometry optimization involves designing parts that maximize the benefits of additive manufacturing, such as complex geometries and internal structures. For example, lattice structures can be designed to achieve specific properties, such as stiffness and strength, while minimizing material usage.

Geometry optimization also involves considering the anisotropic properties of a material. For example, in a part with anisotropic properties, the orientation of the material can be optimized to achieve specific performance goals. This can be achieved through topology optimization, which involves using algorithms to generate optimal geometries based on performance criteria.

Design Considerations

When designing for material properties and anisotropy, several considerations must be taken into account. These include:

- **Material compatibility**: Ensuring that the selected material is compatible with the desired application and processing conditions.

- **Part functionality**: Ensuring that the part design meets the required functional performance criteria.

- **Manufacturing constraints**: Ensuring that the part design can be manufactured within the constraints of the additive manufacturing process.

- **Cost and lead time**: Ensuring that the part design is cost-effective and can be produced within the required lead time.

Case Studies

Several case studies demonstrate the benefits of designing for material properties and anisotropy in additive manufacturing. For example:

- **Aerospace**: A company used additive manufacturing to produce a lightweight, high-strength aircraft component. By optimizing the material properties and anisotropy of the part, the company was able to reduce the weight of the component by 30% while maintaining its structural integrity.

- **Medical**: A researcher used additive manufacturing to produce a custom implant with tailored material properties and anisotropy. The implant was designed to mimic the natural properties of bone, allowing for improved osseointegration and reduced risk of implant failure.

- **Automotive**: A company used additive manufacturing to produce a lightweight, high-performance engine component. By optimizing the material properties and anisotropy of the part, the company was able to reduce the weight of the component by 20% while maintaining its structural integrity.

Challenges and Limitations

While designing for material properties and anisotropy offers many benefits, there are also several challenges and limitations to consider. These include:

- **Material variability**: Material properties can vary depending on the specific batch and processing conditions, making it challenging to achieve consistent results.

- **Process variability**: Processing conditions can also vary, making it challenging to achieve consistent results.

- **Limited material options**: The range of materials available for additive manufacturing is limited, making it challenging to find materials with specific properties and anisotropy.

- **Design complexity**: Designing for material properties and anisotropy can be complex, requiring specialized knowledge and expertise.

Designing for material properties and anisotropy is a critical aspect of additive manufacturing. By understanding the relationships between material composition, processing conditions, and part geometry, designers can create parts and products with unique properties and performance. While there are challenges and limitations to consider, the benefits of designing for material properties and anisotropy make it a valuable tool for designers and engineers.

In summary, designing for material properties and anisotropy involves selecting materials with specific properties and anisotropy for specific applications, optimizing processing conditions to achieve desired material properties and anisotropy, and optimizing part geometry to take advantage of material properties and anisotropy. By considering material compatibility, part functionality, manufacturing constraints, and cost and lead time, designers can create parts and products that meet the required performance criteria while minimizing material usage and reducing weight. Case studies demonstrate the benefits of designing for material properties and anisotropy in additive manufacturing, while challenges and limitations must also be considered. Figure 71, titled "Anisotropy and Material Properties," is divided into three main sections:

Properties

- **Stress-Strain Graphs**: Shows stress (in MPa) against strain for materials like Nickel, Iron, and Copper with different textures labeled 'Cube,' 'Goss,' and 'Brass.'

- **Multi-Axial Stress State**: A graph with axes labeled σ1, σ2, and σ3 representing different stress states in a material.

Single vs. Polycrystalline

- **Single Crystal Structure**: Depicts an orderly arrangement of atoms.

- **Polycrystalline Structure**: Shows clusters of atoms in various orientations.
- **Processing**:

- **Grain Structures**: Illustrates how random grain structures change through processes like drawing and rolling, resulting in elongated grains oriented in one direction.

This image explains how material properties can vary depending on direction (anisotropy) and how processing methods influence these properties by altering grain structures.

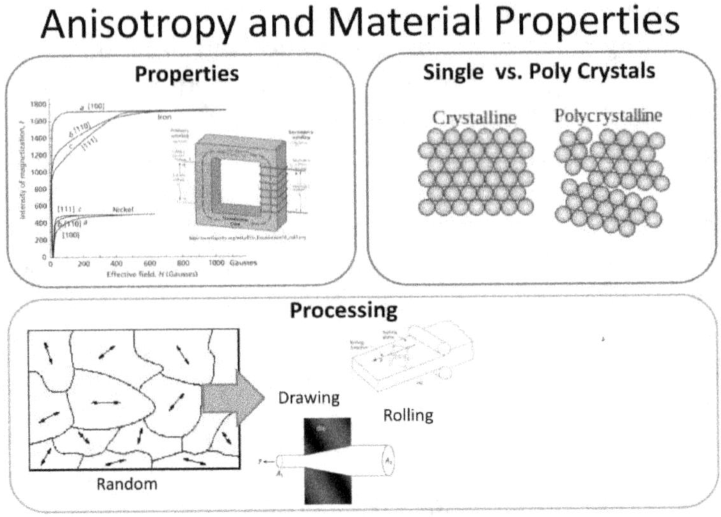

Figure 1: Design for material properties and anisotropy
[Source: https://www.slideserve.com/hoshi/physical-metallurgy-recap-lecture-4]

16.2 Design for Thermal and Mechanical Properties

Additive manufacturing enables the creation of parts and products with tailored thermal and mechanical properties. This includes designing for thermal properties, such as thermal conductivity, thermal expansion, and thermal insulation, as well as mechanical properties, such as strength, stiffness, toughness, and fatigue resistance. By tailoring these properties, designers can create parts and products that meet specific performance requirements, improve efficiency, and reduce weight.

Thermal Properties

Thermal properties are critical in many applications, including aerospace, automotive, and electronics. Additive manufacturing enables the creation of parts with tailored thermal properties, such as:

- **Thermal conductivity**: The ability of a material to conduct heat. High thermal conductivity is desirable in applications such as heat exchangers, while low thermal conductivity is desirable in applications such as thermal insulation.

- **Thermal expansion**: The change in size of a material in response to temperature changes. Low thermal expansion is desirable in applications such as precision instruments, while high thermal expansion is desirable in applications such as thermal actuators.

- **Thermal insulation**: The ability of a material to reduce heat transfer. High thermal insulation is desirable in applications such as building insulation, while low thermal insulation is desirable in applications such as heat sinks.

Designing for thermal properties involves understanding the relationships between material composition, processing conditions, and part geometry. For example, the thermal conductivity of a material can be influenced by the material's composition, the processing conditions used to produce the part, and the part's geometry.

Mechanical Properties

Mechanical properties are critical in many applications, including aerospace, automotive, and consumer products. Additive manufacturing enables the creation of parts with tailored mechanical properties, such as:

- **Strength**: The ability of a material to resist deformation and fracture. High strength is desirable in applications such as structural components, while low strength is desirable in applications such as flexible components.

- **Stiffness**: The ability of a material to resist deformation under load. High stiffness is desirable in applications such as precision instruments, while low stiffness is desirable in applications such as flexible components.

- **Toughness**: The ability of a material to absorb energy and resist fracture. High toughness is desirable in applications such as impact-resistant components, while low toughness is desirable in applications such as brittle components.

- **Fatigue resistance**: The ability of a material to resist cyclic loading and fatigue failure. High fatigue resistance is desirable in applications such as engine components, while low fatigue resistance is desirable in applications such as disposable components.

Designing for mechanical properties involves understanding the relationships between material composition, processing conditions, and part geometry. For example, the strength of a material can be influenced by the material's composition, the processing conditions used to produce the part, and the part's geometry.

Design Considerations

When designing for thermal and mechanical properties, several considerations must be taken into account. These include:

- **Material compatibility**: Ensuring that the selected material is compatible with the desired application and processing conditions.

- **Part functionality**: Ensuring that the part design meets the required functional performance criteria.

- **Manufacturing constraints**: Ensuring that the part design can be manufactured within the constraints of the additive manufacturing process.

- **Cost and lead time**: Ensuring that the part design is cost-effective and can be produced within the required lead time.

Material Selection

Selecting materials with specific thermal and mechanical properties is critical to designing for additive manufacturing. Different materials exhibit unique properties, such as thermal conductivity, strength, and toughness, that can be leveraged to achieve specific performance goals. For example:

- **Metals**: Metals such as aluminum, copper, and steel are often used in applications requiring high thermal conductivity, strength, and stiffness.
- **Polymers**: Polymers such as ABS, PLA, and nylon are often used in applications requiring low thermal conductivity, flexibility, and toughness.
- **Ceramics**: Ceramics such as alumina, zirconia, and silicon carbide are often used in applications requiring high thermal insulation, strength, and stiffness.

Process Optimization

Optimizing processing conditions is critical to achieving desired thermal and mechanical properties. Processing conditions, such as temperature, pressure, and layer thickness, can significantly impact the final properties of a part. For example:

- **Temperature**: The temperature used during processing can influence the thermal conductivity, strength, and toughness of a material.
- **Pressure**: The pressure used during processing can influence the density, strength, and stiffness of a material.
- **Layer thickness**: The layer thickness used during processing can influence the surface roughness, strength, and toughness of a material.

Geometry Optimization

Optimizing part geometry is essential to taking advantage of thermal and mechanical properties. Geometry optimization involves designing parts that maximize the benefits of additive manufacturing, such as complex geometries and internal structures. For example:

- **Topology optimization**: Topology optimization involves using algorithms to generate optimal geometries based on performance criteria, such as thermal conductivity, strength, and stiffness.

- **Lattice structures**: Lattice structures can be designed to achieve specific properties, such as stiffness and strength, while minimizing material usage.

- **Conformal cooling channels**: Conformal cooling channels can be designed to improve the thermal performance of molds and dies.

Case Studies

Several case studies demonstrate the benefits of designing for thermal and mechanical properties in additive manufacturing. For example:

- **Aerospace**: A company used additive manufacturing to produce a lightweight, high-strength aircraft component. By optimizing the thermal and mechanical properties of the part, the company was able to reduce the weight of the component by 30% while maintaining its structural integrity.

- **Medical**: A researcher used additive manufacturing to produce a custom implant with tailored thermal and mechanical properties. The implant was designed to mimic the natural properties of bone, allowing for improved osseointegration and reduced risk of implant failure.

- **Automotive**: A company used additive manufacturing to produce a lightweight, high-performance engine component. By optimizing the thermal and mechanical properties of the part, the company was able to reduce the weight of the component by 20% while maintaining its structural integrity.

Challenges and Limitations

While designing for thermal and mechanical properties offers many benefits, there are also several challenges and limitations to consider. These include:

- **Material variability**: Material properties can vary depending on the specific batch and processing conditions, making it challenging to achieve consistent results.

- **Process variability**: Processing conditions can also vary, making it challenging to achieve consistent results.

- **Limited material options**: The range of materials available for additive manufacturing is limited, making it challenging to find materials with specific thermal and mechanical properties.

- **Design complexity**: Designing for thermal and mechanical properties can be complex, requiring specialized knowledge and expertise.

Designing for thermal and mechanical properties is a critical aspect of additive manufacturing. By understanding the relationships between material composition, processing conditions, and part geometry, designers can create parts and products with tailored thermal and mechanical properties. While there are challenges and limitations to consider, the benefits of designing for thermal and mechanical properties make it a valuable tool for designers and engineers.

In summary, designing for thermal and mechanical properties involves selecting materials with specific thermal and mechanical properties for specific applications, optimizing processing conditions to achieve desired thermal and mechanical properties, and optimizing part geometry to take advantage of thermal and mechanical properties. By considering material compatibility, part functionality, manufacturing constraints, and cost and lead time, designers can create parts and products that meet the required performance criteria while minimizing material usage and reducing weight. Case studies demonstrate the benefits of designing for thermal and mechanical properties in additive manufacturing, while challenges and limitations must also be considered.

16.3 Design for Acoustic and Electromagnetic Properties

Additive manufacturing enables the creation of parts and products with tailored acoustic and electromagnetic properties. This includes designing for acoustic properties, such as sound absorption, sound transmission, and vibration damping, as well as electromagnetic properties, such as electromagnetic shielding, electromagnetic absorption, and electromagnetic transmission. By tailoring these properties, designers can create parts and products that meet specific performance requirements, improve efficiency, and reduce noise and electromagnetic interference.

Acoustic Properties

Acoustic properties are critical in many applications, including aerospace, automotive, and consumer products. Additive manufacturing enables the creation of parts with tailored acoustic properties, such as:

- **Sound absorption**: The ability of a material to absorb sound energy. High sound absorption is desirable in applications such as soundproofing, while low sound absorption is desirable in applications such as loudspeakers.
- **Sound transmission**: The ability of a material to transmit sound energy. High sound transmission is desirable in

applications such as hearing aids, while low sound transmission is desirable in applications such as soundproofing.

- **Vibration damping**: The ability of a material to dampen vibrations. High vibration damping is desirable in applications such as engine mounts, while low vibration damping is desirable in applications such as precision instruments.

Designing for acoustic properties involves understanding the relationships between material composition, processing conditions, and part geometry. For example, the sound absorption of a material can be influenced by the material's composition, the processing conditions used to produce the part, and the part's geometry.

Electromagnetic Properties

Electromagnetic properties are critical in many applications, including aerospace, automotive, and consumer electronics. Additive manufacturing enables the creation of parts with tailored electromagnetic properties, such as:

- **Electromagnetic shielding**: The ability of a material to shield electromagnetic radiation. High electromagnetic shielding is desirable in applications such as electromagnetic interference (EMI) shielding, while low electromagnetic shielding is desirable in applications such as antennas.

- **Electromagnetic absorption**: The ability of a material to absorb electromagnetic radiation. High electromagnetic absorption is desirable in applications such as electromagnetic absorbers, while low electromagnetic absorption is desirable in applications such as radar systems.

- **Electromagnetic transmission**: The ability of a material to transmit electromagnetic radiation. High electromagnetic transmission is desirable in applications such as antennas, while low electromagnetic transmission is desirable in applications such as electromagnetic shielding.

Designing for electromagnetic properties involves understanding the relationships between material composition, processing conditions, and part geometry. For example, the electromagnetic shielding of a material can be influenced by the material's composition, the processing conditions used to produce the part, and the part's geometry.

Design Considerations

When designing for acoustic and electromagnetic properties, several considerations must be taken into account. These include:

- **Material compatibility**: Ensuring that the selected material is compatible with the desired application and processing conditions.

- **Part functionality**: Ensuring that the part design meets the required functional performance criteria.

- **Manufacturing constraints**: Ensuring that the part design can be manufactured within the constraints of the additive manufacturing process.

- **Cost and lead time**: Ensuring that the part design is cost-effective and can be produced within the required lead time.

Material Selection

Selecting materials with specific acoustic and electromagnetic properties is critical to designing for additive manufacturing. Different materials exhibit unique properties, such as sound absorption, electromagnetic shielding, and vibration damping, that can be leveraged to achieve specific performance goals. For example:

- **Metals**: Metals such as aluminum, copper, and steel are often used in applications requiring high electromagnetic shielding, sound absorption, and vibration damping.
- **Polymers**: Polymers such as ABS, PLA, and nylon are often used in applications requiring low electromagnetic shielding, sound transmission, and vibration damping.
- **Ceramics**: Ceramics such as alumina, zirconia, and silicon carbide are often used in applications requiring high electromagnetic shielding, sound absorption, and vibration damping.

Process Optimization

Optimizing processing conditions is critical to achieving desired acoustic and electromagnetic properties. Processing conditions, such as temperature, pressure, and layer thickness, can significantly impact the final properties of a part. For example:

- **Temperature**: The temperature used during processing can influence the sound absorption, electromagnetic shielding, and vibration damping of a material.
- **Pressure**: The pressure used during processing can influence the density, sound transmission, and electromagnetic transmission of a material.
- **Layer thickness**: The layer thickness used during processing can influence the surface roughness, sound absorption, and electromagnetic shielding of a material.

Geometry Optimization

Optimizing part geometry is essential to taking advantage of acoustic and electromagnetic properties. Geometry optimization involves designing parts that maximize the benefits of additive manufacturing, such as complex geometries and internal structures. For example:

- **Topology optimization**: Topology optimization involves using algorithms to generate optimal geometries based on performance criteria, such as sound absorption, electromagnetic shielding, and vibration damping.
- **Lattice structures**: Lattice structures can be designedto achieve specific properties, such as sound absorption, electromagnetic shielding, and vibration damping, while minimizing material usage.
- **Conformal structures**: Conformal structures can be designed to improve the acoustic and electromagnetic performance of parts, such as soundproofing and electromagnetic shielding.

Case Studies

Several case studies demonstrate the benefits of designing for acoustic and electromagnetic properties in additive manufacturing. For example:

- **Aerospace**: A company used additive manufacturing to produce a lightweight, high-performance aircraft component with tailored acoustic and electromagnetic properties. By optimizing the acoustic and electromagnetic properties of the part, the company was able to reduce the weight of the component by 30% while maintaining its structural integrity.
- **Automotive**: A company used additive manufacturing to produce a custom car part with tailored acoustic and electromagnetic properties. The part was designed to reduce noise and electromagnetic interference, improving the overall driving experience.

- **Consumer Electronics**: A company used additive manufacturing to produce a custom phone case with tailored acoustic and electromagnetic properties. The case was designed to improve the sound quality and reduce electromagnetic interference, enhancing the overall user experience.

Challenges and Limitations

While designing for acoustic and electromagnetic properties offers many benefits, there are also several challenges and limitations to consider. These include:

- **Material variability**: Material properties can vary depending on the specific batch and processing conditions, making it challenging to achieve consistent results.

- **Process variability**: Processing conditions can also vary, making it challenging to achieve consistent results.

- **Limited material options**: The range of materials available for additive manufacturing is limited, making it challenging to find materials with specific acoustic and electromagnetic properties.

- **Design complexity**: Designing for acoustic and electromagnetic properties can be complex, requiring specialized knowledge and expertise.

Designing for acoustic and electromagnetic properties is a critical aspect of additive manufacturing. By understanding the relationships between material composition, processing conditions, and part geometry, designers can create parts and products with tailored acoustic and electromagnetic properties. While there are challenges and limitations to consider, the benefits of designing for acoustic and electromagnetic properties make it a valuable tool for designers and engineers.

In summary, designing for acoustic and electromagnetic properties involves selecting materials with specific acoustic and electromagnetic properties for specific applications, optimizing processing conditions to achieve desired acoustic and electromagnetic properties, and optimizing part geometry to take advantage of acoustic and electromagnetic properties. By considering material compatibility, part functionality, manufacturing constraints, and cost and lead time, designers can create parts and products that meet the required performance criteria while minimizing material usage and reducing noise and electromagnetic interference. Case studies demonstrate the benefits of designing for acoustic and electromagnetic properties in additive manufacturing, while challenges and limitations must also be considered. Figure 72 illustrates how materials designed for additive manufacturing can interact with both acoustic and electromagnetic (EM) waves. Here's a breakdown of the key elements:

A. **Incident EM Waves**: These are the waves approaching the shielding material from the left.
B. **Reflected EM Waves**: Some of the incident waves bounce back from the surface of the material.
C. **Absorbed EM Waves**: These waves are absorbed by the material and do not pass through.
D. **Transmitted EM Waves**: These waves pass through the material to the other side.
E. **Multiple Reflected EM Waves**: Within the shielding material, some waves reflect multiple times.
F. **Acoustic Waves**: The bottom part of the diagram shows how acoustic waves are reflected by the material.

This visualization is crucial for understanding how to design materials that can control wave propagation, which is important for applications like electromagnetic interference (EMI) shielding and soundproofing.

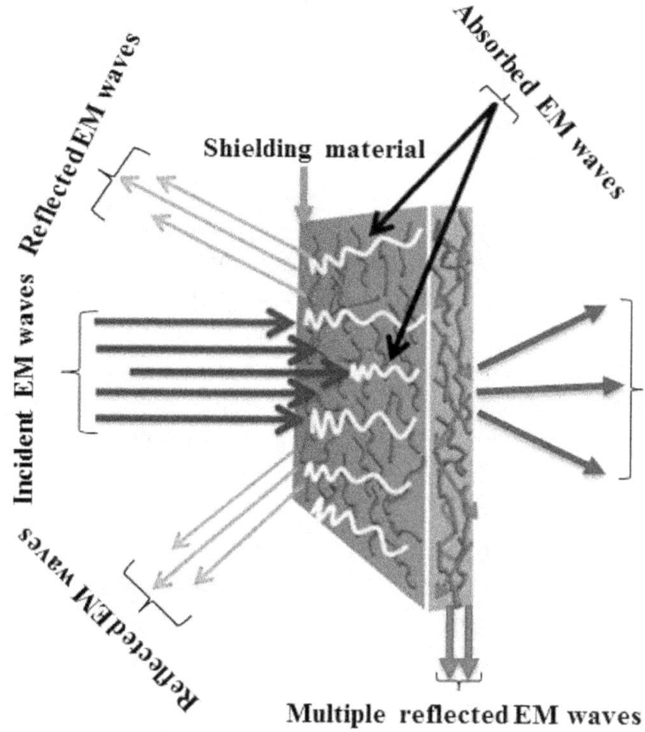

Figure 2: Design for acoustic and electromagnetic properties
[Source: https://link.springer.com/article/10.1007/s00339-020-03927-2]

16.4 Design for Biocompatibility and Biodegradability

Additive manufacturing enables the creation of parts and products with tailored biocompatibility and biodegradability. This includes:

- **Biocompatibility**: Designing for compatibility with living tissues and biological systems. Biocompatible materials and designs minimize adverse reactions, such as inflammation, toxicity, or rejection, when in contact with living tissues.

- **Biodegradability**: Designing for controlled degradation and resorption in biological environments. Biodegradable materials and designs break down over time, eliminating the need for removal or replacement.

- Designing for biocompatibility and biodegradability involves understanding the relationships between material composition, processing conditions, and part geometry. This includes:

- **Material selection**: Selecting materials with specific biocompatibility and biodegradability for specific applications. For example, titanium alloys are commonly used in orthopedic implants due to their high biocompatibility and strength. Polylactic acid (PLA) and polyglycolic acid (PGA) are biodegradable polymers used in tissue engineering and drug delivery applications.

- **Process optimization**: Optimizing processing conditions to achieve desired biocompatibility and biodegradability. For instance, controlling the temperature, pressure, and atmosphere during 3D printing can influence the material's microstructure, porosity, and surface roughness, which in turn affect biocompatibility and biodegradability.

- **Geometry optimization**: Optimizing part geometry to take advantage of biocompatibility and biodegradability. Complex geometries, such as lattice structures, can be designed to promote tissue ingrowth, improve mechanical properties, and enhance biodegradability.

Biocompatibility Considerations

When designing for biocompatibility, several factors must be considered:

I. **Cytotoxicity**: The material's potential to cause cell death or damage.
II. **Immunogenicity**: The material's ability to trigger an immune response.
III. **Inflammation**: The material's potential to cause inflammation or irritation.

IV. **Blood compatibility**: The material's interaction with blood and its components.

V. To ensure biocompatibility, designers can employ various strategies, such as:

VI. **Surface modification**: Modifying the material's surface chemistry or topography to improve biocompatibility.

VII. **Material blending**: Combining materials with different properties to achieve optimal biocompatibility.

VIII. **Porous structures**: Designing porous structures to promote tissue ingrowth and integration.

Biodegradability Considerations

When designing for biodegradability, several factors must be considered:

- **Degradation rate**: The rate at which the material breaks down in a biological environment.

- **Degradation products**: The byproducts of degradation, which should be non-toxic and easily resorbed.

- **Mechanical properties**: The material's mechanical properties during degradation, which should maintain structural integrity.

To ensure biodegradability, designers can employ various strategies, such as:

- **Material selection**: Selecting materials with inherent biodegradability, such as PLA or PGA.

- **Design for degradation**: Designing parts with intentional weaknesses or features that facilitate degradation.

- **Coatings and surface treatments**: Applying coatings or surface treatments to control degradation rates.

Applications of Biocompatible and Biodegradable Designs

The ability to design for biocompatibility and biodegradability has significant implications for various industries, including:

- **Medical devices**: Implants, prosthetics, and surgical instruments can be designed for biocompatibility and biodegradability.
- **Tissue engineering**: Biodegradable scaffolds can be designed to support tissue growth and regeneration.
- **Pharmaceuticals**: Biodegradable materials can be used for controlled drug release and delivery.
- **Cosmetics**: Biocompatible and biodegradable materials can be used in cosmetic products, such as skin care and hair care.

In conclusion, designing for biocompatibility and biodegradability is a critical aspect of additive manufacturing. By understanding the relationships between material composition, processing conditions, and part geometry, designers can create parts and products that interact safely and effectively with living tissues and biological systems. Figure 73 illustrates the design considerations for biocompatibility and biodegradability in materials used for additive manufacturing. Here's a breakdown of the key elements:

- **Biocompatibility**: This circle includes factors such as:
- **Size**: The dimensions of the material, which can affect how it interacts with biological systems.
- **Surface**: The surface properties, which influence cell attachment and interaction.
- **Stability**: How stable the material is within a biological environment.

- **Biodegradability**: This circle highlights:

- **In vitro**: The material's behavior in a controlled laboratory environment.

- **In vivo**: The material's behavior within a living organism.

The overlap between biocompatibility and biodegradability suggests that materials can be designed to be both compatible with biological systems and capable of breaking down safely within those systems.

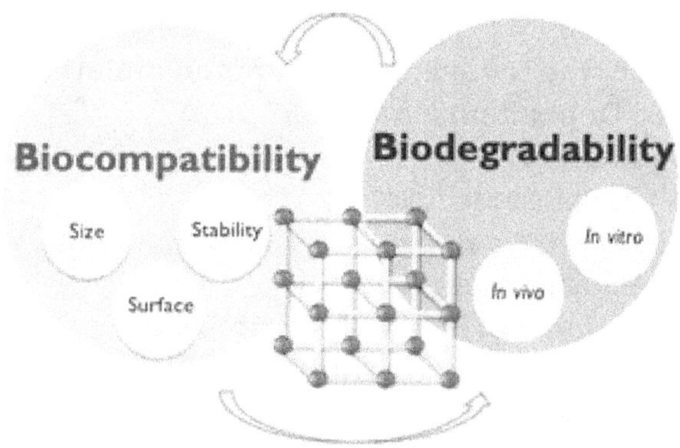

Figure 3: Design for Biocompatibility and Biodegradability
[Source:
https://pubs.rsc.org/en/content/articlelanding/2021/tb/d1tb01044a]

16.5 Questions

16.5.1 Long Response Questions

1. How do material properties and anisotropy impact the performance of additively manufactured parts?
2. What are the key considerations for designing for thermal and mechanical properties?
3. How can acoustic and electromagnetic properties be tailored in additively manufactured parts?
4. What are the key considerations for designing for biocompatibility and biodegradability?

5. How can design for additive manufacturing be used to create complex products with tailored properties?

16.5.2 Multiple Choice Questions

1) What is additive manufacturing?
- a) A subtractive process
- b) A process where material is added layer by layer
- c) A method to mold materials
- d) A chemical reaction process

Solution: - b) A process where material is added layer by layer

2) Which of the following is a common material used in FDM (Fused Deposition Modeling)?
- a) Aluminum
- b) PLA (Polylactic Acid)
- c) Glass
- d) Stainless Steel

Solution: - b) PLA (Polylactic Acid)

3) In additive manufacturing, what does the term 'anisotropy' refer to?
- a) Uniform properties in all directions
- b) Direction-dependent properties
- c) The ability to absorb sound
- d) Process efficiency

Solution: - b) Direction-dependent properties

4) Which type of additive manufacturing process uses a laser to fuse powder materials?
- a) FDM
- b) SLA
- c) SLS
- d) Inkjet Printing

Solution: - c) SLS

5) What is the primary advantage of additive manufacturing over traditional methods?
- a) Higher material waste
- b) Faster production speed
- c) Ability to create complex geometries
- d) Lower material strength

Solution: - c) Ability to create complex geometries

6) **Which property describes a material's ability to resist deformation under load?**
   a) Toughness
   b) Strength
   c) Stiffness
   d) Ductility
   Solution: - c) Stiffness
7) **What is the purpose of process optimization in additive manufacturing?**
   a) To increase material waste
   b) To achieve desired material properties and performance
   c) To slow down production speed
   d) To reduce part complexity
   Solution: - b) To achieve desired material properties and performance
8) **What does a lattice structure in additive manufacturing help to achieve?**
   a) Increase weight
   b) Increase material usage
   c) Reduce material usage while maintaining strength
   d) Decrease part durability
   Solution: - c) Reduce material usage while maintaining strength
9) **Which material property is particularly important for heat exchangers in additive manufacturing?**
   a) Low thermal conductivity
   b) High thermal conductivity
   c) High electrical conductivity
   d) High magnetic permeability
   Solution: - b) High thermal conductivity
10) **What does the term 'process variability' refer to in additive manufacturing?**
   a) Variation in material properties from batch to batch
   b) Consistency of part dimensions
   c) Changes in processing conditions
   d) Standardization of material costs
   Solution: - c) Changes in processing conditions

11) What is 'topology optimization' used for in additive manufacturing?

a) To optimize material costs

b) To design parts with the best possible performance based on criteria

c) To increase machine throughput

d) To decrease part durability

Solution: - b) To design parts with the best possible performance based on criteria

12) How can thermal expansion be advantageous in additive manufacturing?

a) It prevents material from deforming

b) It enhances the material's flexibility

c) It helps in designing thermal actuators

d) It reduces the material's strength

Solution: - c) It helps in designing thermal actuators

13) Which additive manufacturing process involves curing liquid resin with UV light?

a) FDM

b) SLA

c) SLS

d) EBM

Solution: - b) SLA

14) Which factor is least likely to influence the mechanical properties of a part in additive manufacturing?

a) Layer thickness

b) Material composition

c) Ambient temperature

d) Part geometry

Solution: - c) Ambient temperature

15) What is the impact of material anisotropy on the design of a part?

a) It simplifies the design process

b) It requires consideration of direction-dependent properties in the design

c) It eliminates the need for material selection

d) It reduces the overall cost

Solution: - b) It requires consideration of direction-dependent properties in the design

16) **What role do conformal cooling channels play in additive manufacturing?**
   a) They increase part weight
   b) They enhance the thermal performance of molds and dies
   c) They simplify part geometry
   d) They reduce manufacturing speed
   Solution: - b) They enhance the thermal performance of molds and dies

17) **Which technique can be used to simulate the effects of different processing conditions on material properties before actual manufacturing?**
   a) Experimental testing
   b) Finite Element Analysis (FEA)
   c) Simple calculation
   d) Manual inspection
   Solution: - b) Finite Element Analysis (FEA)

18) **In what scenario would a researcher choose a polymer over a metal for additive manufacturing?**
   a) When high thermal conductivity is required
   b) When flexibility and low density are desired
   c) When high strength and stiffness are needed
   d) When electromagnetic shielding is necessary
   Solution: - b) When flexibility and low density are desired

19) **Which of the following represents a challenge in designing for acoustic properties in additive manufacturing?**
   a) Achieving consistent sound absorption across different batches
   b) Designing for high tensile strength
   c) Maximizing part hardness
   d) Improving thermal conductivity
   Solution: - a) Achieving consistent sound absorption across different batches

20) **What does 'grain structure' refer to in the context of material processing in additive manufacturing?**
   a) The distribution of particles in a polymer matrix
   b) The arrangement of crystalline structures in a material

c) The surface texture of the material
d) The type of laser used in the process
Solution: - b) The arrangement of crystalline structures in a material

21) **What is the primary benefit of using topology optimization for part geometry in additive manufacturing?**
a) Reducing the number of parts
b) Optimizing the geometry based on specific performance criteria
c) Increasing the manufacturing cost
d) Simplifying the material selection process
Solution: - b) Optimizing the geometry based on specific performance criteria

22) **Which material property would be most critical for an implant designed to mimic natural bone?**
a) High thermal conductivity
b) High strength
c) Tailored mechanical properties and anisotropy
d) High electrical conductivity
Solution: - c) Tailored mechanical properties and anisotropy

23) **What is the main focus of process optimization in selective laser sintering (SLS)?**
a) Reducing part weight
b) Controlling laser power and scan speed to achieve specific material properties
c) Simplifying the design process
d) Increasing material wastage
Solution: - b) Controlling laser power and scan speed to achieve specific material properties

24) **In additive manufacturing, what does 'electromagnetic shielding' aim to achieve?**
a) Increase the material's electrical conductivity
b) Protect components from electromagnetic interference
c) Enhance thermal insulation
d) Improve sound absorption
Solution: - b) Protect components from electromagnetic interference

25) What is a key factor in ensuring the consistent quality of parts produced through additive manufacturing?

a) Variability in material composition
b) Consistency in processing conditions
c) Inconsistent part design
d) Variability in machine calibration

Solution: - b) Consistency in processing conditions

26) How does the layer-by-layer deposition process in FDM impact the material properties?

a) It enhances isotropy
b) It creates anisotropic properties
c) It reduces part strength
d) It increases material density

Solution: - b) It creates anisotropic properties

27) Which of the following materials is typically chosen for high thermal insulation applications in additive manufacturing?

a) Copper
b) Steel
c) Ceramic
d) Titanium

Solution: - c) Ceramic

28) What does 'vibration damping' in additive manufacturing refer to?

a) The material's ability to resist deformation
b) The material's ability to absorb sound
c) The material's ability to reduce vibrations
d) The material's ability to conduct electricity

Solution: - c) The material's ability to reduce vibrations

29) Which property would be most important for a material used in a high-performance engine component?

a) Low thermal conductivity
b) High fatigue resistance
c) Low strength
d) High sound absorption

Solution: - b) High fatigue resistance

30) Which of the following best describes 'conformal cooling channels'?

a) Channels designed to optimize airflow

b) Cooling channels integrated into the mold design to follow the part's geometry

c) Channels used to increase part weight

d) Channels designed for electrical insulation

Solution: - b) Cooling channels integrated into the mold design to follow the part's geometry

31) What is a common challenge when using lattice structures in additive manufacturing?

a) Reduced part accuracy

b) Increased material consumption

c) Complexity in design and optimization

d) Simplified manufacturing process

Solution: - c) Complexity in design and optimization

32) Which process improvement strategy is used to predict the impact of various conditions on material properties before manufacturing?

a) Trial-and-error experimentation

b) Simulation and modeling techniques

c) Basic calculations

d) Material selection alone

Solution: - b) Simulation and modeling techniques

33) In which additive manufacturing technique is a binder selectively deposited onto powder material?

a) Stereolithography (SLA)

b) Fused Deposition Modeling (FDM)

c) Binder Jetting

d) Direct Metal Laser Sintering (DMLS)

Solution: - c) Binder Jetting

34) What is the main advantage of using a high-density lattice structure in additive manufacturing?

a) Increased material weight

b) Improved thermal insulation

c) Enhanced mechanical strength with reduced material use

d) Simplified part design

Solution: - c) Enhanced mechanical strength with reduced material use

35) What does 'response surface methodology' help achieve in the context of additive manufacturing?

a) Simple part design
b) Optimization of process parameters for better quality
c) Increased material wastage
d) Reduced part complexity

Solution: - b) Optimization of process parameters for better quality

36) Which of the following factors would not typically be considered in the optimization of mechanical properties for an additive manufacturing process?

a) Material type
b) Layer thickness
c) Part color
d) Print orientation

Solution: - c) Part color

37) What is a primary consideration when designing parts for additive manufacturing to ensure structural integrity?

a) The color of the part
b) The layer alignment and bonding strength
c) The printer's brand
d) The type of post-processing used

Solution: - b) The layer alignment and bonding strength

38) Which material property is critical for parts subjected to high temperatures in additive manufacturing?

a) High electrical conductivity
b) High thermal stability
c) Low mechanical strength
d) High sound absorption

Solution: - b) High thermal stability

39) How does 'design for manufacturability' influence the additive manufacturing process?

a) It focuses on simplifying part complexity to fit manufacturing constraints
b) It increases material costs
c) It eliminates the need for design optimization
d) It reduces part strength

Solution: - a) It focuses on simplifying part complexity to fit manufacturing constraints

40) What is a significant benefit of using 3D printing for prototyping?

a) Increased production lead times

b) Higher prototyping costs

c) Faster design iteration and testing

d) Limited design flexibility

Solution: - c) Faster design iteration and testing

Chapter 17

Computer-Aided Design (CAD) for 3D Printing

17.1 Introduction to CAD for 3D Printing

Computer-Aided Design (CAD) is a crucial step in the 3D printing process, as it enables designers to create digital models of parts and products that can be directly manufactured using additive manufacturing technologies. CAD for 3D printing involves the use of specialized software and techniques to create models that are optimized for 3D printing.

The Importance of CAD in 3D Printing

CAD plays a vital role in the 3D printing process, as it allows designers to:

- **Create complex geometries**: CAD software enables designers to create complex shapes and geometries that would be difficult or impossible to produce using traditional manufacturing methods.

- **Optimize designs for 3D printing**: CAD software can be used to optimize designs for 3D printing, taking into account factors such as material properties, print orientation, and support structures.

- **Reduce errors and iterations**: CAD software can help reduce errors and iterations by allowing designers to test and refine their designs virtually before sending them to print.

- **Improve communication and collaboration**: CAD software can facilitate communication and collaboration between designers, engineers, and manufacturers by providing a common language and platform for design and production.

CAD Software for 3D Printing

There are several CAD software options available for 3D printing, each with its own strengths and weaknesses. Some popular CAD software options include:

- **Autodesk Fusion 360**: A powerful and feature-rich CAD software that is widely used in the 3D printing industry.
- **SolidWorks**: A popular CAD software that is known for its ease of use and powerful modeling capabilities.
- **Blender**: A free and open-source CAD software that is popular among hobbyists and professionals alike.
- **Tinkercad**: A web-based CAD software that is great for beginners and hobbyists.

Key Features of CAD Software for 3D Printing

When selecting a CAD software for 3D printing, there are several key features to look for, including:

- **Parametric modeling**: The ability to create models using parameters and equations, which allows for easy modification and iteration.
- **Freeform modeling**: The ability to create complex, organic shapes using tools such as NURBS (Non-uniform rational B-spline) and subdivision surfaces.
- **Mesh repair and optimization**: The ability to repair and optimize mesh models for 3D printing, including tools for fixing holes, gaps, and other defects.
- **Support generation**: The ability to generate support structures for 3D printing, including tools for customizing support density and pattern.

- **Slicing and toolpath generation**: The ability to generate slicing and toolpath data for 3D printing, including tools for customizing layer height, infill density, and print speed.

Design Considerations for 3D Printing

When designing for 3D printing, there are several key considerations to keep in mind, including:

- **Material properties**: The physical properties of the material being used, including its strength, stiffness, and melting point.

- **Print orientation**: The orientation of the part during printing, which can affect the strength and accuracy of the final product.

- **Support structures**: The use of support structures to prevent warping, sagging, and other defects during printing.

- **Layer height and infill density**: The thickness of each layer and the density of the infill material, which can affect the strength, weight, and cost of the final product.

- **Post-processing and finishing**: The steps taken after printing to remove supports, sand, and finish the part, which can affect the final appearance and performance of the product.

Best Practices for CAD Design for 3D Printing

To get the most out of CAD design for 3D printing, it's essential to follow best practices, including:

- **Design for manufacturability**: Designing parts and products with the limitations and capabilities of 3D printing in mind.

- **Use simple and robust designs**: Avoiding complex and fragile designs that may be prone to failure or defects.

- **Optimize designs for material properties**: Designing parts and products that take advantage of the unique properties of 3D printing materials.

- **Use simulation and analysis tools**: Using simulation and analysis tools to test and refine designs virtually before sending them to print.

- **Collaborate with manufacturers and engineers**: Working closely with manufacturers and engineers to ensure that designs are feasible and manufacturable.

By following these best practices and using specialized CAD software and techniques, designers can create optimized designs for 3D printing that take advantage of the unique capabilities of additive manufacturing technologies. Figure 74 shows a CAD drawing for a 3D printing project. It features a mechanical part with various views, including top, front, and side perspectives. The drawing includes dimensions and specifications necessary for manufacturing the part, providing a detailed blueprint for creating a precise physical object through 3D printing technology.

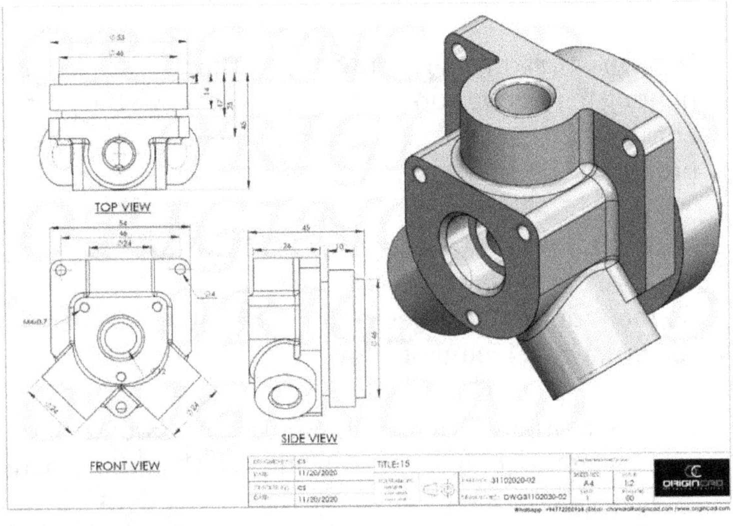

Figure 1: CAD for 3D printing
[Source: https://ferhire.weebly.com/blog/free-3d-printing-cad]

17.2 CAD Software and Tools

There are many CAD software and tools available for 3D printing, each with its own strengths and weaknesses. Some popular CAD software for 3D printing includes:

Autodesk Fusion 360

A powerful CAD software that offers advanced modeling and simulation capabilities. Fusion 360 is widely used in the 3D printing industry due to its ability to create complex models, simulate real-world performance, and collaborate with others in real-time. Its features include:

Parametric modeling: Create models using parameters and equations, allowing for easy modification and iteration.

Freeform modeling: Create complex, organic shapes using tools such as NURBS (Non-uniform rational B-spline) and subdivision surfaces.

Simulation: Analyze and simulate the performance of designs under various loads and conditions.

Collaboration: Collaborate with others in real-time, using cloud-based collaboration tools.

SolidWorks

A popular CAD software that offers a range of tools and features for 3D modeling and simulation. SolidWorks is known for its ease of use and powerful modeling capabilities, making it a popular choice among engineers and designers. Its features include:

Parametric modeling: Create models using parameters and equations, allowing for easy modification and iteration.

Assembly modeling: Create complex assemblies using a variety of tools and features.

Simulation: Analyze and simulate the performance of designs under various loads and conditions.

Rendering: Create high-quality renderings and animations of designs.

Blender

A free and open-source CAD software that offers advanced modeling and animation capabilities. Blender is popular among hobbyists and professionals alike, due to its flexibility and customizability. Its features include:

3D modeling: Create complex models using a variety of tools and features.

Animation: Create animations and simulations using keyframe animation and physics engines.

Rendering: Create high-quality renderings and animations of designs.

Scripting: Customize and automate tasks using Python scripting.

Tinkercad

A web-based CAD software that offers a simple and intuitive interface for beginners. Tinkercad is a great introduction to CAD design, allowing users to create simple models and designs using a variety of tools and features. Its features include:

Block-based modeling: Create models using a block-based system, allowing for easy modification and iteration.

Shape generators: Create complex shapes using a variety of shape generators and tools.

Export: Export designs in a variety of formats, including STL and OBJ.

Other CAD Software and Tools

In addition to these popular CAD software options, there are many other tools and software available for 3D printing, including:

A. **Onshape**: A cloud-based CAD software that offers real-time collaboration and advanced modeling capabilities.

B. **OpenSCAD**: A free and open-source CAD software that allows users to create models using a programming language.

C. **SketchUp**: A popular CAD software that offers a simple and intuitive interface for architectural and interior design applications.

D. **Rhino**: A powerful CAD software that offers advanced modeling and simulation capabilities for industrial design and engineering applications.

CAD Software for Specific 3D Printing Applications

Different CAD software may be better suited for specific 3D printing applications, such as:

- **Jewelry design**: Software such as Rhino and Blender are popular for jewelry design due to their ability to create complex, organic shapes.

- **Architectural design**: Software such as SketchUp and Revit are popular for architectural design due to their ability to create complex buildings and structures.

- **Product design**: Software such as SolidWorks and Fusion 360 are popular for product design due to their ability to create complex models and simulate real-world performance.

- **Artistic design**: Software such as Blender and Tinkercad are popular for artistic design due to their ability to create complex, organic shapes and animations.

Choosing the Right CAD Software

When choosing a CAD software for 3D printing, it's essential to consider the following factors:

- **Level of expertise**: Choose a software that matches your level of expertise, from beginner-friendly options like Tinkercad to advanced software like Fusion 360.

- **Specific application**: Choose a software that is well-suited for your specific 3D printing application, such as jewelry design or architectural design.

- **Cost**: Consider the cost of the software, including subscription fees and one-time purchases.

- **Collaboration**: Consider the collaboration features of the software, including real-time collaboration and cloud-based storage.

By choosing the right CAD software and tools, designers and engineers can create optimized designs for 3D printing that take advantage of the unique capabilities of additive manufacturing technologies. Figure 75 showcases a CAD software interface featuring a detailed 3D model of a jet engine. The model is displayed in a split view, with one half in wireframe and the other half smoothly shaded. This demonstrates the software's capability to visualize models in different ways. The interface includes various tools and options, such as:

- Wireframe View: Displays the model as a network of lines, showing the underlying structure.

- Smoothly Shaded View: Renders the model with smooth surfaces, providing a realistic appearance.

- Shaded with Edges View: Combines shading with visible edges to highlight the model's geometry.
- Color Palettes and Axis Indicators: Located at the bottom right, these help in orienting and customizing the model.

This setup is crucial for creating precise and complex designs, such as engine parts, which are essential in engineering and manufacturing processes.

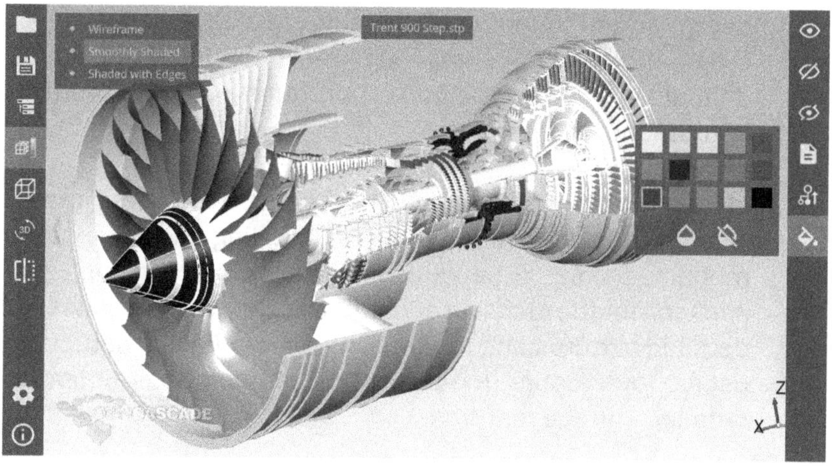

Figure 2: CAD software and tools
[Source: https://www.sculpteo.com/blog/2018/03/21/top-13-of-the-best-open-source-cad-software/]

17.3 CAD Modeling Techniques for 3D Printing

CAD modeling techniques for 3D printing involve the use of specialized techniques and strategies to create models that are optimized for 3D printing. These techniques allow designers to create complex shapes and models that take advantage of the unique capabilities of additive manufacturing technologies. Some key techniques include:

- **Parametric modeling**: Using parameters and equations to create complex shapes and models. Parametric modeling involves defining a model's shape and features using mathematical equations and parameters, allowing for easy modification and iteration. This technique is particularly useful for creating models with complex geometries and precise tolerances.

- **Freeform modeling**: Using curves and surfaces to create organic and complex shapes. Freeform modeling involves using curves and surfaces to create models with smooth, flowing shapes, often used in applications such as product design and architecture. This technique allows designers to create models with complex, organic shapes that would be difficult or impossible to produce using traditional manufacturing methods.

- **Mesh modeling**: Using polygon meshes to create models with complex geometries. Mesh modeling involves creating models using a network of interconnected polygons, allowing for the creation of complex shapes and models with detailed textures and features. This technique is particularly useful for creating models with complex geometries and detailed surface features.

- **Lattice modeling**: Using lattice structures to create models with complex internal geometries. Lattice modeling involves creating models using a network of interconnected lattice structures, allowing for the creation of models with complex internal geometries and optimized material properties. This technique is particularly useful for creating models with complex internal structures, such as those used in aerospace and biomedical applications.

Other CAD Modeling Techniques

In addition to these key techniques, there are several other CAD modeling techniques that can be used for 3D printing, including:

- **Subdivision surface modeling**: Using subdivision surfaces to create models with smooth, curved shapes. Subdivision surface modeling involves creating models using a network of interconnected curves and surfaces, allowing for the creation of models with smooth, curved shapes and detailed features.

- **NURBS modeling**: Using NURBS (Non-uniform rational B-spline) curves and surfaces to create models with complex shapes and geometries. NURBS modeling involves creating models using a network of interconnected curves and surfaces, allowing for the creation of models with complex shapes and geometries.

- **Sweep and loft modeling**: Using sweep and loft operations to create models with complex shapes and geometries. Sweep and loft modeling involves creating models by sweeping or lofting curves and surfaces along a path, allowing for the creation of models with complex shapes and geometries.

- **Boolean operations**: Using Boolean operations to create models with complex shapes and geometries. Boolean operations involve combining or subtracting shapes and models using logical operations, allowing for the creation of models with complex shapes and geometries.

Design Considerations for 3D Printing

When using CAD modeling techniques for 3D printing, it's essential to consider several key design considerations, including:

- **Material properties**: The physical properties of the material being used, including its strength, stiffness, and melting point.

- **Print orientation**: The orientation of the part during printing, which can affect the strength and accuracy of the final product.

- **Support structures**: The use of support structures to prevent warping, sagging, and other defects during printing.
- **Layer height and infill density**: The thickness of each layer and the density of the infill material, which can affect the strength, weight, and cost of the final product.
- **Post-processing and finishing**: The steps taken after printing to remove supports, sand, and finish the part, which can affect the final appearance and performance of the product.

Best Practices for CAD Modeling for 3D Printing

To get the most out of CAD modeling for 3D printing, it's essential to follow best practices, including:

- **Design for manufacturability**: Designing parts and products with the limitations and capabilities of 3D printing in mind.
- **Use simple and robust designs**: Avoiding complex and fragile designs that may be prone to failure or defects.
- **Optimize designs for material properties**: Designing parts and products that take advantage of the unique properties of 3D printing materials.
- **Use simulation and analysis tools**: Using simulation and analysis tools to test and refine designs virtually before sending them to print.
- **Collaborate with manufacturers and engineers**: Working closely with manufacturers and engineers to ensure that designs are feasible and manufacturable.

By following these best practices and using specialized CAD modeling techniques, designers and engineers can create optimized designs for 3D printing that take advantage of the unique capabilities of additive manufacturing technologies. Figure 76 illustrates various

3D CAD modeling techniques using a semi-transparent car model. This allows viewers to see the internal components, such as the engine and other mechanisms, highlighting the detailed structure and layout. The title "3D CAD Modeling Techniques" suggests that this image is used for educational purposes, possibly by an institution like the CADD Centre, which specializes in teaching CAD skills.

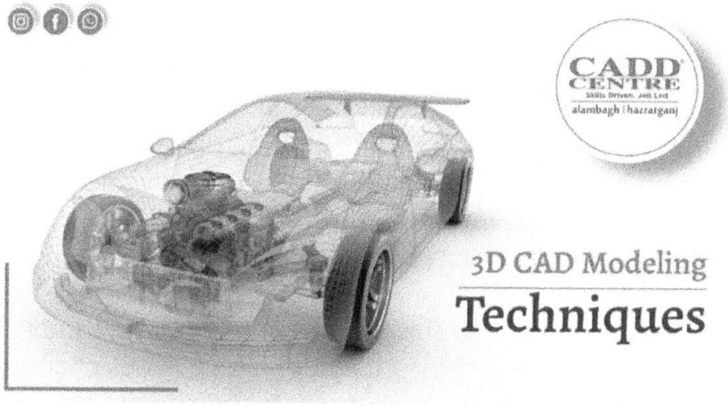

Figure 3: CAD modeling techniques
[Source: https://www.sicces.co.in/cadd/effective-3d-cad-modeling-techniques/]

17.4 CAD File Formats and Conversion

CAD file formats and conversion are critical steps in the 3D printing process, as they enable designers to export and import models between different software and systems. The ability to seamlessly transfer designs between different CAD software, 3D modeling tools, and 3D printing systems is essential for efficient and effective design-to-print workflows. In this section, we will explore some popular CAD file formats and conversion techniques used in 3D printing.

Popular CAD File Formats

There are several popular CAD file formats used in 3D printing, each with its own strengths and weaknesses. Some of the most widely used formats include:

- **STL (STereoLithography)**: A widely used file format for 3D printing that represents models as a series of triangles. STL files are simple, lightweight, and easy to work with, making them a popular choice for 3D printing applications. However, they can be limited in terms of precision and accuracy, and may not support complex geometries or textures.

- **OBJ (Object File Format)**: A file format that represents models as a series of polygons and vertices. OBJ files are more detailed and precise than STL files, and can support complex geometries, textures, and materials. They are commonly used in computer-aided design (CAD), computer-aided manufacturing (CAM), and 3D modeling applications.

- **STEP (STandard for the Exchange of Product model data)**: A file format that represents models as a series of curves and surfaces. STEP files are highly detailed and precise, and can support complex geometries, textures, and materials. They are commonly used in CAD, CAM, and product lifecycle management (PLM) applications.

- **IGES (Initial Graphics Exchange Specification)**: A file format that represents models as a series of curves and surfaces. IGES files are similar to STEP files but are more focused on 2D and 3D geometry exchange. They are commonly used in CAD, CAM, and engineering applications.

- **Collada (COLLAborative Design Activity)**: A file format that represents models as a series of polygons, vertices, and materials. Collada files are highly detailed and precise, and can support complex geometries, textures, and materials. They are commonly used in 3D modeling, animation, and gaming applications.

CAD File Conversion

CAD file conversion is the process of translating a CAD file from one format to another. This is often necessary when working with different CAD software, 3D modeling tools, or 3D printing systems, as each may have its own proprietary file format. There are several ways to convert CAD files, including:

- **Native conversion**: Many CAD software and 3D modeling tools offer native conversion capabilities, allowing users to export and import files in different formats.
- **Third-party conversion software**: There are several third-party software tools available that specialize in CAD file conversion, such as Autodesk's FBX Converter and Okino's PolyTrans.
- **Online conversion services**: There are several online services available that offer CAD file conversion, such as Convertio and Online-Convert.com.

Challenges and Limitations

While CAD file formats and conversion are essential steps in the 3D printing process, they can also present several challenges and limitations. Some common issues include:

- **File compatibility**: Different CAD software and 3D printing systems may not support the same file formats, making it difficult to transfer designs between systems.
- **Data loss**: CAD file conversion can result in data loss or corruption, particularly when converting between different file formats.
- **Geometry and topology**: CAD file formats may not accurately represent the geometry and topology of complex

models, leading to errors or inaccuracies in the 3D printing process.

- **Material and texture information**: CAD file formats may not support material and texture information, making it difficult to accurately represent these properties in the 3D printing process.

Best Practices for CAD File Formats and Conversion

To overcome these challenges and limitations, it's essential to follow best practices for CAD file formats and conversion, including:

- **Use standardized file formats**: Use standardized file formats such as STL, OBJ, and STEP to ensure compatibility between different CAD software and 3D printing systems.

- **Verify file integrity**: Verify the integrity of CAD files before conversion to ensure that they are accurate and complete.

- **Use high-quality conversion software**: Use high-quality conversion software that is specifically designed for CAD file conversion to minimize data loss and corruption.

- **Test and validate designs**: Test and validate designs after conversion to ensure that they are accurate and complete.

By following these best practices and understanding the different CAD file formats and conversion techniques available, designers and engineers can ensure seamless design-to-print workflows and produce high-quality 3D printed parts and products. Figure 77 illustrates various CAD file formats commonly used in 3D modeling and printing. Here are the formats displayed:

- **STP**
- **MAX**
- **FBX**
- **OBJ**
- **X3D**
- **VRML**

- **3DS**
- **3MF**
- **STL**
- **DAE**

Each format has its own unique features and applications. For instance, **STL** is widely used for 3D printing due to its simplicity and compatibility with most 3D printers, while **FBX** is popular in animation and game development for its ability to store complex models and animations.

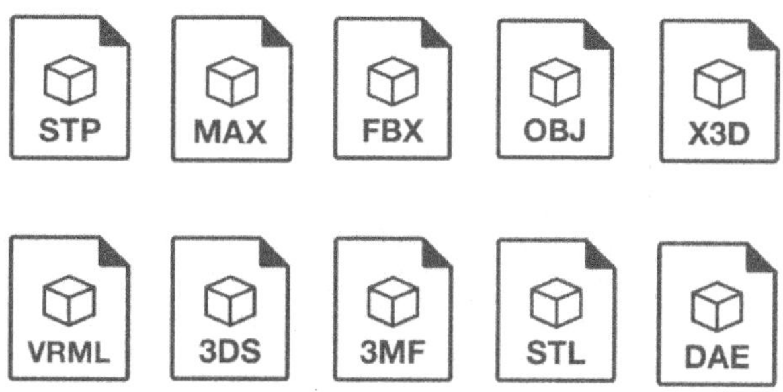

Figure 4: CAD file formats and conversion
[Source: https://pshdesign.com/the-most-common-cad-file-formats/]

17.5 Questions

17.5.1 Long Response Questions

1. What are the key benefits of using CAD software for 3D printing?
2. How do parametric modeling and freeform modeling differ in CAD for 3D printing?
3. What are the advantages and disadvantages of using mesh modeling and lattice modeling in CAD for 3D printing?
4. How do CAD file formats and conversion impact the 3D printing process?
5. What are some best practices for designing and modeling parts and products for 3D printing using CAD software?

17.5.2 Multiple Choice Questions

1) **What does CAD stand for in 3D printing?**
   a) Computer-Aided Drafting
   b) Computer-Aided Design
   c) Computer-Assisted Design
   d) Computer-Aided Drawing
   Solution: - b) Computer-Aided Design
2) **Which of the following is a common CAD file format used in 3D printing?**
   a) .jpg
   b) .pdf
   c) .stl
   d) .mp4
   Solution: - c) .stl
3) **What is the primary function of CAD software in 3D printing?**
   a) To print the model
   b) To design the model
   c) To repair the printer
   d) To apply colors
   Solution: - b) To design the model
4) **Which CAD tool is used to create a 3D model from a 2D drawing?**
   a) Loft
   b) Extrude
   c) Sweep
   d) Revolve
   Solution: - b) Extrude
5) **In CAD, what is the purpose of a 'Mesh'?**
   a) To add texture to the model
   b) To define the surface of the model
   c) To color the model
   d) To enhance model size
   Solution: - b) To define the surface of the model
6) **What does 'Boolean Union' accomplish in CAD modeling?**
   a) Removes overlapping parts of models
   b) Merges two or more models into one
   c) Creates holes in the model

d) Divides a model into parts

Solution: - b) Merges two or more models into one

7) **Which CAD feature allows for viewing the inside of a model without cutting it?**

a) Section View

b) Isometric View

c) Orthographic View

d) Perspective View

Solution: - a) Section View

8) **In CAD, what is 'Scaling' used for?**

a) To change the model's color

b) To adjust the model's size

c) To rotate the model

d) To add texture

Solution: - b) To adjust the model's size

9) **What does 'Mesh Repair' involve in CAD modeling?**

a) Adding color to the model

b) Correcting errors in the mesh structure

c) Optimizing the print speed

d) Creating textures

Solution: - b) Correcting errors in the mesh structure

10) **Which technique in CAD involves pulling or pushing a surface to change a model's shape?**

a) Lofting

b) Sculpting

c) Extruding

d) Sweeping

Solution: - b) Sculpting

11) **What is the role of 'Slicing' in the 3D printing process?**

a) To convert the model into layers for printing

b) To add color to the model

c) To assemble the print

d) To clean the printer

Solution: - a) To convert the model into layers for printing

12) **How does 'Topology Optimization' benefit CAD designs for 3D printing?**

a) Reduces the model's weight while maintaining strength

b) Simplifies color application

c) Increases the print speed

d) Improves the model's visual appearance

Solution: - a) Reduces the model's weight while maintaining strength

13) What does 'Parametric Design' in CAD allow you to do?

a) Create freeform shapes

b) Design with adjustable parameters

c) Improve color accuracy

d) Increase print speed

Solution: - b) Design with adjustable parameters

14) What is the purpose of 'Support Structures' in 3D printing?

a) To stabilize the printer

b) To support overhangs and complex features during printing

c) To color the printed model

d) To speed up the printing process

Solution: - b) To support overhangs and complex features during printing

15) Which CAD feature helps in optimizing the path for reducing material use in 3D printing?

a) Path Optimization

b) Mesh Repair

c) Slicing

d) Print Optimization

Solution: - a) Path Optimization

16) What is 'Lofting' used for in CAD modeling?

a) Creating a model from two or more cross-sectional shapes

b) Adding textures

c) Rotating the model

d) Coloring the model

Solution: - a) Creating a model from two or more cross-sectional shapes

17) What does 'Boolean Subtraction' achieve in CAD?

a) Merges two models

b) Subtracts one shape from another

c) Adds a new shape to the existing model

d) Divides a model into sections

Solution: - b) Subtracts one shape from another

18) In CAD, what is the use of 'Revolve'?

a) To rotate a 2D shape around an axis to create a 3D model
b) To add color to the model
c) To change the model's size
d) To create texture maps

Solution: - a) To rotate a 2D shape around an axis to create a 3D model

19) Which CAD tool is used to create complex geometries from simple shapes?

a) Sweeping
b) Lofting
c) Extruding
d) Revolving

Solution: - b) Lofting

20) What does 'Mesh Smoothing' do to a 3D model in CAD?

a) Adds textures
b) Refines the surface to reduce jagged edges
c) Changes the model's color
d) Increases the model's size

Solution: - b) Refines the surface to reduce jagged edges

21) What type of CAD view helps in understanding how parts of a model fit together?

a) Section View
b) Isometric View
c) Perspective View
d) Orthographic View

Solution: - a) Section View

22) What does 'Mesh Decimation' do in CAD modeling?

a) Adds more detail to the model
b) Reduces the number of polygons to simplify the mesh
c) Changes the model's color
d) Increases the model's resolution

Solution: - b) Reduces the number of polygons to simplify the mesh

23) In CAD, what does 'Texturing' refer to?

a) Adding physical textures to the model
b) Applying images or patterns to the model's surface

c) Changing the model's size
d) Rotating the model

Solution: - b) Applying images or patterns to the model's surface

24) Which CAD operation is used to combine multiple shapes into one?

a) Boolean Union
b) Boolean Subtraction
c) Extrusion
d) Revolution

Solution: - a) Boolean Union

25) What is the use of 'Subdivision Surface Modeling' in CAD?

a) To increase the polygon count for smoother surfaces
b) To reduce the polygon count for faster processing
c) To apply textures
d) To change the color of the model

Solution: - a) To increase the polygon count for smoother surfaces

26) What does the 'Boolean Difference' operation do in CAD?

a) Adds material to the model
b) Subtracts one shape from another
c) Merges two shapes
d) Divides a model into sections

Solution: - b) Subtracts one shape from another

27) In CAD, what is a 'Parametric Constraint'?

a) A constraint that fixes the model's size
b) A variable that defines relationships between model features
c) A feature that colors the model
d) A tool that rotates the model

Solution: - b) A variable that defines relationships between model features

28) What is 'Drafting' in the context of CAD for 3D printing?

a) Creating detailed engineering drawings
b) Adding textures to the model
c) Adjusting the model's color
d) Printing a prototype

Solution: - a) Creating detailed engineering drawings

29) What does the 'Mirror' function do in CAD modeling?

a) Flips the model along a plane
b) Adds details to the model
c) Changes the model's size
d) Colors the model
Solution: - a) Flips the model along a plane

30) In CAD, what is the purpose of 'Patterning'?
a) To create repeating patterns of features in the model
b) To apply textures
c) To change the model's size
d) To rotate the model
Solution: - a) To create repeating patterns of features in the model

31) Which CAD tool is used to create complex 3D shapes by sweeping a 2D profile along a path?
a) Lofting
b) Sweeping
c) Extruding
d) Revolving
Solution: - b) Sweeping

32) What is the significance of 'Resolution' in CAD models for 3D printing?
a) Determines the print speed
b) Affects the level of detail in the model
c) Sets the color of the print
d) Controls the printer's temperature
Solution: - b) Affects the level of detail in the model

33) Which CAD feature would you use to create a new shape by rotating a 2D profile around an axis?
a) Extruding
b) Revolving
c) Sweeping
d) Lofting
Solution: - b) Revolving

34) In CAD, what does the 'Offset' tool do?
a) Creates a copy of the shape at a different distance
b) Colors the model
c) Changes the model's size
d) Adds textures

Solution: - a) Creates a copy of the shape at a different distance

35) **What does the term 'Topology' refer to in CAD for 3D printing?**
   a) The model's surface texture
   b) The geometric arrangement of the model's features
   c) The color of the model
   d) The model's internal structure
   Solution: - b) The geometric arrangement of the model's features

36) **Which CAD tool is used to edit the surface of a 3D model for smoother transitions?**
   a) Mesh Smoothing
   b) Mesh Repair
   c) Patterning
   d) Sculpting
   Solution: - a) Mesh Smoothing

37) **What does 'Boolean Intersection' do in CAD modeling?**
   a) Creates a new shape from the overlapping part of two models
   b) Adds a new shape to the model
   c) Subtracts a part of the model
   d) Merges two shapes
   Solution: - a) Creates a new shape from the overlapping part of two models

38) **What is the purpose of 'Mesh Analysis' in CAD?**
   a) To check the structural integrity of the model
   b) To apply colors
   c) To rotate the model
   d) To create textures
   Solution: - a) To check the structural integrity of the model

39) **In CAD modeling, what is the benefit of using 'Adaptive Mesh Refinement'?**
   a) Increases the resolution in areas with complex details
   b) Simplifies the model's texture
   c) Reduces the model's file size
   d) Colors the model

Solution: - a) Increases the resolution in areas with complex details

40) Which CAD function is crucial for creating complex and intricate designs by manipulating shapes?

a) Parametric Constraints
b) Mesh Repair
c) Boolean Operations
d) Texturing

Solution: - c) Boolean Operations

Chapter 18

3D Scanning and Reverse Engineering

18.1 Introduction to 3D Scanning and Reverse Engineering

3D scanning and reverse engineering are critical technologies that enable the creation of digital models from physical objects. These technologies have revolutionized the way we design, prototype, and manufacture products, and have opened up new possibilities for innovation and creativity. In the context of 3D printing, 3D scanning and reverse engineering are essential tools for creating accurate digital models of parts and products, allowing designers and engineers to rapidly prototype, test, and refine their designs.

What is 3D Scanning?

3D scanning is the process of capturing the shape and geometry of an object using specialized equipment. This equipment, known as 3D scanners, uses various technologies such as laser, structured light, or photogrammetry to capture detailed information about the object's surface. The resulting data is then used to create a digital model of the object, which can be used for a variety of purposes such as design, analysis, and simulation.

There are several types of 3D scanning technologies, each with its own strengths and limitations. Some common types of 3D scanning technologies include:

- **Laser scanning**: Uses a laser to capture detailed information about the object's surface. Laser scanning is highly accurate and can capture complex geometries, but can be slow and expensive.

- **Structured light scanning**: Uses a projector to project a pattern of light onto the object, which is then captured by a camera. Structured light scanning is fast and accurate, but can be limited by the complexity of the object's surface.

- **Photogrammetry**: Uses multiple cameras to capture images of the object from different angles, which are then used to create a digital model. Photogrammetry is fast and cost-effective but can be limited by the quality of the images and the complexity of the object's surface.

What is Reverse Engineering?

Reverse engineering is the process of using a digital model to recreate or modify an existing object. This involves analyzing the digital model to understand the object's design, materials, and manufacturing processes, and then using this information to create a new version of the object. Reverse engineering is commonly used in a variety of industries, including aerospace, automotive, and consumer products.

In the context of 3D printing, reverse engineering is an essential tool for creating accurate digital models of parts and products. By scanning an existing object and creating a digital model, designers and engineers can rapidly prototype, test, and refine their designs, reducing the time and cost associated with traditional prototyping methods.

Applications of 3D Scanning and Reverse Engineering

3D scanning and reverse engineering have a wide range of applications in various industries, including:

- **Product design and development**: 3D scanning and reverse engineering can be used to create accurate digital models of existing products, allowing designers and engineers to rapidly prototype, test, and refine their designs.

- **Manufacturing and production**: 3D scanning and reverse engineering can be used to create digital models of manufacturing tools and equipment, allowing for rapid prototyping and production of new parts and products.

- **Quality control and inspection**: 3D scanning and reverse engineering can be used to inspect and analyze the quality of manufactured parts and products, allowing for rapid detection of defects and imperfections.

- **Archaeology and cultural heritage**: 3D scanning and reverse engineering can be used to create digital models of historical artifacts and cultural heritage objects, allowing for preservation and analysis of these valuable resources.

Benefits of 3D Scanning and Reverse Engineering

3D scanning and reverse engineering offer a range of benefits, including:

- **Increased accuracy**: 3D scanning and reverse engineering allow for the creation of highly accurate digital models, reducing the risk of errors and inaccuracies.

- **Reduced time and cost**: 3D scanning and reverse engineering can reduce the time and cost associated with traditional prototyping methods, allowing designers and engineers to rapidly test and refine their designs.

- **Improved collaboration**: 3D scanning and reverse engineering enable designers and engineers to collaborate more effectively, by providing a common digital language and platform for design and development.

- **Increased innovation**: 3D scanning and reverse engineering enable designers and engineers to push the boundaries of innovation, by allowing them to create complex geometries and shapes that would be difficult or impossible to produce using traditional manufacturing methods.

By understanding the principles and applications of 3D scanning and reverse engineering, designers and engineers can unlock the full potential of these technologies and create innovative products and solutions that were previously impossible to produce. Figure 78 illustrates the process of 3D scanning and reverse engineering. It shows a person using a handheld 3D scanner to scan what appears to be a car part, possibly an engine block. The object is partially covered with green and blue film, likely to aid in the scanning process. The person is also holding a laptop that displays the 3D model of the scanned object.

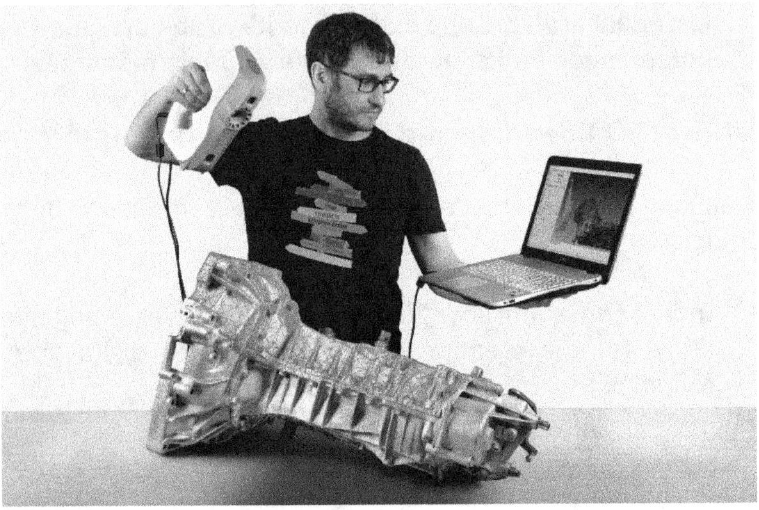

Figure 1: 3D scanning and reverse engineering
[Source: https://www.javelin-tech.com/3d/services/3d-scanning-service/]

18.2 Scanning Technologies and Techniques

There are several 3D scanning technologies and techniques available, each with its own strengths and weaknesses. The choice of 3D scanning technology depends on the specific application, the type of object being scanned, and the desired level of accuracy and detail. In this section, we will explore some popular 3D scanning technologies and techniques, including laser scanning, structured light scanning, photogrammetry, and computed tomography (CT) scanning.

Laser Scanning

Laser scanning is a popular 3D scanning technology that uses a laser to capture the shape and geometry of an object. The laser is directed at the object, and the reflected light is measured to create a point cloud of data. This data is then used to create a 3D model of the object.

There are several types of laser scanning technologies, including:

- **Time-of-Flight (ToF) laser scanning**: Measures the time it takes for the laser to bounce back from the object, allowing for accurate distance measurements.

- **Phase Shift laser scanning**: Measures the phase shift of the laser as it bounces back from the object, allowing for accurate distance measurements.

- **Frequency Modulated Continuous Wave (FMCW) laser scanning**: Measures the frequency shift of the laser as it bounces back from the object, allowing for accurate distance measurements.

Laser scanning is highly accurate and can capture complex geometries, making it ideal for applications such as industrial inspection, reverse engineering, and architecture. However, it can be slow and expensive, and may not be suitable for scanning large or complex objects.

Structured Light Scanning

Structured light scanning is a 3D scanning technology that uses a projector to project patterns onto an object and capture its shape and geometry. The projector projects a series of patterns, such as stripes or dots, onto the object, and a camera captures images of the object from different angles. The images are then used to create a 3D model of the object.

Structured light scanning is fast and accurate, and can capture complex geometries, making it ideal for applications such as product design, prototyping, and quality control. However, it may not be suitable for scanning objects with complex textures or reflective surfaces.

Photogrammetry

Photogrammetry is a 3D scanning technology that uses multiple cameras to capture images of an object from different angles and create a 3D model. The cameras are positioned around the object, and the images are then used to create a point cloud of data. This data is then used to create a 3D model of the object.

Photogrammetry is fast and cost-effective, and can capture complex geometries, making it ideal for applications such as architecture, surveying, and cultural heritage preservation. However, it may not be suitable for scanning objects with complex textures or reflective surfaces.

Computed Tomography (CT) Scanning

Computed Tomography (CT) scanning is a 3D scanning technology that uses X-rays to capture the internal structure of an object. The X-rays are directed at the object, and the resulting data is used to create a 3D model of the object's internal structure.

CT scanning is highly accurate and can capture complex internal geometries, making it ideal for applications such as medical imaging, industrial inspection, and materials science. However, it can be slow and expensive and may not be suitable for scanning large or complex objects.

Other 3D Scanning Technologies

In addition to laser scanning, structured light scanning, photogrammetry, and CT scanning, there are several other 3D scanning technologies available, including:

- **Stereo vision**: Uses two cameras to capture images of an object from different angles and create a 3D model.

- **Shape from shading**: Uses a single camera to capture images of an object and create a 3D model based on the shading and texture of the object.

- **Moire fringe analysis**: Uses a projector to project patterns onto an object and capture its shape and geometry.

- Each of these technologies has its own strengths and weaknesses, and the choice of technology depends on the specific application and the desired level of accuracy and detail.

Challenges and Limitations

While 3D scanning technologies have revolutionized the way we capture and analyze data, there are still several challenges and limitations to consider. Some common challenges and limitations include:

- **Data accuracy and quality**: 3D scanning data can be affected by factors such as noise, distortion, and resolution, which can impact the accuracy and quality of the resulting 3D model.

- **Object complexity**: 3D scanning technologies may struggle to capture complex geometries or objects with complex textures or reflective surfaces.

- **Scanning time and cost**: 3D scanning can be a time-consuming and expensive process, particularly for large or complex objects.

- **Data processing and analysis**: 3D scanning data can be complex and require specialized software and expertiseto process and analyze.

By understanding the different 3D scanning technologies and techniques available, designers and engineers can choose the best approach for their specific application and overcome the challenges and limitations associated with 3D scanning. Figure 79 demonstrates the use of 3D scanning technologies and techniques. It shows a person sitting at a desk with a laptop displaying a 3D scanned image of a human face. To the right of the laptop, there's a 3D scanning device aimed at a classical bust placed on the desk. The person appears to be gesturing towards the bust or the scanning device. This setup highlights how 3D scanning can be used to capture detailed digital representations of physical objects. Such technology is valuable in various fields, including art restoration, medical imaging, virtual reality, and manufacturing.

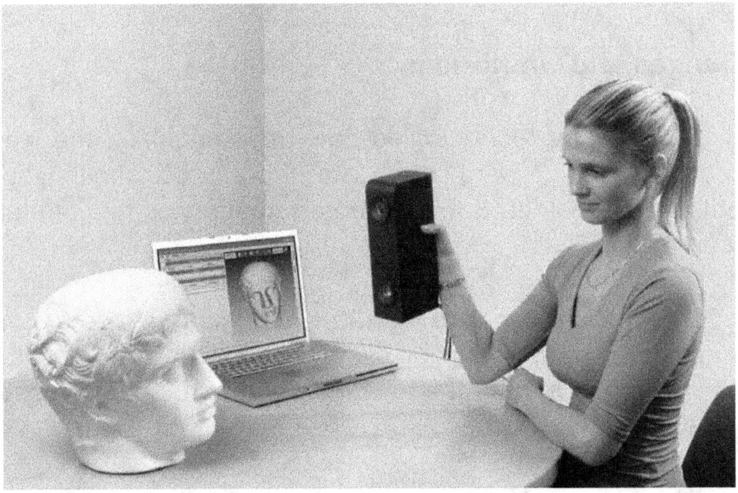

Figure 2: 3D scanning technologies and techniques
[Source: https://www.aniwaa.com/guide/3d-scanners/3d-scanning-technologies-and-the-3d-scanning-process/]

18.3 Reverse Engineering Methods and Applications

Reverse engineering involves the use of 3D scanning data to create a digital model of an object. This process is essential in various industries, including aerospace, automotive, healthcare, and consumer products, where the need to replicate or improve existing designs is crucial. There are several reverse engineering methods and applications, each with its own strengths and weaknesses.

Parametric Modeling

Parametric modeling is a reverse engineering method that uses 3D scanning data to create a parametric model of an object. This approach involves creating a digital model using parameters such as dimensions, shapes, and features. Parametric models are ideal for objects with complex geometries, as they can be easily modified and updated by changing the underlying parameters.

In parametric modeling, the 3D scanning data is used to create a point cloud, which is then converted into a parametric model using software such as Computer-Aided Design (CAD) or Computer-Aided Engineering (CAE). The resulting model can be used for various applications, including design optimization, simulation, and analysis.

Freeform Modeling

Freeform modeling is another reverse engineering method that uses 3D scanning data to create a freeform model of an object. This approach involves creating a digital model using curves and surfaces, rather than parameters. Freeform models are ideal for objects with complex, organic shapes, such as those found in aerospace or biomedical applications.

In freeform modeling, the 3D scanning data is used to create a mesh, which is then converted into a freeform model using software such as CAD or CAE. The resulting model can be used for various applications, including design visualization, prototyping, and manufacturing.

Surface Modeling

Surface modeling is a reverse engineering method that uses 3D scanning data to create a surface model of an object. This approach involves creating a digital model using surfaces, rather than solids.

Surface models are ideal for objects with complex, curved surfaces, such as those found in automotive or aerospace applications.

In surface modeling, the 3D scanning data is used to create a mesh, which is then converted into a surface model using software such as CAD or CAE. The resulting model can be used for various applications, including design visualization, prototyping, and manufacturing.

Inspection and Analysis

Inspection and analysis is a reverse engineering method that uses 3D scanning data to inspect and analyze the shape and geometry of an object. This approach involves using software to analyze the 3D scanning data and identify defects, deviations, or other issues with the object.

In inspection and analysis, the 3D scanning data is used to create a point cloud, which is then analyzed using software such as Geometric Dimensioning and Tolerancing (GD&T) or Computer-Aided Inspection (CAI). The resulting analysis can be used to identify areas for improvement, optimize designs, and ensure quality control.

Applications of Reverse Engineering

Reverse engineering has numerous applications across various industries, including:

- **Aerospace**: Reverse engineering is used to replicate or improve existing aircraft designs, reducing development time and costs.

- **Automotive**: Reverse engineering is used to design and manufacture car parts, improving performance and reducing production costs.

- **Healthcare**: Reverse engineering is used to create custom implants, prosthetics, and surgical models, improving patient outcomes and reducing recovery times.

- **Consumer Products**: Reverse engineering is used to design and manufacture consumer products, such as toys, appliances, and furniture, improving performance and reducing production costs.

Benefits of Reverse Engineering

Reverse engineering offers numerous benefits, including:

- **Reduced Development Time**: Reverse engineering reduces the time and effort required to design and develop new products, allowing companies to get to market faster.

- **Improved Accuracy**: Reverse engineering ensures accurate digital models, reducing errors and improving product quality.

- **Cost Savings**: Reverse engineering reduces production costs by minimizing prototyping and testing, and improving manufacturing efficiency.

- **Increased Innovation**: Reverse engineering enables companies to improve existing designs, leading to increased innovation and competitiveness.

In conclusion, reverse engineering is a powerful tool that enables companies to create digital models of existing objects, improving design, development, and manufacturing processes. By understanding the different reverse engineering methods and applications, companies can unlock the full potential of 3D scanning technology and stay ahead of the competition. Figure 80 is a circular infographic highlighting the six advantages of reverse engineering methods.

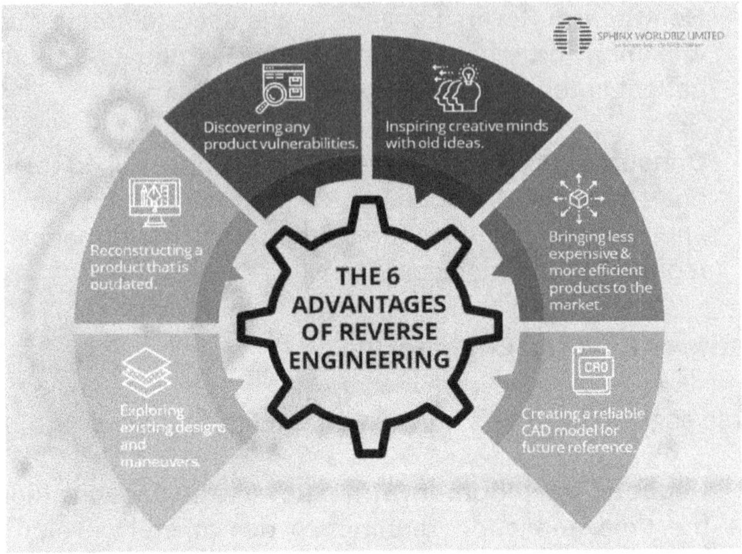

Figure 3: Advantages of reverse engineering methods
[Source: https://www.sphinxworldbiz.com/blog/the-6-advantages-of-reverse-engineering/]

18.4 3D Scanning and Reverse Engineering for 3D Printing

3D scanning and reverse engineering are essential tools for 3D printing, as they enable the creation of accurate digital models of parts and products. These technologies have revolutionized the 3D printing industry, allowing for rapid prototyping, production, and innovation. Some key applications of 3D scanning and reverse engineering for 3D printing include:

Part Replication

Part replication is a critical application of 3D scanning and reverse engineering for 3D printing. This process involves creating exact replicas of parts and products using 3D scanning data. The scanned data is used to create a digital model, which is then 3D printed using various materials and technologies.

Part replication is useful in various industries, including aerospace, automotive, and healthcare. For example, in aerospace, 3D scanning and reverse engineering can be used to replicate complex aircraft parts, reducing production time and costs. In healthcare, 3D scanning and reverse engineering can be used to create custom implants and prosthetics, improving patient outcomes and reducing recovery times.

Design Modification

Design modification is another key application of 3D scanning and reverse engineering for 3D printing. This process involves using 3D scanning data to modify existing designs and create new ones. The scanned data is used to create a digital model, which is then modified using computer-aided design (CAD) software.

Design modification is useful in various industries, including consumer products, industrial equipment, and automotive. For example, in consumer products, 3D scanning, and reverse engineering can be used to modify existing product designs, improving performance and reducing production costs. In industrial equipment, 3D scanning, and reverse engineering can be used to modify existing machinery, improving efficiency and reducing downtime.

Inspection and Quality Control

Inspection and quality control is a critical application of 3D scanning and reverse engineering for 3D printing. This process involves using 3D scanning data to inspect and analyze the shape and geometry of 3D printed parts. The scanned data is used to identify defects, deviations, or other issues with the part, ensuring that it meets the required specifications and standards.

Inspection and quality control is useful in various industries, including aerospace, automotive, and healthcare. For example, in aerospace, 3D scanning, and reverse engineering can be used to inspect and analyze aircraft parts, ensuring that they meet the required safety and performance standards. In healthcare, 3D scanning, and reverse engineering can be used to inspect and analyze medical implants, ensuring that they meet the required safety and performance standards.

Benefits of 3D Scanning and Reverse Engineering for 3D Printing

3D scanning and reverse engineering offer numerous benefits for 3D printing, including:

- **Improved Accuracy**: 3D scanning and reverse engineering enable the creation of accurate digital models, reducing errors and improving product quality.

- **Reduced Production Time**: 3D scanning and reverse engineering reduce the time and effort required to design and develop new products, allowing companies to get to market faster.

- **Cost Savings**: 3D scanning and reverse engineering reduce production costs by minimizing prototyping and testing, and improving manufacturing efficiency.

- **Increased Innovation**: 3D scanning and reverse engineering enable companies to improve existing designs, leading to increased innovation and competitiveness.

Challenges and Limitations

While 3D scanning and reverse engineering offer numerous benefits for 3D printing, there are also some challenges and limitations to consider. These include:

- **Data Accuracy**: 3D scanning data may not always be accurate, leading to errors and inaccuracies in the digital model.

- **Complexity**: 3D scanning and reverse engineering can be complex and time-consuming, requiring specialized skills and expertise.
- **Cost**: 3D scanning and reverse engineering can be expensive, especially for high-end equipment and software.

Future Directions

The future of 3D scanning and reverse engineering for 3D printing looks promising, with ongoing advancements in technology and software. Some potential future directions include:

- **Increased Automation**: Increased automation of 3D scanning and reverse engineering processes, reducing the need for human intervention and improving efficiency.
- **Improved Data Accuracy**: Improved data accuracy and resolution, enabling the creation of more accurate digital models and reducing errors.
- **New Materials and Technologies**: The development of new materials and technologies, enabling the creation of complex and innovative products.

In conclusion, 3D scanning and reverse engineering are essential tools for 3D printing, enabling the creation of accurate digital models of parts and products. These technologies have numerous applications, including part replication, design modification, and inspection and quality control. While there are some challenges and limitations to consider, the benefits of 3D scanning and reverse engineering for 3D printing are clear, and the future looks promising for these technologies. Figure 81 outlines the key steps involved in reverse engineering and 3D scanning for 3D printing. Here's a breakdown of the process:

- Original Part: The process starts with the physical object, such as a turbine blade.

- 3D Scanning: A handheld 3D scanner is used to capture the geometry of the original part.
- Scan-to-CAD: The scanned data is converted into a CAD model, which can be edited and refined.
- Engineering or Design: The CAD model is further developed, with detailed wireframe and surface models.
- Production: The final product is created, often starting with a prototype and leading to the finished item.

This process is essential for creating accurate digital models from physical objects, which can then be used for manufacturing, prototyping, or further design work.

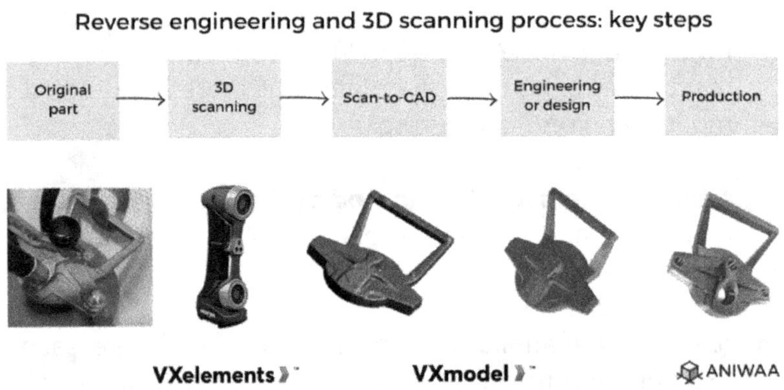

Figure 4: 3D scanning and reverse engineering for 3D printing
[Source: https://www.aniwaa.com/]

18.5 Questions

18.5.1 Long Response Questions

1. What are the key benefits of using 3D scanning and reverse engineering in the context of 3D printing?

2. How do laser scanning, structured light scanning, photogrammetry, and CT scanning differ in terms of their applications and benefits?
3. What are the advantages and disadvantages of using parametric modeling, freeform modeling, and surface modeling in reverse engineering?
4. How can 3D scanning and reverse engineering be used for inspection and quality control in 3D printing?
5. What are some best practices for using 3D scanning and reverse engineering in the context of 3D printing?

18.5.2 Multiple Choice Questions

1) **What is the primary purpose of 3D scanning?**
   a) To create physical objects
   b) To capture the geometry of a physical object
   c) To enhance the color of objects
   d) To increase the size of objects
   Solution: - b) To capture the geometry of a physical object
2) **Which technology is commonly used for 3D scanning?**
   a) Laser Scanning
   b) Inkjet Printing
   c) CNC Machining
   d) Ultrasonic Testing
   Solution: - a) Laser Scanning
3) **Which of the following is a type of 3D scanner?**
   a) Handheld Scanner
   b) Projection Printer
   c) Extrusion Printer
   d) Inkjet Scanner
   Solution: - a) Handheld Scanner
4) **What does 'reverse engineering' primarily involve?**
   a) Designing new products from scratch
   b) Disassembling and analyzing an existing product
   c) Enhancing the color of products
   d) Printing prototypes
   Solution: - b) Disassembling and analyzing an existing product

5) **Which of the following is an advantage of 3D scanning?**
   a) Reduces material waste
   b) Provides accurate digital models
   c) Increases the weight of objects
   d) Requires manual measurements
   Solution: - b) Provides accurate digital models

6) **Which of the following is a common file format for 3D scanning data?**
   a) STL
   b) JPEG
   c) PDF
   d) DOCX
   Solution: - a) STL

7) **What is a key benefit of using laser scanners in reverse engineering?**
   a) High-speed color detection
   b) High precision in capturing complex geometries
   c) Low-cost operation
   d) Improved material strength
   Solution: - b) High precision in capturing complex geometries

8) **In 3D scanning, what does 'point cloud' refer to?**
   a) A series of colors
   b) A collection of data points in space
   c) A type of 3D printer
   d) A method of material extrusion
   Solution: - b) A collection of data points in space

9) **Which software is commonly used to process 3D scanning data?**
   a) AutoCAD
   b) Photoshop
   c) Excel
   d) Word
   Solution: - a) AutoCAD

10) **What is 'mesh reconstruction' in the context of 3D scanning?**
   a) Converting a 2D image into a 3D model
   b) Creating a 3D model from a point cloud
   c) Improving the texture of a scanned object

d) Scaling the size of a 3D model

Solution: - b) Creating a 3D model from a point cloud

11) Which factor is crucial for ensuring accuracy in 3D scanning?

a) Scanner resolution
b) Scanner color
c) Scanner brand
d) Scanner weight

Solution: - a) Scanner resolution

12) What does 'registration' refer to in 3D scanning?

a) Aligning multiple scans to create a complete model
b) Creating a new product design
c) Converting 3D data to 2D
d) Printing the scanned object

Solution: - a) Aligning multiple scans to create a complete model

13) Which of the following is NOT a type of 3D scanning technology?

a) Structured Light Scanning
b) Contact Scanning
c) Photogrammetry
d) Thermal Imaging

Solution: - d) Thermal Imaging

14) In reverse engineering, why is it important to analyze the existing product?

a) To create a marketing strategy
b) To understand and replicate design features
c) To improve product aesthetics
d) To increase production speed

Solution: - b) To understand and replicate design features

15) What is a common application of 3D scanning in manufacturing?

a) Quality control
b) Texturing materials
c) Material extrusion
d) Increasing product weight

Solution: - a) Quality control

16) **What challenge is associated with scanning highly reflective surfaces?**

a) Inaccurate color reproduction
b) Data loss due to reflections
c) Increased scanning speed
d) Enhanced resolution

Solution: - b) Data loss due to reflections

17) **Which of the following methods can be used to improve the accuracy of 3D scanning in the presence of noise?**

a) Filtering and smoothing algorithms
b) Increasing scan speed
c) Reducing the resolution of the scanner
d) Ignoring noise in post-processing

Solution: - a) Filtering and smoothing algorithms

18) **In reverse engineering, what is 'feature extraction'?**

a) Identifying key geometric features from a scanned model
b) Removing unnecessary data from a scan
c) Enhancing the color of a 3D model
d) Reducing the size of the scanned object

Solution: - a) Identifying key geometric features from a scanned model

19) **What is the main difference between active and passive 3D scanning technologies?**

a) Active technologies project a signal, while passive technologies capture ambient light

b) Passive technologies project a signal, while active technologies capture ambient light

c) Active technologies use lasers, while passive technologies use sound

d) Passive technologies use lasers, while active technologies use sound

Solution: - a) Active technologies project a signal, while passive technologies capture ambient light

20) **Which software feature is critical for working with large 3D scan data sets?**

a) Data compression
b) Basic color adjustment
c) Texturing
d) Data visualization

Solution: - a) Data compression

21) How does photogrammetry differ from laser scanning in 3D scanning?

a) Photogrammetry uses photos and software for reconstruction, while laser scanning uses lasers

b) Laser scanning uses photos, while photogrammetry uses lasers

c) Photogrammetry is faster, while laser scanning is slower

d) Laser scanning is cheaper, while photogrammetry is more expensive

Solution: - a) Photogrammetry uses photos and software for reconstruction, while laser scanning uses lasers

22) What is 'polygon mesh' in 3D modeling?

a) A network of polygons representing the surface of a 3D object

b) A method for colorizing a 3D model

c) A tool for smoothing 3D surfaces

d) A technique for scaling 3D objects

Solution: - a) A network of polygons representing the surface of a 3D object

23) Why is texture mapping used in 3D scanning and modeling?

a) To apply surface details and color to a 3D model

b) To increase the model's resolution

c) To improve the model's structural integrity

d) To simplify the model's geometry

Solution: - a) To apply surface details and color to a 3D model

24) What is 'geometric distortion' in the context of 3D scanning?

a) Deformation of the scanned object's geometry due to inaccuracies in scanning

b) Enhancement of the scanned object's color details

c) Increase in scanning speed

d) Reduction of data size during scanning

Solution: - a) Deformation of the scanned object's geometry due to inaccuracies in scanning

25) **In reverse engineering, what is the role of CAD (Computer-Aided Design) software?**
   a) To create and modify digital models based on scanned data
   b) To increase the scanning speed
   c) To improve the texture of the scanned object
   d) To manage the physical scanning equipment
   Solution: - a) To create and modify digital models based on scanned data
26) **What is 'scan fusion' in 3D scanning?**
   a) Combining data from multiple scans to create a complete model
   b) Reducing the number of scans needed
   c) Improving the color accuracy of a single scan
   d) Converting 3D data to 2D images
   Solution: - a) Combining data from multiple scans to create a complete model
27) **What is the main advantage of using a structured light scanner over a laser scanner?**
   a) Higher speed in capturing detailed textures
   b) Better for scanning large objects
   c) More accurate color reproduction
   d) Lower cost of operation
   Solution: - a) Higher speed in capturing detailed textures
28) **In reverse engineering, what does 'surface fitting' involve?**
   a) Creating smooth surfaces from scanned data to improve model quality
   b) Adjusting the color of scanned data
   c) Increasing the resolution of the scan
   d) Reducing the size of the scanned object
   Solution: - a) Creating smooth surfaces from scanned data to improve model quality
29) **What challenge does 'data alignment' address in 3D scanning?**
   a) Merging multiple scans to create a unified model
   b) Enhancing the color accuracy of the scan
   c) Speeding up the scanning process

d) Improving the mechanical strength of the scanned object

Solution: - a) Merging multiple scans to create a unified model

30) Which type of data is primarily used for 'model reconstruction' from 3D scans?

a) Point cloud data

b) Textual data

c) Color data

d) Audio data

Solution: - a) Point cloud data

31) In which field is 3D scanning used to preserve historical artifacts?

a) Archaeology

b) Manufacturing

c) Medicine

d) Automotive

Solution: - a) Archaeology

32) What does 'noise reduction' refer to in 3D scanning?

a) Removing unwanted artifacts and errors from the scan data

b) Enhancing the texture of the scanned object

c) Increasing the scanning speed

d) Improving the color accuracy of the scan

Solution: - a) Removing unwanted artifacts and errors from the scan data

33) What is a common limitation of photogrammetry compared to laser scanning?

a) Lower precision in capturing surface details

b) Higher cost

c) Slower processing speed

d) Greater complexity in data processing

Solution: - a) Lower precision in capturing surface details

34) What is the purpose of 'feature-based registration' in 3D scanning?

a) Aligning scans based on identifiable features or landmarks

b) Reducing the number of scans needed

c) Enhancing the color of scanned data

d) Increasing the resolution of the scan

Solution: - a) Aligning scans based on identifiable features or landmarks

35) Which of the following technologies can be used for scanning internal structures?

a) Computed Tomography (CT) Scanning
b) Structured Light Scanning
c) Photogrammetry
d) Laser Scanning

Solution: - a) Computed Tomography (CT) Scanning

36) What is 'point cloud density' in the context of 3D scanning?

a) The number of points collected per unit area or volume
b) The color accuracy of the scanned object
c) The speed of the scanning process
d) The size of the scanned object

Solution: - a) The number of points collected per unit area or volume

37) What does 'model simplification' involve in reverse engineering?

a) Reducing the complexity of a 3D model while preserving essential features
b) Increasing the detail of a 3D model
c) Changing the color of a 3D model
d) Enhancing the texture of a 3D model

Solution: - a) Reducing the complexity of a 3D model while preserving essential features

38) Which factor does NOT typically affect the quality of a 3D scan?

a) Ambient light conditions
b) Object surface texture
c) Scanner color
d) Scanner resolution

Solution: - c) Scanner color

39) In 3D scanning, what does 'resampling' refer to?

a) Adjusting the density of points in a point cloud
b) Changing the color of the scan data
c) Enhancing the texture of the scan
d) Increasing the size of the scanned object

Solution: - a) Adjusting the density of points in a point cloud

40) Which of the following is a common method for improving the quality of 3D scanned models?

a) Post-processing techniques such as smoothing and filtering

b) Increasing the color contrast of the scanned object

c) Reducing the number of scans taken

d) Changing the scanning speed

Solution: - a) Post-processing techniques such as smoothing and filtering

Part 4

Applications and Industries

Chapter 19

Additive Manufacturing in Aerospace and Defense

19.1 Introduction to AM in Aerospace and Defense

Additive manufacturing (AM) has revolutionized the aerospace and defense industries by enabling the rapid production of complex, lightweight, and high-performance components. AM offers a range of benefits, including reduced lead times, increased design flexibility, and improved part performance. In this chapter, we will explore the applications, benefits, challenges, and limitations of AM in aerospace and defense.

History of AM in Aerospace and Defense

The use of AM in aerospace and defense dates back to the 1990s, when the first 3D printing technologies were developed. Initially, AM was used for prototyping and model making, but as the technology improved, it began to be used for production parts. Today, AM is used in a wide range of aerospace and defense applications, from aircraft and spacecraft components to missile guidance systems and satellite parts.

Benefits of AM in Aerospace and Defense

AM offers a range of benefits for aerospace and defense applications, including:

- **Reduced Lead Times**: AM enables rapid production of parts, reducing lead times from months to weeks or even days.

- **Increased Design Flexibility**: AM allows for the creation of complex geometries and internal structures that cannot be produced with traditional manufacturing methods.

- **Improved Part Performance**: AM enables the creation of lightweight, high-strength parts with improved thermal and mechanical properties.
- **Reduced Material Waste**: AM is a net-shape process, reducing material waste and the environmental impact of production.
- **Increased Supply Chain Efficiency**: AM enables on-demand production, reducing inventory costs and supply chain complexity.

Applications of AM in Aerospace and Defense

AM is used in a wide range of aerospace and defense applications, including:

- **Aircraft Components**: AM is used to produce aircraft components such as engine parts, satellite components, and structural components.
- **Spacecraft Components**: AM is used to produce spacecraft components such as satellite parts, propulsion systems, and habitat modules.
- **Missile Guidance Systems**: AM is used to produce missile guidance systems, including seeker heads and navigation systems.
- **Satellite Parts**: AM is used to produce satellite parts, including antennas, reflectors, and structural components.
- **Unmanned Aerial Vehicles (UAVs)**: AM is used to produce UAV components, including airframes, wings, and propulsion systems.

Challenges and Limitations of AM in Aerospace and Defense

While AM offers many benefits for aerospace and defense applications, there are also some challenges and limitations to consider, including:

- **Material Properties**: AM materials may not have the same properties as traditionally manufactured materials, which can affect part performance.
- **Scalability**: AM is still a relatively slow process, which can limit its scalability for large production runs.
- **Regulatory Framework**: There is a need for a regulatory framework to govern the use of AM in aerospace and defense applications.
- **Cybersecurity**: AM systems and data are vulnerable to cyber threats, which can compromise part quality and security.

Future Directions

The future of AM in aerospace and defense looks promising, with ongoing advancements in technology and materials. Some potential future directions include:

- **Increased Use of Metal AM**: Metal AM is expected to play a larger role in aerospace and defense applications, enabling the production of high-strength, lightweight components.
- **Development of New Materials**: New materials are being developed specifically for AM, offering improved properties and performance.
- **Integration with Traditional Manufacturing**: AM is expected to be integrated with traditional manufacturing methods, enabling the production of complex, hybrid components.

In conclusion, AM has revolutionized the aerospace and defense industries by enabling the rapid production of complex, lightweight, and high-performance components. While there are some challenges and limitations to consider, the benefits of AM are clear, and the future looks promising for this technology. Figure 82 showcases various aerospace and defense components produced using Additive Manufacturing (AM) technologies. Here's a brief overview of each labeled part:

(a) **Hinge Holder**: Produced by Airbus using AM technology, featuring a complex, geometrically intricate design.

(b) **Rolls Royce Trent XWB Engine**: Manufactured with the Electron Beam Melting (EBM) method, showing detailed internal structures.

(c) **Liquid Oxygen Flange**: Created using AM, with a smooth surface and precise edges.

(d) **Aerospace Hinge Part**: Produced with the laser powder bed method, featuring a lightweight and open lattice structure.

(e) **Gas Turbine Engine Component**: Created by 3D printing, demonstrating the capability of AM to produce functional engine parts.

(f) **Bracket Connector**: Produced by Airbus, optimized for weight reduction while maintaining robustness.

(g) **LEAP Engine**: Fabricated with AM, highlighting complex internal channels achievable with this technology.

(h) **Bald Bracket**: Manufactured by EBM, showing a solid and durable design.

(i) **Complex Shape Bracket**: Designed with AM, exhibiting an organic form optimized for stress distribution.

These examples highlight the advanced applications of AM in the aerospace and defense industries, where components benefit from being lighter, more complex, and potentially more efficient than those made with traditional manufacturing methods.

Figure 1: AM in aerospace and defense: (a) Hing holder produced by Airbus with AM technology, (b) Rolls Royce Trent XWB engine manufactured with EBM method, (c) liquid oxygen flange produced by AM, (d) aerospace hinge part produced with laser powder bed method, (e) gas turbine engine by 3D printing, (f) bracket connector produced by Airbus, (g) leap engine fabricated with AM, (h) bald bracket manufactured by EBM, (i) complex shape Bracket designed with AM
[Source: https://www.sciencedirect.com/science/article/pii/S223878542031646X?via%3Dihub]

19.2 Applications and Benefits of AM in Aerospace

Additive manufacturing (AM) has numerous applications in aerospace, including the production of aircraft components, spacecraft components, missile and rocket components, and unmanned aerial vehicles (UAVs). The benefits of AM in aerospace are numerous, including weight reduction, increased design flexibility, improved part performance, and reduced lead times.

Aircraft Components

AM is used to produce lightweight, complex components for aircraft, such as:

- **Engine Parts**: AM is used to produce engine components, such as turbine blades, compressor blades, and engine mounts. These components require high strength, stiffness, and resistance to high temperatures, making AM an ideal production method.

- **Satellite Components**: AM is used to produce satellite components, such as antennas, reflectors, and structural components. These components require high precision, low weight, and high strength, making AM an ideal production method.

- **Aircraft Structures**: AM is used to produce aircraft structures, such as fuselage components, wing components, and control surfaces. These components require high strength, stiffness, and resistance to fatigue, making AM an ideal production method.

Spacecraft Components

AM is used to produce components for spacecraft, such as:

- **Satellite Components**: AM is used to produce satellite components, such as propulsion systems, habitat modules, and life support systems. These components require high precision, low weight, and high strength, making AM an ideal production method.

- **Propulsion Systems**: AM is used to produce propulsion systems, such as rocket engines, fuel tanks, and thrusters. These components require high strength, stiffness, and resistance to high temperatures, making AM an ideal production method.

- **Habitat Modules**: AM is used to produce habitat modules, such as living quarters, laboratories, and life support systems. These components require high precision, low weight, and high strength, making AM an ideal production method.

Missile and Rocket Components

AM is used to produce components for missiles and rockets, such as:

- **Guidance Systems**: AM is used to produce guidance systems, such as seeker heads, navigation systems, and control systems. These components require high precision, low weight, and high strength, making AM an ideal production method.

- **Propulsion Systems**: AM is used to produce propulsion systems, such as rocket engines, fuel tanks, and thrusters. These components require high strength, stiffness, and resistance to high temperatures, making AM an ideal production method.

- **Structural Components**: AM is used to produce structural components, such as fuselage components, wing components, and control surfaces. These components require high strength, stiffness, and resistance to fatigue, making AM an ideal production method.

Unmanned Aerial Vehicles (UAVs)

AM is used to produce components for UAVs, such as:

- **Airframes**: AM is used to produce airframes, including fuselage components, wing components, and control surfaces. These components require high strength, stiffness, and resistance to fatigue, making AM an ideal production method.

- **Propulsion Systems**: AM is used to produce propulsion systems, such as engines, fuel tanks, and thrusters. These components require high strength, stiffness, and resistance to high temperatures, making AM an ideal production method.

- **Sensors**: AM is used to produce sensors, such as cameras, radar systems, and communication systems. These components require high precision, low weight, and high strength, making AM an ideal production method.

Benefits of AM in Aerospace

The benefits of AM in aerospace are numerous, including:

I. **Weight Reduction**: AM enables the production of lightweight components, reducing fuel consumption and increasing aircraft performance.
II. **Increased Design Flexibility**: AM allows for the production of complex geometries and internal structures, enabling the creation of innovative designs.
III. **Improved Part Performance**: AM enables the production of parts with improved mechanical properties, such as strength, stiffness, and fatigue resistance.
IV. **Reduced Lead Times**: AM enables rapid production, reducing lead times and enabling just-in-time production.
V. **Increased Supply Chain Efficiency**: AM enables on-demand production, reducing inventory costs and supply chain complexity.

VI. **Improved Sustainability**: AM enables the production of parts with reduced material waste, reducing the environmental impact of production.

In conclusion, AM has numerous applications in aerospace, including the production of aircraft components, spacecraft components, missile and rocket components, and UAVs. The benefits of AM in aerospace are numerous, including weight reduction, increased design flexibility, improved part performance, and reduced lead times. As the technology continues to evolve, we can expect to see even more innovative applications of AM in aerospace. Figure 83 illustrates the benefits of Additive Manufacturing (AM) in the aerospace industry. AM, also known as 3D printing, offers several advantages:

- **Cost Efficiency**: By streamlining the supply chain and enabling on-demand production of spare parts, AM reduces costs significantly. Notably, removing just one pound from an aircraft can save a staggering 14,000 gallons of fuel annually.

- **Lightweight & Innovative Materials**: AM allows for weight-saving techniques, such as lattice structures and conformal features. Unique materials, like Titanium Ti64 and Aluminum A710, meet FAA regulations while enhancing performance.

- **Reduced Lead Times**: AM slashes lead times from six months (with traditional manufacturing) to just six weeks. Thinner walls and custom material development contribute to faster production.

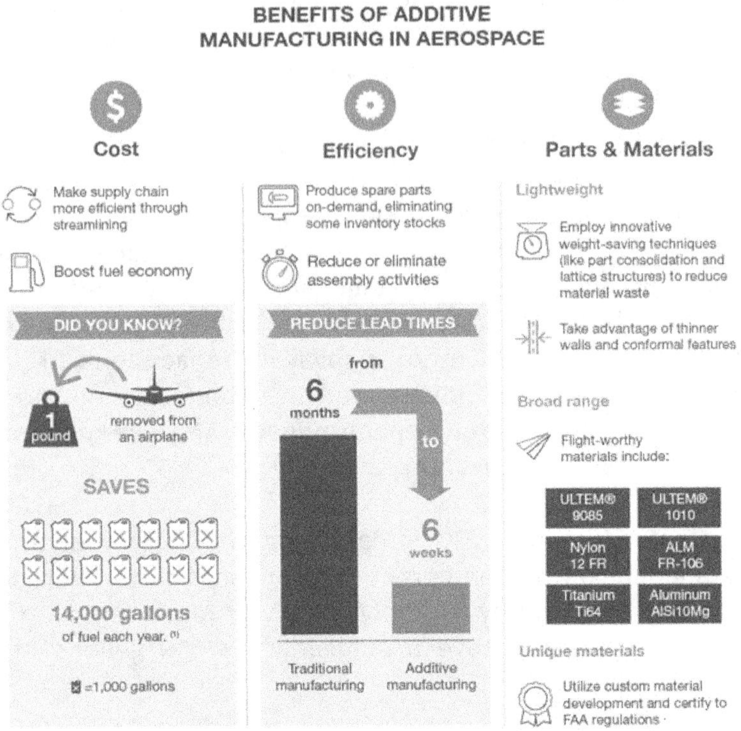

Figure 2: Benefits of AM in aerospace
[Source: https://www.makepartsfast.com/a-closer-look-at-the-use-of-3d-printing-on-planes/]

19.3 Challenges and Limitations of AM in Aerospace

Despite the benefits of additive manufacturing (AM) in aerospace, there are several challenges and limitations that need to be addressed. These challenges and limitations can be categorized into four main areas: material properties, scalability, certification and regulation, and cost.

Material Properties

One of the significant challenges of AM in aerospace is the properties of the materials used. The properties of AM materials may not meet the required specifications for aerospace applications, which can affect the performance and safety of the components. For example:

- **Mechanical Properties**: AM materials may not have the same mechanical properties as traditionally manufactured materials, such as strength, stiffness, and fatigue resistance.

- **Thermal Properties**: AM materials may not have the same thermal properties as traditionally manufactured materials, such as thermal conductivity, specific heat capacity, and thermal expansion.

- **Chemical Properties**: AM materials may not have the same chemical properties as traditionally manufactured materials, such as corrosion resistance, reactivity, and toxicity.

To address these challenges, researchers and manufacturers are working to develop new AM materials with improved properties. For example, some companies are developing AM materials with enhanced mechanical properties, such as high-strength aluminum and titanium alloys. Others are developing AM materials with improved thermal properties, such as thermal insulation materials and heat-resistant materials.

Scalability

Another challenge of AM in aerospace is scalability. AM machines may not be able to produce large components, limiting the scalability of AM in aerospace. For example:

- **Build Volume**: AM machines have limited build volumes, which can restrict the size of the components that can be produced.

- **Production Rate**: AM machines may not be able to produce components at a rate that meets the demands of large-scale production.

- **Material Handling**: AM machines may not be able to handle large quantities of material, which can limit the scalability of AM in aerospace.

To address these challenges, manufacturers are working to develop larger AM machines with increased build volumes and production rates. For example, some companies are developing AM machines with build volumes of up to 1 cubic meter, which can produce large components such as aircraft wings and fuselage sections.

Certification and Regulation

A third challenge of AM in aerospace is certification and regulation. There is a need for standardized certification and regulation of AM components in aerospace, which can ensure the safety and performance of the components. For example:

- **Standards and Specifications**: There is a need for standardized standards and specifications for AM components in aerospace, which can ensure consistency and quality.

- **Regulatory Framework**: There is a need for a regulatory framework that governs the use of AM in aerospace, which can ensure safety and performance.

- **Certification and Testing**: There is a need for certification and testing procedures that can ensure the safety and performance of AM components in aerospace.

To address these challenges, regulatory bodies and industry organizations are working to develop standardized certification and regulation procedures for AM components in aerospace. For example, the Federal Aviation Administration (FAA) has developed guidelines for the certification of AM components in aerospace, which include requirements for material properties, design, and testing.

Cost

A final challenge of AM in aerospace is cost. AM can be a costly process, particularly for large-scale production. For example:

- **Machine Cost**: AM machines can be expensive, particularly for large-scale production.
- **Material Cost**: AM materials can be expensive, particularly for high-performance materials.
- **Labor Cost**: AM requires skilled labor, which can be expensive.

To address these challenges, manufacturers are working to reduce the cost of AM in aerospace. For example, some companies are developing more affordable AM machines and materials, while others are working to improve the efficiency of the AM process.

In conclusion, despite the benefits of AM in aerospace, there are several challenges and limitations that need to be addressed. These challenges and limitations can be categorized into four main areas: material properties, scalability, certification and regulation, and cost. By addressing these challenges and limitations, manufacturers and researchers can ensure the safe and effective use of AM in aerospace. Let's explore the challenges associated with Additive Manufacturing (AM) in the aerospace industry, as depicted in Figure 84. AM, also known as 3D printing, has ushered in a new era of possibilities within aerospace engineering. Here are some key challenges:

- Inconsistent Accuracy and Quality: Safety is always the top priority in aerospace. For AM to be seriously considered, it must improve accuracy and replicability. Ensuring consistent quality across different designs, machines, and facilities is crucial.

- Surface Texture Repeatability: Achieving consistent surface quality remains a challenge. AM-produced parts may exhibit variations in texture, which can impact performance and reliability.

- Inspection of Components: Validating the quality and integrity of AM parts requires robust inspection methods. Detecting defects early in the process is essential for safety and reliability.

- Residual Stresses and Distortion: AM processes can introduce residual stresses and distortions in manufactured components. Managing these effects is critical for structural reliability.

- Material Properties and Databases: Understanding the properties of AM alloys and having reliable material databases are essential. Engineers need accurate data to make informed decisions during design and production.

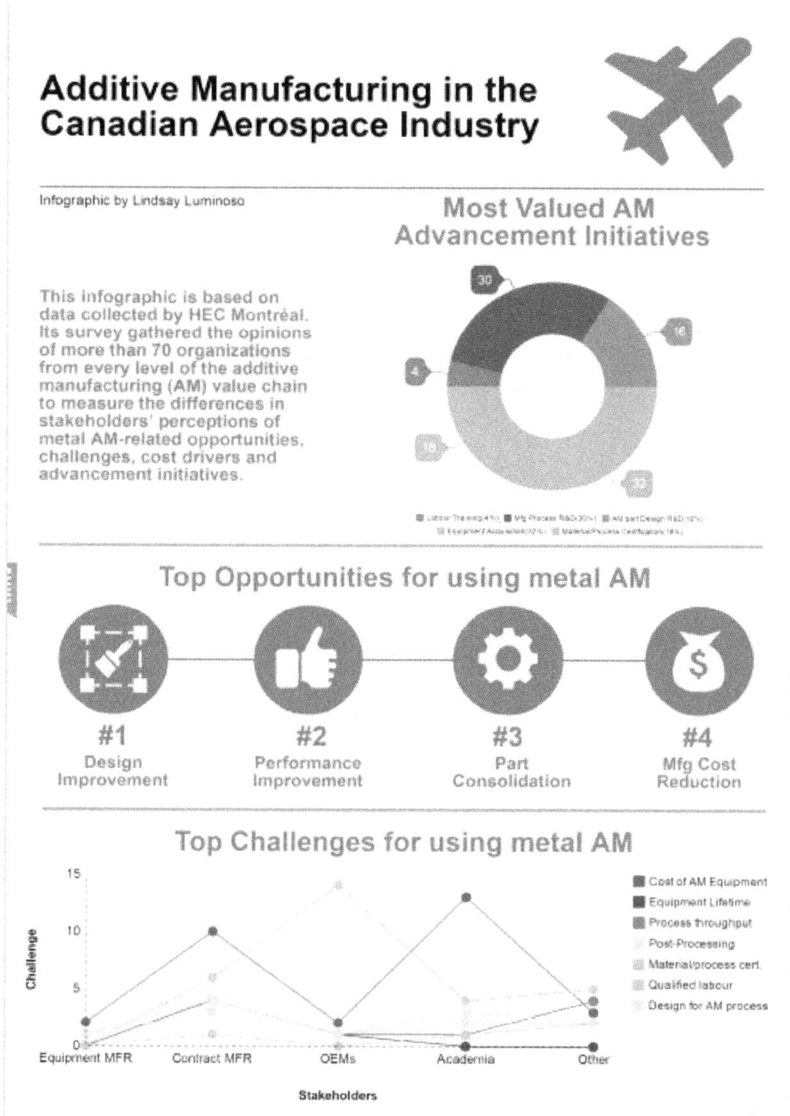

Figure 3: Challenges of AM in aerospace
[Source: https://www.design-engineering.com/features/metal-3d-print-1004025136/aerospace-am-infographic/]

19.4 Case Studies and Examples

Several companies and organizations are already leveraging additive manufacturing (AM) in aerospace, demonstrating its potential to transform the industry. Here are some examples of how AM is being used in aerospace:

NASA

NASA is a pioneer in the use of AM in aerospace. The agency has been using AM to produce components for spacecraft and aircraft, including:

- **Satellite Components**: NASA has used AM to produce satellite components, such as antennas and reflectors, which require high precision and low weight.

- **Engine Parts**: NASA has used AM to produce engine parts, such as turbine blades and compressor blades, which require high strength and resistance to high temperatures.

- **Spacecraft Components**: NASA has used AM to produce spacecraft components, such as habitat modules and life support systems, which require high precision and low weight.

NASA's use of AM has enabled the agency to reduce production time and cost, while also improving the performance and safety of its spacecraft and aircraft.

Boeing

Boeing is another major player in the aerospace industry that is leveraging AM. The company is using AM to produce components for aircraft, including:

- **Engine Parts**: Boeing is using AM to produce engine parts, such as turbine blades and compressor blades, which require high strength and resistance to high temperatures.

- **Structural Components**: Boeing is using AM to produce structural components, such as fuselage sections and wing components, which require high strength and stiffness.
- **Interior Components**: Boeing is using AM to produce interior components, such as seat frames and cabin components, which require high precision and low weight.

Boeing's use of AM has enabled the company to reduce production time and cost, while also improving the performance and safety of its aircraft.

Lockheed Martin

Lockheed Martin is a leading defense contractor that is also leveraging AM in aerospace. The company is using AM to produce components for spacecraft and aircraft, including:

- **Satellite Components**: Lockheed Martin is using AM to produce satellite components, such as antennas and reflectors, which require high precision and low weight.
- **Propulsion Systems**: Lockheed Martin is using AM to produce propulsion systems, such as rocket engines and fuel tanks, which require high strength and resistance to high temperatures.
- **Structural Components**: Lockheed Martin is using AM to produce structural components, such as fuselage sections and wing components, which require high strength and stiffness.

Lockheed Martin's use of AM has enabled the company to reduce production time and cost, while also improving the performance and safety of its spacecraft and aircraft.

Airbus

Airbus is a leading aircraft manufacturer that is also leveraging AM in aerospace. The company is using AM to produce components for aircraft, including:

- **Engine Parts**: Airbus is using AM to produce engine parts, such as turbine blades and compressor blades, which require high strength and resistance to high temperatures.

- **Structural Components**: Airbus is using AM to produce structural components, such as fuselage sections and wing components, which require high strength and stiffness.

- **Interior Components**: Airbus is using AM to produce interior components, such as seat frames and cabin components, which require high precision and low weight.

Airbus's use of AM has enabled the company to reduce production time and cost, while also improving the performance and safety of its aircraft.

Other Examples

In addition to these major players, there are many other companies and organizations that are leveraging AM in aerospace. For example:

- **United Launch Alliance**: United Launch Alliance is using AM to produce components for its rockets, including engine parts and structural components.

- **Blue Origin**: Blue Origin is using AM to produce components for its spacecraft, including engine parts and structural components.

- **SpaceX**: SpaceX is using AM to produce components for its rockets and spacecraft, including engine parts and structural components.

These examples demonstrate the potential of AM to transform the aerospace industry, enabling the production of complex components with improved performance, safety, and efficiency. As the technology continues to evolve, we can expect to see even more innovative applications of AM in aerospace. Let's explore the innovative applications of Additive Manufacturing (AM) in aerospace engineering and beyond. Figure 85 provides a glimpse into the diverse fields that benefit from this transformative technology:

- **Aerospace**: AM has revolutionized aircraft and spacecraft design and production. Notable applications include:

- **Improved Cooling for Engine Components**: 3D printing allows intricate geometries and high-strength materials, leading to advancements in engine efficiency.

- **Lightweight Engine Parts**: AM enables the creation of lightweight yet robust components, enhancing overall performance.

- **Improved Structural Components:** Complex geometries enhance structural integrity while reducing weight.

- **Better Tooling Through Additive Manufacturing:** Customized tooling aids in efficient manufacturing processes.

Other Fields

- **Automotive**: AM is used for rapid prototyping, customized parts, and lightweight components.

- **Biomedical**: Custom implants, prosthetics, and medical devices benefit from AM.

- **Energy**: AM contributes to efficient energy production and storage.

- **Retail**: Customized consumer products and electronics.

- **Academic Institutions**: Research and education leverage AM for innovation.

- **Government/Military**: Applications range from defense to logistics.

In summary, AM transcends boundaries, enabling groundbreaking solutions across various industries. Its flexibility, cost-effectiveness, and design freedom continue to drive innovation worldwide.

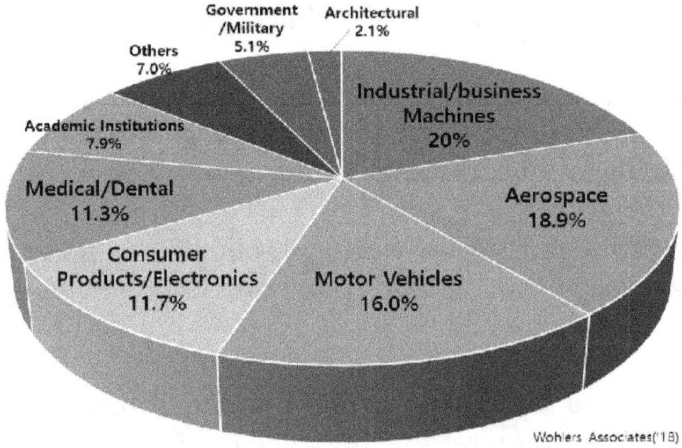

Figure 4: Case studies and examples along with other fields
[Source: https://link.springer.com/article/10.1007/s42405-020-00283-6]

19.5 Questions

19.5.1 Long Response Questions

1. What are the key benefits of using AM in aerospace and defense?
2. How do the properties of AM materials impact their use in aerospace applications?
3. What are the challenges and limitations of scaling up AM production in aerospace?
4. How can certification and regulation of AM components in aerospace be improved?

5. What are some examples of successful AM applications in aerospace and defense?

19.5.2 Multiple Choice Questions

1) **What is additive manufacturing commonly referred to as?**
   a) Subtractive manufacturing
   b) 3D printing
   c) Injection molding
   d) Forging
   Solution: - b) 3D printing
2) **Which material is often used in additive manufacturing for aerospace components?**
   a) Concrete
   b) Aluminum
   c) Wood
   d) Glass
   Solution: - b) Aluminum
3) **What is the primary benefit of using additive manufacturing in aerospace?**
   a) Increased weight
   b) Reduced complexity
   c) Higher costs
   d) More manual labor
   Solution: - b) Reduced complexity
4) **Which additive manufacturing technology is frequently used for creating metal parts?**
   a) Fused Deposition Modeling (FDM)
   b) Stereolithography (SLA)
   c) Selective Laser Melting (SLM)
   d) Digital Light Processing (DLP)
   Solution: - c) Selective Laser Melting (SLM)
5) **Additive manufacturing can help in reducing what in the aerospace industry?**
   a) Production time
   b) Energy consumption
   c) Labor costs
   d) All of the above
   Solution: - d) All of the above

6) **Which of the following is a challenge of additive manufacturing in aerospace?**
   a) High production speed
   b) Limited material options
   c) Reduced precision
   d) Increased waste
   Solution: - b) Limited material options
7) **How does additive manufacturing contribute to the reduction of waste in aerospace manufacturing?**
   a) By using excess material
   b) By recycling failed parts
   c) By adding only the necessary amount of material
   d) By increasing the use of scrap metal
   Solution: - c) By adding only the necessary amount of material
8) **What type of design can be more easily implemented using additive manufacturing in aerospace?**
   a) Complex geometries
   b) Simple geometries
   c) Large mass production designs
   d) Traditional designs
   Solution: - a) Complex geometries
9) **In which phase of aerospace component development does additive manufacturing show significant advantages?**
   a) Mass production
   b) Prototyping
   c) Final assembly
   d) Painting
   Solution: - b) Prototyping
10) **Which additive manufacturing technology is known for its high-resolution capabilities and is used for creating detailed aerospace components?**
   a) Fused Deposition Modeling (FDM)
   b) Stereolithography (SLA)
   c) Electron Beam Melting (EBM)
   d) Laminated Object Manufacturing (LOM)
   Solution: - b) Stereolithography (SLA)

11) Which company is known for using additive manufacturing to produce rocket engine parts?

a) Boeing
b) SpaceX
c) Lockheed Martin
d) Northrop Grumman

Solution: - b) SpaceX

12) What is one advantage of using polymer-based additive manufacturing in defense applications?

a) Increased weight
b) Higher cost
c) Rapid production of lightweight parts
d) Limited durability

Solution: - c) Rapid production of lightweight parts

13) Which aspect of additive manufacturing is crucial for ensuring the structural integrity of aerospace parts?

a) Layer thickness
b) Material color
c) Printer brand
d) Build volume

Solution: - a) Layer thickness

14) What type of additive manufacturing is particularly useful for creating strong, lightweight parts in aerospace?

a) Binder Jetting
b) Fused Deposition Modeling (FDM)
c) Direct Energy Deposition (DED)
d) Material Jetting

Solution: - c) Direct Energy Deposition (DED)

15) Which additive manufacturing process uses a laser to melt and fuse powder particles together?

a) Fused Deposition Modeling (FDM)
b) Selective Laser Sintering (SLS)
c) Stereolithography (SLA)
d) Laminated Object Manufacturing (LOM)

Solution: - b) Selective Laser Sintering (SLS)

16) What is one of the main benefits of using additive manufacturing for customized aerospace parts?

a) Uniformity in all parts
b) Reduced need for design iterations

c) Faster production of standard parts
d) Decreased design flexibility

Solution: - b) Reduced need for design iterations

17) **Which material is commonly used in additive manufacturing to create parts with high thermal resistance for aerospace applications?**

a) Titanium alloys
b) PLA
c) Polyethylene
d) ABS

Solution: - a) Titanium alloys

18) **What is one major limitation of additive manufacturing when applied to aerospace structures?**

a) High flexibility
b) Large build volumes
c) Surface finish and resolution
d) High material cost

Solution: - c) Surface finish and resolution

19) **How does the use of additive manufacturing influence the lifecycle of aerospace components?**

a) Increases lifecycle duration
b) Reduces lifecycle duration
c) No impact on lifecycle
d) Varies depending on material

Solution: - d) Varies depending on material

20) **Which aerospace company has successfully employed additive manufacturing to create a titanium part used in its jet engines?**

a) Airbus
b) Rolls-Royce
c) General Electric
d) Pratt & Whitney

Solution: - c) General Electric

21) **What role does advanced simulation play in additive manufacturing for aerospace applications?**

a) Reduces accuracy
b) Enhances design optimization
c) Increases production time
d) Limits material choices

Solution: - b) Enhances design optimization

22) **What is a critical factor to ensure the reliability of additive manufacturing parts in aerospace?**
    a) Aesthetic design
    b) Process repeatability
    c) Material color consistency
    d) Printer type
    Solution: - b) Process repeatability

23) **Which additive manufacturing technique involves the use of an electron beam to melt metal powders for aerospace applications?**
    a) Selective Laser Melting (SLM)
    b) Electron Beam Melting (EBM)
    c) Direct Metal Laser Sintering (DMLS)
    d) Laser Engineered Net Shaping (LENS)
    Solution: - b) Electron Beam Melting (EBM)

24) **What is one significant challenge in using additive manufacturing for high-temperature aerospace environments?**
    a) Insufficient material strength
    b) High material cost
    c) Limited printer technology
    d) Inadequate cooling systems
    Solution: - a) Insufficient material strength

25) **Which organization is known for setting standards and certifications for additive manufacturing in aerospace?**
    a) ASTM International
    b) ISO
    c) SAE International
    d) ASME
    Solution: - a) ASTM International

26) **How does additive manufacturing support the concept of 'on-demand production' in aerospace?**
    a) By requiring large inventory storage
    b) By allowing production of parts as needed
    c) By increasing lead times
    d) By focusing on mass production
    Solution: - b) By allowing production of parts as needed

27) What advantage does additive manufacturing provide in terms of repairing and retrofitting aerospace components?

a) Decreased repair time
b) Limited repair options
c) Increased repair costs
d) Standardized repair methods

Solution: - a) Decreased repair time

28) Which additive manufacturing technology is best suited for creating intricate cooling channels within aerospace components?

a) Direct Energy Deposition (DED)
b) Stereolithography (SLA)
c) Selective Laser Sintering (SLS)
d) Fused Deposition Modeling (FDM)

Solution: - c) Selective Laser Sintering (SLS)

29) What is a primary concern when integrating additive manufacturing parts into existing aerospace systems?

a) Compatibility with existing systems
b) Increased production speed
c) Uniformity of parts
d) Simplified design

Solution: - a) Compatibility with existing systems

30) Which aerospace application benefits from additive manufacturing's capability to produce parts with complex internal structures?

a) Aircraft fuselage
b) Rocket nozzles
c) Airplane wings
d) Engine housings

Solution: - b) Rocket nozzles

31) What impact does additive manufacturing have on the supply chain for aerospace components?

a) Increases complexity
b) Simplifies inventory management
c) Requires more suppliers
d) Reduces part availability

Solution: - b) Simplifies inventory management

32) **Which property of additive manufacturing materials is critical for ensuring durability in aerospace applications?**
   a) Surface finish
   b) Density
   c) Aesthetic quality
   d) Print speed
   Solution: - b) Density

33) **What is one of the major advantages of additive manufacturing in producing prototypes for aerospace applications?**
   a) Higher cost
   b) Longer lead times
   c) Faster design iteration
   d) Limited design options
   Solution: - c) Faster design iteration

34) **Which technique is often employed to improve the mechanical properties of additive manufactured parts?**
   a) Post-processing heat treatment
   b) Increased layer height
   c) Reduced build volume
   d) Lower material density
   Solution: - a) Post-processing heat treatment

35) **What role does advanced materials science play in the future of additive manufacturing for aerospace?**
   a) Reducing the need for new technologies
   b) Limiting material choices
   c) Enhancing material properties and performance
   d) Simplifying design processes
   Solution: - c) Enhancing material properties and performance

36) **Which additive manufacturing process is known for producing high-strength parts with minimal porosity?**
   a) Direct Metal Laser Sintering (DMLS)
   b) Fused Deposition Modeling (FDM)
   c) Stereolithography (SLA)
   d) Laminated Object Manufacturing (LOM)
   Solution: - a) Direct Metal Laser Sintering (DMLS)

37) What is a key consideration for ensuring that additive manufactured parts meet aerospace certification standards?

a) Aesthetic design
b) Material consistency and quality
c) Print speed
d) Design complexity

Solution: - b) Material consistency and quality

38) Which aerospace company used additive manufacturing to create a titanium part for its jet engines?

a) Boeing
b) Airbus
c) Rolls-Royce
d) Northrop Grumman

Solution: - c) Rolls-Royce

39) How does additive manufacturing influence the overall cost of producing aerospace parts?

a) Increases cost due to machine expense
b) Reduces cost due to lower material waste
c) Has no impact on cost
d) Increases cost due to longer production times

Solution: - b) Reduces cost due to lower material waste

40) Which additive manufacturing technology is known for its ability to create high-precision aerospace components with complex geometries?

a) Fused Deposition Modeling (FDM)
b) Stereolithography (SLA)
c) Selective Laser Melting (SLM)
d) Binder Jetting

Solution: - c) Selective Laser Melting (SLM)

Chapter 20

Additive Manufacturing in Healthcare and Biomedical

20.1 Introduction to AM in Healthcare and Biomedical

Additive manufacturing (AM) has revolutionized the healthcare and biomedical industries by enabling the creation of customized, complex, and personalized medical devices, implants, and models. AM offers a range of benefits, including improved patient outcomes, reduced recovery times, and increased accessibility to medical care. The technology has transformed the way medical professionals approach patient care, enabling them to create tailored solutions that meet the unique needs of individual patients.

Personalized Medicine

One of the most significant advantages of AM in healthcare is its ability to facilitate personalized medicine. By creating customized implants, prosthetics, and devices, medical professionals can provide patients with tailored solutions that meet their specific needs. For example, AM can be used to create customized hip and knee implants that are designed to fit individual patients' anatomy, reducing the risk of complications and improving outcomes.

Complex Geometries

AM also enables the creation of complex geometries that cannot be produced using traditional manufacturing techniques. This is particularly useful in the development of medical devices, such as stents, implants, and surgical guides, which require intricate designs and precise tolerances. By creating complex geometries, medical professionals can develop devices that are more effective, efficient, and safe.

Customized Models

AM can also be used to create customized models of organs and tissues, enabling medical professionals to practice and plan complex surgeries. These models can be used to simulate surgical procedures, reducing the risk of complications and improving outcomes. Additionally, AM can be used to create models of patient-specific anatomy, enabling medical professionals to develop personalized treatment plans.

Benefits of AM in Healthcare

The benefits of AM in healthcare are numerous and far-reaching. Some of the most significant advantages include:

- **Improved Patient Outcomes**: AM enables the creation of customized medical devices and implants that are tailored to individual patients' needs, improving outcomes and reducing the risk of complications.

- **Reduced Recovery Times**: AM can be used to create devices and implants that are designed to promote faster healing and reduce recovery times.

- **Increased Accessibility**: AM can be used to create medical devices and implants that are more accessible to patients in remote or underserved areas.

- **Cost Savings**: AM can be used to reduce the cost of medical devices and implants, making them more affordable for patients and healthcare providers.

Challenges and Limitations

While AM has the potential to revolutionize the healthcare and biomedical industries, there are several challenges and limitations that must be addressed. Some of the most significant challenges include:

- **Regulatory Framework**: There is a need for a regulatory framework that governs the use of AM in healthcare, ensuring that devices and implants meet safety and efficacy standards.

- **Material Properties**: There is a need for materials that are biocompatible, durable, and suitable for use in medical devices and implants.

- **Scalability**: There is a need for AM machines that can produce devices and implants at scale, reducing production time and cost.

- **Training and Education**: There is a need for training and education programs that enable medical professionals to design and use AM devices and implants effectively.

Future Directions

As AM continues to evolve, we can expect to see even more innovative applications in healthcare and biomedical. Some of the most promising areas of research include:

- **Bioprinting**: The use of AM to create living tissues and organs, enabling the development of personalized organ transplants and regenerative medicine.

- **Nanotechnology**: The use of AM to create devices and implants at the nanoscale, enabling the development of targeted therapies and personalized medicine.

- **Artificial Intelligence**: The use of AM in conjunction with artificial intelligence (AI) to develop personalized treatment plans and optimize medical device design.

In conclusion, AM has the potential to transform the healthcare and biomedical industries, enabling the creation of customized, complex, and personalized medical devices, implants, and models. While there are challenges and limitations that must be addressed, the benefits of AM in healthcare are clear, and we can expect to see even more innovative applications in the future. AM is reshaping healthcare and biomedical fields by offering patient-centric solutions, advancing research, and transforming how we approach medical challenges. And, Figure 86 shows some examples of AM in healthcare and biomedical engineering.

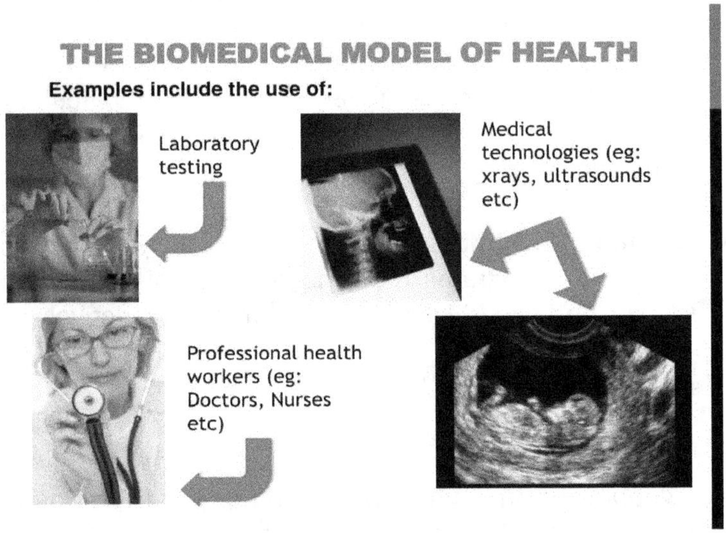

Figure 1: AM in healthcare and biomedical
[Source: https://www.slideserve.com/jada/approaches-to-health-care]

20.2 Applications and Benefits of AM in Healthcare

Additive manufacturing (AM) has numerous applications in healthcare, transforming the way medical professionals approach patient care. From customized implants to pharmaceuticals and drug delivery, AM is revolutionizing the healthcare industry.

Customized Implants

One of the most significant applications of AM in healthcare is the production of customized implants. Traditional implants are often designed to fit a wide range of patients, but they may not provide the best fit for individual patients. AM enables the production of customized implants that are tailored to individual patients' needs, improving fit, function, and patient outcomes.

For example, AM can be used to create customized hip and knee implants that are designed to fit individual patients' anatomy. This can reduce the risk of complications, improve mobility, and enhance patient outcomes. Additionally, AM can be used to create customized implants for patients with rare or complex conditions, such as craniofacial implants for patients with facial deformities.

Prosthetics and Orthotics

AM is also being used to create customized prosthetics and orthotics, improving mobility and quality of life for patients. Traditional prosthetics and orthotics are often designed to fit a wide range of patients, but they may not provide the best fit for individual patients. AM enables the production of customized prosthetics and orthotics that are tailored to individual patients' needs, improving fit, function, and patient outcomes.

For example, AM can be used to create customized prosthetic limbs that are designed to fit individual patients' anatomy. This can improve mobility, reduce discomfort, and enhance patient outcomes. Additionally, AM can be used to create customized orthotics, such as braces and splints, that are designed to provide support and stability for patients with musculoskeletal conditions.

Surgical Models and Guides

AM is also being used to create accurate surgical models and guides, improving surgical planning and reducing complications. Traditional surgical models and guides are often created using traditional manufacturing techniques, which can be time-consuming and expensive. AM enables the rapid production of accurate surgical models and guides, improving surgical planning and reducing complications.

For example, AM can be used to create customized surgical models of organs and tissues, enabling surgeons to practice and plan complex surgeries. This can reduce the risk of complications, improve patient outcomes, and enhance surgical training. Additionally, AM can be used to create customized surgical guides, such as templates and jigs, that are designed to facilitate precise surgical procedures.

Dental Restorations

AM is also being used to create customized dental restorations, such as crowns, bridges, and dentures. Traditional dental restorations are often created using traditional manufacturing techniques, which can be time-consuming and expensive. AM enables the rapid production of customized dental restorations, improving fit, function, and patient outcomes.

For example, AM can be used to create customized dental crowns that are designed to fit individual patients' teeth. This can improve fit, reduce discomfort, and enhance patient outcomes. Additionally, AM can be used to create customized dental bridges and dentures, improving oral health and quality of life for patients.

Pharmaceuticals and Drug Delivery

AM is also being used to create customized pharmaceuticals and drug delivery systems, improving patient outcomes and reducing side effects. Traditional pharmaceuticals and drug delivery systems are often designed to fit a wide range of patients, but they may not provide the best fit for individual patients. AM enables the production of customized pharmaceuticals and drug delivery systems that are tailored to individual patients' needs, improving patient outcomes and reducing side effects.

For example, AM can be used to create customized pills and capsules that are designed to release medication at a specific rate, improving patient outcomes and reducing side effects. Additionally, AM can be used to create customized drug delivery systems, such as implants and patches, that are designed to provide targeted therapy and reduce side effects.

Benefits of AM in Healthcare

The benefits of AM in healthcare are numerous and far-reaching. Some of the most significant benefits include:

- **Improved Patient Outcomes**: AM enables the production of customized medical devices and implants that are tailored to individual patients' needs, improving patient outcomes and reducing complications.

- **Reduced Recovery Times**: AM enables the production of medical devices and implants that are designed to promote rapid healing and reduce recovery times.

- **Increased Accessibility**: AM enables the production of medical devices and implants that are more accessible to patients in remote or underserved areas.

- **Cost Savings**: AM enables the production of medical devices and implants at a lower cost, reducing healthcare costs and improving affordability.

In conclusion, AM has numerous applications in healthcare, transforming the way medical professionals approach patient care. From customized implants to pharmaceuticals and drug delivery, AM is revolutionizing the healthcare industry. The benefits of AM in healthcare are clear, and we can expect to see even more innovative applications in the future. Figure 87 offers several advantages in this field:

- **Customization & Personalization**: AM allows for patient-specific designs, such as customized implants, prosthetics, and dental devices. Each patient's unique needs can be addressed.

- **On-demand Manufacturing:** Healthcare professionals can quickly produce medical tools, surgical guides, and patient-specific models using 3D printing. This agility improves patient care.

- **Improved Accuracy & Productivity:** Precise anatomical models aid surgical planning, leading to better outcomes. Additionally, AM streamlines production processes.

- **Utilization of Different Materials:** AM accommodates a wide range of materials, including biocompatible ones. This versatility supports various medical applications.

- **Surgeons as Designers:** Surgeons can actively participate in designing patient-specific implants and surgical tools, tailoring them to individual cases.

- **Better Understanding of Anatomy:** Anatomical models created through AM enhance medical education and training. Students and practitioners gain hands-on experience.

- **Flexibility & Ease of Use:** AM allows rapid iteration and adjustments, ensuring optimal fit and function for medical devices.

- **Antibiotic Incorporation:** Researchers explore 3D-printed implants with built-in antibiotic properties to prevent infections.

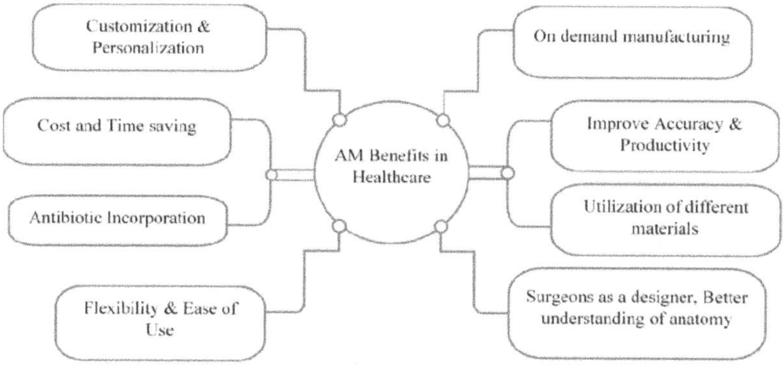

Figure 2: Benefits of AM in healthcare
[Source: https://iopscience.iop.org/article/10.1088/1757-899X/1201/1/012041]

20.3 Challenges and Limitations of AM in Healthcare

Despite the numerous benefits of additive manufacturing (AM) in healthcare, there are several challenges and limitations that must be addressed. These challenges and limitations can impact the adoption and implementation of AM in healthcare, and must be overcome to fully realize the potential of AM in this industry.

Regulatory Frameworks

One of the most significant challenges facing AM in healthcare is the need for standardized regulatory frameworks. Currently, there is a lack of clear guidelines and regulations governing the use of AM in healthcare, which can make it difficult for manufacturers to ensure the safety and efficacy of AM-produced medical devices and implants.

Regulatory agencies such as the US Food and Drug Administration (FDA) have begun to issue guidance documents and regulations specific to AM in healthcare, but more work is needed to establish a comprehensive regulatory framework. This framework must balance the need to ensure the safety and efficacy of AM-produced medical devices and implants with the need to encourage innovation and entrepreneurship in the industry.

Material Properties

Another challenge facing AM in healthcare is the properties of AM materials. While AM materials have improved significantly in recent years, they may not always meet the required specifications for medical devices and implants. For example, AM materials may not have the same strength, durability, or biocompatibility as traditional materials, which can impact their performance and safety.

To address this challenge, researchers and manufacturers are working to develop new AM materials that meet the required specifications for medical devices and implants. This includes the development of new metal alloys, polymers, and ceramics that can be used in AM applications.

Scalability

Scalability is another challenge facing AM in healthcare. While AM machines are capable of producing small batches of medical devices and implants, they may not be able to produce large quantities. This can limit the scalability of AM in healthcare, making it difficult to meet the demand for AM-produced medical devices and implants.

To address this challenge, manufacturers are working to develop larger AM machines that can produce larger quantities of medical devices and implants. This includes the development of industrial-scale AM machines that can produce hundreds or thousands of units per day.

Cost and Reimbursement

The cost of AM-produced medical devices and implants is another challenge facing the industry. While AM can reduce the cost of production in some cases, the cost of AM machines and materials can be high. Additionally, the cost of AM-produced medical devices and implants may not be reimbursed by insurance providers, which can limit accessibility to these technologies.

To address this challenge, manufacturers and healthcare providers are working to develop new business models that can reduce the cost of AM-produced medical devices and implants. This includes the development of subscription-based models, where healthcare providers pay a monthly fee for access to AM machines and materials.

Training and Education

Training and education are also critical challenges facing AM in healthcare. Healthcare professionals may not have the necessary training and education to design and use AM-produced medical devices and implants effectively. This can impact the safety and efficacy of these devices and limit their adoption in healthcare.

To address this challenge, manufacturers and healthcare providers are working to develop training and education programs that can teach healthcare professionals how to design and use AM-produced medical devices and implants. This includes the development of online courses, workshops, and certification programs.

Cybersecurity

Cybersecurity is another challenge facing AM in healthcare. AM machines and systems can be vulnerable to cyber-attacks, which can impact the safety and efficacy of AM-produced medical devices and implants. This can also lead to data breaches and intellectual property theft.

To address this challenge, manufacturers and healthcare providers are working to develop cybersecurity protocols and standards that can protect AM machines and systems from cyber-attacks. This includes the development of encryption technologies, firewalls, and intrusion detection systems.

Standardization

Standardization is also a critical challenge facing AM in healthcare. There is a need for standardized protocols and standards for AM in healthcare, which can ensure the safety and efficacy of AM-produced medical devices and implants. This includes the development of standards for AM machine performance, material properties, and device design.

To address this challenge, industry organizations and regulatory agencies are working to develop standardized protocols and standards for AM in healthcare. This includes the development of standards for AM machine performance, material properties, and device design.

In conclusion, while AM has the potential to transform the healthcare industry, there are several challenges and limitations that must be addressed. These challenges and limitations include regulatory frameworks, material properties, scalability, cost and reimbursement, training and education, cybersecurity, and standardization. By addressing these challenges and limitations, we can ensure the safe and effective adoption of AM in healthcare, and realize the full potential of this technology. Let's explore the challenges and limitations associated with Additive Manufacturing (AM) in healthcare which is shown in Figure 88. While AM offers significant benefits, it also faces certain obstacles:

- **High Initial Capital Requirement**: Implementing AM technology requires substantial upfront investment in equipment, materials, and training.

- **Potential for Increased Unemployment**: As AM automates certain manufacturing processes, there may be concerns about job displacement in traditional manufacturing sectors.

- **Difficulty in Deployment**: Integrating AM into existing healthcare systems can be complex due to infrastructure changes and workflow adjustments.

- **Reluctance Among Medical Practitioners to Adopt AI**: Healthcare professionals may be hesitant to embrace new technologies, including AM, without robust evidence of safety and efficacy.

- **Ambiguous Regulatory Guidelines for Medical Software**: Clear regulations are essential to ensure the safety and effectiveness of AM-produced medical devices.

- **Lack of Curated Healthcare Data**: High-quality data specific to AM applications in healthcare is essential for evidence-based decision-making.

- **Concerns Regarding Privacy and Security**: Protecting patient data and ensuring cybersecurity are critical when using AM technology.

- **Lack of Interoperability Between AI Solutions**: Ensuring seamless communication and compatibility between different AM systems and software remains a challenge.

- **State and Federal Regulations**: The regulatory landscape varies across regions, impacting the adoption and implementation of AM in healthcare.

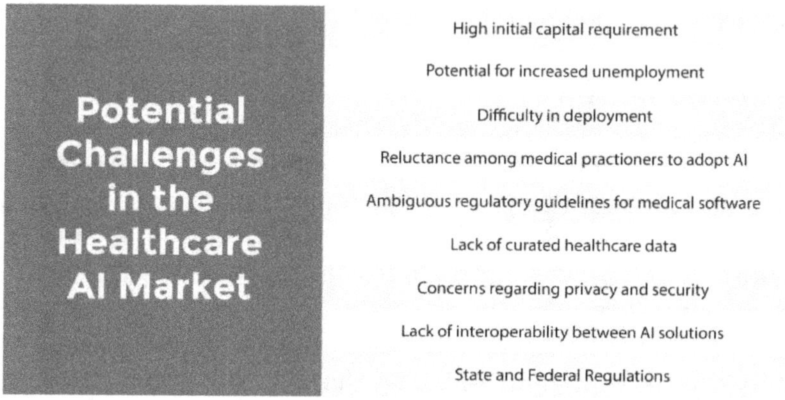

Figure 3: Challenges and limitations of AM in healthcare
[Source: https://www.chthealthcare.com/blog/artificial-intelligence-in-healthcare]

20.4 Case Studies and Examples

Several companies and organizations are already leveraging additive manufacturing (AM) in healthcare, demonstrating the potential of this technology to transform the industry. Here are some examples of companies and organizations that are using AM in healthcare:

Stryker

Stryker, a leading medical technology company, is using AM to produce customized spinal implants and surgical guides. By using AM, Stryker is able to create implants that are tailored to individual patients' needs, improving fit, function, and patient outcomes. For example, Stryker's AM-produced spinal implants can be designed to match the unique anatomy of each patient, reducing the risk of complications and improving recovery times.

Stryker is also using AM to produce surgical guides, which are used to help surgeons navigate complex procedures. These guides are created using patient-specific data, such as CT scans and MRI images, and are designed to provide surgeons with a precise and accurate roadmap for the procedure. By using AM-produced surgical guides, surgeons can reduce the risk of complications and improve patient outcomes.

Johnson & Johnson

Johnson & Johnson, a multinational healthcare company, is using AM to produce customized orthopedic implants and surgical models. By using AM, Johnson & Johnson is able to create implants that are tailored to individual patients' needs, improving fit, function, and patient outcomes. For example, Johnson & Johnson's AM-produced hip implants can be designed to match the unique anatomy of each patient, reducing the risk of complications and improving recovery times.

Johnson & Johnson is also using AM to produce surgical models, which are used to help surgeons plan and practice complex procedures. These models are created using patient-specific data, such as CT scans and MRI images, and are designed to provide surgeons with a precise and accurate representation of the patient's anatomy. By using AM-produced surgical models, surgeons can reduce the risk of complications and improve patient outcomes.

Medtronic

Medtronic, a leading medical technology company, is using AM to produce customized cardiovascular devices and implants. By using AM, Medtronic is able to create devices and implants that are tailored to individual patients' needs, improving fit, function, and patient outcomes. For example, Medtronic's AM-produced heart valves can be designed to match the unique anatomy of each patient, reducing the risk of complications and improving recovery times.

Medtronic is also using AM to produce customized surgical models, which are used to help surgeons plan and practice complex cardiovascular procedures. These models are created using patient-specific data, such as CT scans and MRI images, and are designed to provide surgeons with a precise and accurate representation of the patient's anatomy. By using AM-produced surgical models, surgeons can reduce the risk of complications and improve patient outcomes.

University of California, Los Angeles (UCLA)

The University of California, Los Angeles (UCLA) is using AM to produce customized prosthetics and orthotics for patients. By using AM, UCLA is able to create prosthetics and orthotics that are tailored to individual patients' needs, improving fit, function, and patient outcomes. For example, UCLA's AM-produced prosthetic limbs can be designed to match the unique anatomy of each patient, reducing the risk of complications and improving recovery times.

UCLA is also using AM to produce customized orthotics, such as braces and splints, which are used to support and stabilize patients' joints and muscles. This orthotics are created using patient-specific data, such as CT scans and MRI images, and are designed to provide patients with a precise and accurate fit. By using AM-produced orthotics, patients can reduce the risk of complications and improve their overall quality of life.

Other Examples

In addition to these examples, there are many other companies and organizations that are using AM in healthcare. For example, the National Institutes of Health (NIH) is using AM to produce customized models of organs and tissues, which are used to help researchers understand the underlying causes of diseases and develop new treatments.

The US Department of Veterans Affairs (VA) is also using AM to produce customized prosthetics and orthotics for veterans. By using AM, the VA is able to provide veterans with customized devices that are tailored to their individual needs, improving their overall quality of life.

In conclusion, these case studies and examples demonstrate the potential of AM to transform the healthcare industry. By using AM, companies and organizations can create customized medical devices and implants that are tailored to individual patients' needs, improving fit, function, and patient outcomes. As the technology continues to evolve, we can expect to see even more innovative applications of AM in healthcare. Let's delve into the case studies and examples of how Additive Manufacturing (AM) is transforming various sectors, including healthcare which is displayed in Figure 89:

- **Telemedicine in Rural Areas:** Telemedicine leverages technology to expand access to care for patients in remote regions. It allows consultations and follow-ups without the need for patients to travel long distances.

- **Electronic Health Records (EHRs):** Transitioning to a new EHR system can be challenging. In one case study, involving physicians in planning and implementation led to successful integration, improving efficiency and patient care.

- **Customized Implants and Prosthetics**: AM enables patient-specific designs for implants and prosthetics. Whether it's a dental crown or a custom-fitted prosthetic limb, 3D printing ensures precise fit and better outcomes.

- **Bioprinting and Tissue Engineering**: Researchers are using AM to create functional tissues and organs. Imagine a future where replacement organs are printed on demand.

- **Anatomical Models for Surgical Planning**: Surgeons use 3D-printed models to plan complex procedures. These models provide tangible representations of patient-specific anatomy, enhancing precision and outcomes.

- **Drug Delivery Devices**: AM allows intricate designs for drug delivery systems. Customized capsules, implants, and microneedles can release medications at specific rates, improving treatment efficacy.

- **Orthotics and Wearables**: From customized orthotic insoles to wearable medical devices, AM offers personalized solutions for patients with various conditions.

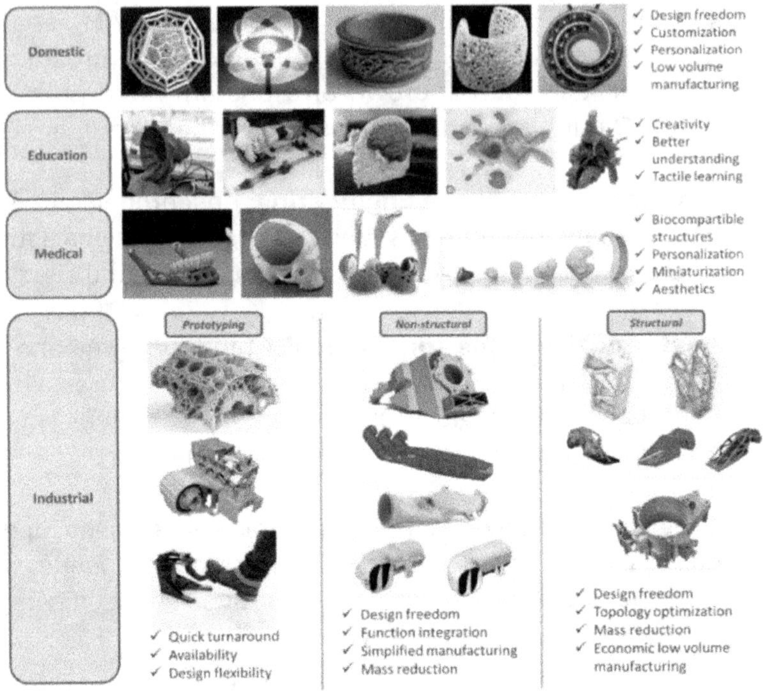

Figure 4: Case studies and examples AM in healthcare and other areas [Source: https://papers.ssrn.com/sol3/papers.cfm?abstract_id=4169003]

20.5 Questions

20.5.1 Long Response Questions

1. What are the key benefits of using AM in healthcare and biomedical?
2. How do the properties of AM materials impact their use in medical devices and implants?
3. What are the challenges and limitations of scaling up AM production in healthcare?
4. How can regulatory frameworks for AM in healthcare be improved?
5. What are some examples of successful AM applications in healthcare and biomedical?

20.5.2 Multiple Choice Questions

1) **What is a common application of 3D printing in healthcare?**
   a) Automotive parts
   b) Aerospace components
   c) Prosthetics
   d) Consumer electronics
   Solution: - c) Prosthetics
2) **Which of the following materials is often used in 3D printing for biomedical applications?**
   a) Steel
   b) Plastic
   c) Titanium
   d) Wood
   Solution: - c) Titanium
3) **What is the main benefit of using 3D printing for prosthetics?**
   a) Faster production
   b) Lower cost of raw materials
   c) Increased weight
   d) Limited customization
   Solution: - a) Faster production
4) **Which 3D printing technique is commonly used to create customized dental implants?**
   a) Fused Deposition Modeling (FDM)
   b) Stereolithography (SLA)
   c) Selective Laser Sintering (SLS)
   d) Digital Light Processing (DLP)
   Solution: - b) Stereolithography (SLA)
5) **3D printing can be used to create models for which purpose in medical training?**
   a) Vehicle simulations
   b) Structural engineering
   c) Surgical planning
   d) Aerospace design
   Solution: - c) Surgical planning
6) **Which technology is used in 3D printing to produce biocompatible scaffolds for tissue engineering?**

a) Inkjet printing
b) Laser sintering
c) Stereolithography (SLA)
d) Direct Ink Writing (DIW)
Solution: - d) Direct Ink Writing (DIW)

7) **What is a primary challenge associated with 3D printing of human organs?**
a) Lack of printing materials
b) High cost of technology
c) Biocompatibility issues
d) Limited software options
Solution: - c) Biocompatibility issues

8) **Which of the following is a key advantage of using 3D printing for creating surgical instruments?**
a) Increased manual labor
b) Longer lead times
c) Customization for specific procedures
d) High material wastage
Solution: - c) Customization for specific procedures

9) **What is the purpose of using 3D-printed anatomical models in medical education?**
a) To improve software development
b) To enhance hands-on learning and practice
c) To test new manufacturing processes
d) To design consumer products
Solution: - b) To enhance hands-on learning and practice

10) **In which process is bioprinting commonly used?**
a) Printing car parts
b) Creating customized jewelry
c) Tissue engineering and organ printing
d) Manufacturing household items
Solution: - c) Tissue engineering and organ printing

11) **Which bioprinting technique uses living cells as ink to create tissue structures?**
a) Fused Deposition Modeling (FDM)
b) Inkjet-based bioprinting
c) Laser Sintering
d) Direct Metal Laser Sintering (DMLS)

Solution: - b) Inkjet-based bioprinting

12) **What role does CAD (Computer-Aided Design) play in 3D printing of medical devices?**

a) CAD is used to create the software interface for the printer.

b) CAD helps in designing the precise model of the medical device.

c) CAD manages the raw material inventory.

d) CAD operates the mechanical components of the printer.

Solution: - b) CAD helps in designing the precise model of the medical device.

13) **How does 3D printing contribute to personalized medicine?**

a) By providing one-size-fits-all solutions

b) By creating customized implants and prosthetics

c) By standardizing drug dosages

d) By reducing the need for patient-specific data

Solution: - b) By creating customized implants and prosthetics

14) **Which of the following is a challenge faced by 3D-printed implants in long-term applications?**

a) Low strength of the material

b) High cost of production

c) Limited availability of printing technology

d) Degradation or biocompatibility over time

Solution: - d) Degradation or biocompatibility over time

15) **In the context of 3D printing in healthcare, what is the significance of "bio-inks"?**

a) They are used to print electronic circuits for medical devices.

b) They are materials containing living cells used in bioprinting.

c) They are inks used for marking surgical tools.

d) They are non-biological materials used to create medical implants.

Solution: - b) They are materials containing living cells used in bioprinting.

16) What is a potential future application of 3D printing in regenerative medicine?

a) Mass production of over-the-counter drugs

b) Customization of wearable electronics

c) Printing of complex tissue structures and organs

d) Standardized production of surgical gloves

Solution: - c) Printing of complex tissue structures and organs

17) Which of the following best describes the process of "bioprinting"?

a) Printing biological samples for research

b) Creating 3D structures using biomaterials and living cells

c) Printing labels for medical packaging

d) Producing prototypes for medical devices

Solution: - b) Creating 3D structures using biomaterials and living cells

18) What is the primary goal of using 3D-printed scaffolds in tissue engineering?

a) To create structural supports for building construction

b) To provide temporary support for tissue regeneration

c) To print decorative elements for medical facilities

d) To manufacture cost-effective surgical tools

Solution: - b) To provide temporary support for tissue regeneration

19) Which method is frequently used to ensure the accuracy of 3D-printed medical implants?

a) X-ray diffraction

b) Mechanical stress testing

c) Post-printing imaging and inspection

d) Thermal analysis

Solution: - c) Post-printing imaging and inspection

20) How does 3D printing help in the development of patient-specific surgical plans?

a) By using standardized models for all patients

b) By creating customizable anatomical models for pre-surgical planning

c) By providing generic surgical procedures

d) By limiting the options available to surgeons

Solution: - b) By creating customizable anatomical models for pre-surgical planning

21) **Which technology is often used to create high-resolution medical models for pre-surgical planning?**
    a) Fused Deposition Modeling (FDM)
    b) Stereolithography (SLA)
    c) Selective Laser Sintering (SLS)
    d) Digital Light Processing (DLP)
    Solution: - b) Stereolithography (SLA)

22) **What is one advantage of using 3D printing over traditional manufacturing in the medical field?**
    a) It requires rawer materials
    b) It allows for on-demand production and customization
    c) It increases the time required for production
    d) It limits the design options available
    Solution: - b) It allows for on-demand production and customization

23) **Which factor is critical in selecting materials for 3D-printed biomedical devices?**
    a) Color variety
    b) Cost of raw materials
    c) Biocompatibility and mechanical properties
    d) Availability of materials in the local market
    Solution: - c) Biocompatibility and mechanical properties

24) **What is one limitation of current 3D-printed organ models in clinical applications?**
    a) They are not customizable
    b) They cannot replicate complex tissue structures
    c) They are too inexpensive
    d) They have low resolution
    Solution: - b) They cannot replicate complex tissue structures

25) **How does 3D printing contribute to the field of regenerative medicine?**
    a) By producing non-biological materials
    b) By creating patient-specific biomaterials and scaffolds
    c) By replacing traditional surgical methods
    d) By reducing the need for tissue engineering

Solution: - b) By creating patient-specific biomaterials and scaffolds

26) Which aspect of 3D printing is crucial for the development of functional tissue constructs?

a) Material cost
b) Printing speed
c) Accuracy and resolution of the printing process
d) Size of the printer

Solution: - c) Accuracy and resolution of the printing process

27) What is one of the primary challenges when integrating 3D-printed implants with human tissue?

a) Incompatibility with medical imaging
b) Difficulty in achieving the desired shape
c) Ensuring long-term biocompatibility
d) Limited availability of printing technologies

Solution: - c) Ensuring long-term biocompatibility

28) Which type of 3D printing technology is most suitable for creating detailed anatomical models for surgical planning?

a) Fused Deposition Modeling (FDM)
b) Stereolithography (SLA)
c) Selective Laser Sintering (SLS)
d) Direct Metal Laser Sintering (DMLS)

Solution: - b) Stereolithography (SLA)

29) In what way does 3D printing enhance the field of personalized medicine?

a) By providing generic solutions for all patients
b) By enabling the production of patient-specific medical devices and implants
c) By reducing the number of available treatment options
d) By standardizing patient treatment plans

Solution: - b) By enabling the production of patient-specific medical devices and implants

30) What is the significance of "bio-ink" in the context of bioprinting?

a) It is used to create non-living medical models
b) It contains living cells and biomaterials for tissue construction

c) It is a standard ink used in medical imaging

d) It is a material used to print electronic components

Solution: - b) It contains living cells and biomaterials for tissue construction

31) Which material is commonly used in the 3D printing of bone scaffolds due to its biocompatibility and mechanical strength?

a) Polyethylene

b) Ceramic

c) Nylon

d) Glass

Solution: - b) Ceramic

32) What is a significant advantage of using 3D printing for creating patient-specific surgical implants?

a) Decreased precision

b) Increased production time

c) Enhanced fit and functionality tailored to individual patient needs

d) Standardization of implant sizes

Solution: - c) Enhanced fit and functionality tailored to individual patient needs

33) In the context of 3D printing, what does "layer-by-layer manufacturing" refer to?

a) The process of applying multiple coats of paint on an object

b) The method of building a 3D object by adding material layer by layer

c) The technique of layering different types of plastics

d) The process of combining different technologies in one printer

Solution: - b) The method of building a 3D object by adding material layer by layer

34) What is one of the key considerations when designing 3D-printed implants for complex surgeries?

a) The color of the implant

b) The material's resistance to high temperatures

c) The material's biocompatibility and mechanical properties

d) The size of the implant relative to standard models

Solution: - c) The material's biocompatibility and mechanical properties

35) **Which bioprinting technique involves the use of a laser to deposit cells and biomaterials onto a substrate?**

a) Inkjet bioprinting
b) Laser-assisted bioprinting
c) Extrusion-based bioprinting
d) Stereolithography

Solution: - b) Laser-assisted bioprinting

36) **What is a primary application of 3D printing in developing medical training aids?**

a) Manufacturing electronic devices
b) Creating realistic anatomical models for practice and education
c) Producing non-medical prototypes
d) Designing consumer products

Solution: - b) Creating realistic anatomical models for practice and education

37) **What factor limits the current use of 3D-printed organs in clinical settings?**

a) The complexity of creating functional organ tissues
b) The cost of 3D printing equipment
c) The availability of raw materials
d) The simplicity of the printing process

Solution: - a) The complexity of creating functional organ tissues

38) **Which 3D printing technology is often employed for creating durable and functional orthopedic implants?**

a) Fused Deposition Modeling (FDM)
b) Selective Laser Sintering (SLS)
c) Digital Light Processing (DLP)
d) Stereolithography (SLA)

Solution: - b) Selective Laser Sintering (SLS)

39) **What role do computational models play in 3D printing for medical applications?**

a) They manage the printer's mechanical parts
b) They optimize the design and predict the performance of printed objects
c) They handle the material supply chain

d) They control the color of the printed items

Solution: - b) They optimize the design and predict the performance of printed objects

40) Which challenge is associated with integrating 3D-printed components into existing medical devices?

a) Lack of compatibility with medical imaging

b) Difficulty in achieving required precision and fit

c) Increased material cost

d) Shortage of 3D printing technologies

Solution: - b) Difficulty in achieving required precision and fit

Chapter 21

Additive Manufacturing in Automotive and Transportation

21.1 Introduction to AM in Automotive and Transportation

Additive manufacturing (AM) has revolutionized the automotive and transportation industries by enabling the rapid production of complex, lightweight, and customized components. AM offers a range of benefits, including improved fuel efficiency, reduced production times, and increased design flexibility. In this chapter, we will explore the applications, benefits, challenges, and limitations of AM in automotive and transportation.

The Need for Innovation

The automotive and transportation industries are facing significant challenges in the 21st century. With growing concerns about climate change, air pollution, and traffic congestion, there is a pressing need for innovative solutions that can reduce the environmental impact of transportation while improving safety, efficiency, and performance. AM has emerged as a key technology that can help address these challenges by enabling the production of lightweight, fuel-efficient, and customized components.

Benefits of AM in Automotive and Transportation

AM offers a range of benefits in automotive and transportation, including:

- **Improved Fuel Efficiency**: AM enables the production of lightweight components that can reduce the weight of vehicles, improving fuel efficiency and reducing greenhouse gas emissions.

- **Reduced Production Times**: AM allows for rapid prototyping and production, reducing the time and cost associated with traditional manufacturing methods.

- **Increased Design Flexibility**: AM enables the production of complex geometries and customized components that cannot be produced using traditional manufacturing methods.

- **Improved Safety**: AM enables the production of components with complex internal structures that can improve safety and reduce the risk of injury in the event of a crash.

- **Reduced Material Waste**: AM is an additive process that builds components layer by layer, reducing material waste and the environmental impact of production.

Applications of AM in Automotive and Transportation

AM is being used in a range of applications in automotive and transportation, including:

- **Aerospace**: AM is being used to produce lightweight components for aircraft and spacecraft, improving fuel efficiency and reducing emissions.

- **Automotive**: AM is being used to produce customized components for cars, trucks, and motorcycles, improving performance, safety, and fuel efficiency.

- **Rail**: AM is being used to produce components for trains and rail infrastructure, improving safety, efficiency, and reducing maintenance costs.

- **Marine**: AM is being used to produce components for ships and boats, improving fuel efficiency, reducing emissions, and improving performance.

Challenges and Limitations of AM in Automotive and Transportation

While AM offers a range of benefits in automotive and transportation, there are also challenges and limitations that must be addressed, including:

- **Material Properties**: AM materials may not have the same properties as traditional materials, which can impact performance and safety.

- **Scalability**: AM machines may not be able to produce large quantities of components, which can limit the scalability of AM in automotive and transportation.

- **Regulatory Frameworks**: There is a need for regulatory frameworks that can ensure the safety and efficacy of AM-produced components in automotive and transportation.

- **Cost and Reimbursement**: The cost of AM machines and materials can be high, which can limit accessibility to this technology.

In conclusion, AM has the potential to revolutionize the automotive and transportation industries by enabling the rapid production of complex, lightweight, and customized components. While there are challenges and limitations that must be addressed, the benefits of AM in automotive and transportation are clear. As the technology continues to evolve, we can expect to see even more innovative applications of AM in these industries. Let's explore the impact of Additive Manufacturing (AM) in the automotive and transportation industry. AM has been gradually transforming the way end-use parts are designed, manufactured, and distributed. Here are some key aspects:

Pre-production Benefits: During new product development, 3D printing accelerates the product design phase. Designers can cost-effectively iterate through various designs before finalizing the product.

Volkswagen, for instance, has been using 3D printing technologies for 25 years, investing in additional printers to optimize the design process.

Tooling Optimization: AM allows the production of specific tooling parts (such as grips, jigs, and fixtures) at a lower cost. These tools play a crucial role in ensuring repeatable and high-quality assembly line processes.

End-Use Parts: Beyond rapid prototyping, several original equipment manufacturers (OEMs) are interested in 3D printing for end-use parts. This quest for product innovation drives weight reduction and customization.

Luxury car manufacturers, in particular, explore customization through AM, enhancing performance and aesthetics.

Growing Revenue: By 2029, automotive 3D printing is projected to generate $9 billion in revenue from end-use parts alone, compared to $1.39 billion in 2019.

The focus is on improving part performance, reducing weight, and achieving product innovation.

In summary, AM is reshaping automotive manufacturing by offering flexibility, efficiency, and novel design possibilities. Figure 90 shows the various application of AM in Automotive car.

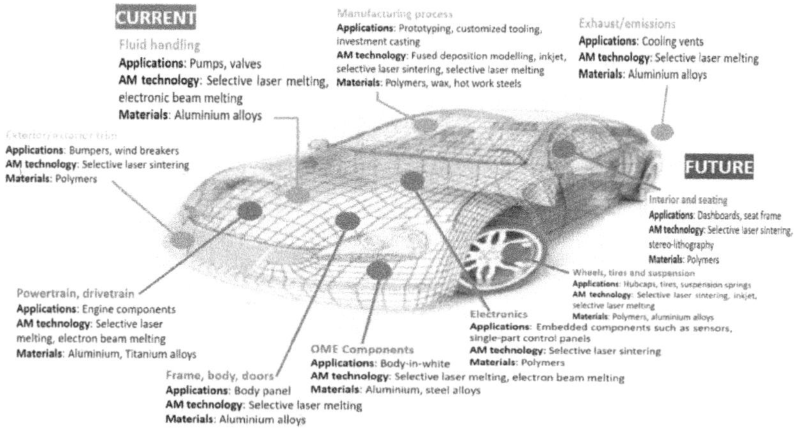

Figure 1: AM in automotive and transportation
[Source: https://link.springer.com/article/10.1007/s00170-021-08569-z]

21.2 Applications and Benefits of AM in Automotive

Additive manufacturing (AM) has numerous applications in the automotive industry, transforming the way vehicles are designed, produced, and maintained. From engine components to interior trim, AM is being used to produce a wide range of parts and components that are lighter, stronger, and more efficient.

Engine Components

AM is being used to produce lightweight, complex engine components, such as:

a. **Cylinder Heads**: AM enables the production of cylinder heads with complex internal structures, improving engine performance and reducing weight.

b. **Engine Blocks**: AM allows for the production of engine blocks with optimized cooling channels, reducing engine weight and improving thermal management.

c. **Turbochargers**: AM enables the production of turbochargers with complex geometries, improving engine performance and reducing emissions.

d. AM is being used to produce customized vehicle structures, such as:

e. **Chassis**: AM enables the production of chassis components with optimized strength-to-weight ratios, improving vehicle performance and reducing weight.

f. **Frames**: AM allows for the production of frames with complex geometries, improving vehicle safety and reducing production costs.

g. **Body Panels**: AM enables the production of body panels with optimized aerodynamics, improving vehicle performance and reducing wind noise.

Interior Components

AM is being used to produce customized interior components, such as:

i. **Dashboards**: AM enables the production of dashboards with complex geometries, improving ergonomics and reducing production costs.

ii. **Trim**: AM allows for the production of trim components with optimized aesthetics, improving vehicle interior design and reducing material waste.

iii. **Seats**: AM enables the production of seats with optimized ergonomics, improving passenger comfort and reducing production costs.

Aerodynamic Components

AM is being used to produce complex aerodynamic components, such as:

1. **Spoilers**: AM enables the production of spoilers with optimized aerodynamics, improving vehicle performance and reducing wind noise.

2. **Airfoils**: AM allows for the production of airfoils with optimized aerodynamics, improving vehicle performance and reducing fuel consumption.

3. **Diffusers**: AM enables the production of diffusers with optimized aerodynamics, improving vehicle performance and reducing wind noise.

Electric Vehicle Components

AM is being used to produce components for electric vehicles, such as:

- **Battery Housings**: AM enables the production of battery housings with optimized thermal management, improving battery performance and reducing weight.

- **Electric Motors**: AM allows for the production of electric motors with optimized performance, improving vehicle range and reducing energy consumption.

- **Charging Systems**: AM enables the production of charging systems with optimized ergonomics, improving user experience and reducing production costs.

Benefits of AM in Automotive

The benefits of AM in automotive include:

- **Weight Reduction**: AM enables the production of lightweight components, reducing fuel consumption and improving vehicle performance.

- **Increased Design Flexibility**: AM allows for the production of complex geometries and internal structures, enabling the creation of innovative designs.
- **Improved Part Performance**: AM enables the production of parts with improved mechanical properties, such as strength, stiffness, and fatigue resistance.

- **Reduced Production Times**: AM enables rapid production, reducing lead times and enabling just-in-time production.

- **Improved Sustainability**: AM enables the production of parts with reduced material waste, improving sustainability and reducing environmental impact.

- **Customization**: AM enables the production of customized parts and components, improving vehicle performance and reducing production costs.

Case Studies

Several companies are already leveraging AM in automotive, including:

- **General Motors**: General Motors is using AM to produce engine components, such as cylinder heads and engine blocks, improving engine performance and reducing weight.

- **Ford Motor Company**: Ford is using AM to produce vehicle structures, such as chassis and frames, improving vehicle safety and reducing production costs.

- **Tesla**: Tesla is using AM to produce electric vehicle components, such as battery housings and electric motors, improving vehicle range and reducing energy consumption.

In conclusion, AM has the potential to transform the automotive industry by enabling the production of lightweight, complex, and customized components. As the technology continues to evolve, we can expect to see even more innovative applications of AM in automotive.

21.3 Challenges and Limitations of AM in Automotive

Despite the benefits of additive manufacturing (AM) in automotive, there are several challenges and limitations that must be addressed to fully realize the potential of this technology. These challenges and limitations include:

Material Properties

One of the significant challenges of AM in automotive is the properties of AM materials. While AM materials have improved significantly in recent years, they may not meet the required specifications for automotive applications. For example:

- **Mechanical Properties**: AM materials may not have the same mechanical properties as traditional materials, such as strength, stiffness, and fatigue resistance.

- **Thermal Properties**: AM materials may not have the same thermal properties as traditional materials, such as thermal conductivity and heat resistance.

- **Chemical Properties**: AM materials may not have the same chemical properties as traditional materials, such as corrosion resistance and chemical reactivity.

To address these challenges, researchers and manufacturers are working to develop new AM materials with improved properties. For example, some companies are developing AM materials with enhanced mechanical properties, such as strength and stiffness, while others are developing materials with improved thermal properties, such as thermal conductivity and heat resistance.

Scalability

Another challenge of AM in automotive is scalability. While AM machines can produce complex components with high accuracy and precision, they may not be able to produce large quantities of components, limiting the scalability of AM in automotive. This is because AM machines are typically designed for small-batch production, rather than large-scale production.

To address this challenge, manufacturers are working to develop new AM machines that can produce large quantities of components quickly and efficiently. For example, some companies are developing AM machines with multiple print heads, allowing for faster production times and increased scalability.

Cost

AM can be a costly process, particularly for large-scale production. The cost of AM machines, materials, and labor can be prohibitively expensive for many manufacturers, limiting the adoption of AM in automotive.

To address this challenge, manufacturers are working to reduce the cost of AM production. For example, some companies are developing new AM machines that are more affordable and efficient, while others are developing new AM materials that are less expensive and more sustainable.

Regulatory Frameworks

There is a need for standardized regulatory frameworks for AM in automotive, ensuring the safety and efficacy of AM-produced components. Currently, there is a lack of regulatory frameworks for AM in automotive, which can make it difficult for manufacturers to ensure compliance with safety and performance standards.

To address this challenge, regulatory bodies and industry organizations are working to develop standardized regulatory frameworks for AM in automotive. For example, the International

Organization for Standardization (ISO) is developing standards for AM in automotive, while the Society of Automotive Engineers (SAE) is developing guidelines for the use of AM in automotive.

Other Challenges and Limitations

In addition to these challenges and limitations, there are several other issues that must be addressed to fully realize the potential of AM in automotive. These include:

- **Design for AM**: Designing components for AM production can be challenging, particularly for complex components with multiple features and geometries.

- **Post-processing**: Post-processing of AM components can be time-consuming and expensive, particularly for components with complex geometries and features.

- **Supply Chain Management**: Managing the supply chain for AM production can be challenging, particularly for components with complex geometries and features.

- **Cybersecurity**: Ensuring the cybersecurity of AM production systems and components is critical, particularly in the automotive industry where safety and security are paramount.

Addressing the Challenges and Limitations

To address the challenges and limitations of AM in automotive, manufacturers, researchers, and regulatory bodies must work together to develop new technologies, materials, and processes that can overcome these challenges. This includes:

- **Investing in Research and Development**: Investing in research and development to improve the properties of AM materials, increase the scalability of AM production, and reduce the cost of AM production.

- **Developing New Technologies and Processes**: Developing new technologies and processes that can overcome the challenges and limitations of AM in automotive, such as new AM machines, materials, and software.

- **Establishing Regulatory Frameworks**: Establishing standardized regulatory frameworks for AM in automotive, ensuring the safety and efficacy of AM-produced components.

- **Collaborating and Partnering**: Collaborating and partnering with other manufacturers, researchers, and regulatory bodies to share knowledge, expertise, and resources.

By addressing the challenges and limitations of AM in automotive, manufacturers can fully realize the potential of this technology and create innovative, sustainable, and efficient vehicles that meet the needs of the 21st century. Let's explore some effective cost reduction techniques in the manufacturing industry. These strategies can enhance efficiency and profitability:

- **Reduce Direct Material Costs:**

Move to Competitive Suppliers or Materials: Evaluate alternative suppliers or materials to reduce costs.

Negotiate with Long-Term Contractors: Establish favorable long-term contracts with suppliers.

- **Evaluate Production Process:**

Keep All Machines in Good Condition: Regular maintenance minimizes downtime and ensures optimal performance.

- **Cut Out Surplus:**

Avoid Excess Production: Oversupply decreases market value. Align production with actual consumption.

- **Cut Shipping Costs:**

Adjust Delivery Routes: Optimize travel time and fuel expenses by refining delivery routes.

- **Audit Outsourcing Costs:**

Regularly assess outsourcing processes for cost-effectiveness, adherence to standards, and efficiency.

Remember, a combination of these techniques can lead to significant cost savings and improved competitiveness in the manufacturing sector. Figure 91 shows the Cost Reduction Techniques in Manufacturing Industries.

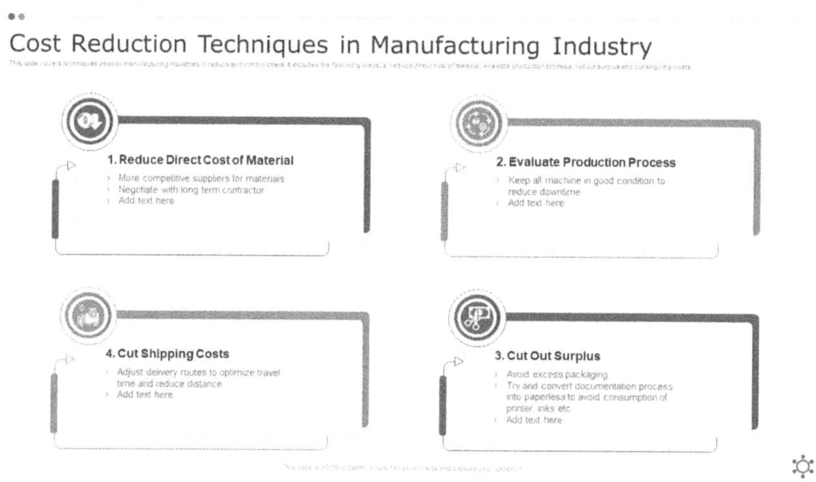

Figure 2: Cost reduction techniques in manufacturing industry [Source: https://www.slideteam.net/cost-reduction-techniques-in-manufacturing-industry.html]

21.4 Case Studies and Examples

Several companies and organizations are already leveraging additive manufacturing (AM) in automotive, demonstrating the potential of this technology to transform the industry. Here are some examples of companies that are using AM in innovative ways:

General Motors

General Motors is a pioneer in the use of AM in automotive. The company has been using AM to produce a range of components, including engine components, vehicle structures, and interior components. For example, General Motors has used AM to produce cylinder heads, engine blocks, and turbochargers, which have improved engine performance and reduced weight.

In addition, General Motors has used AM to produce vehicle structures, such as chassis and frames, which have improved vehicle safety and reduced production costs. The company has also used AM to produce interior components, such as dashboards and trim, which have improved ergonomics and reduced material waste.

Ford Motor Company

Ford Motor Company is another company that is leveraging AM in automotive. The company has used AM to produce engine components, such as cylinder heads and engine blocks, which have improved engine performance and reduced weight. Ford has also used AM to produce vehicle structures, such as chassis and frames, which have improved vehicle safety and reduced production costs.

In addition, Ford has used AM to produce aerodynamic components, such as spoilers and airfoils, which have improved vehicle performance and reduced wind noise. The company has also used AM to produce customized vehicle components, such as dashboards and trim, which have improved ergonomics and reduced material waste.

BMW

BMW is a luxury vehicle manufacturer that is using AM to produce customized vehicle components. The company has used AM to produce dashboards, trim, and other interior components that are tailored to individual customers' preferences. This has improved customer satisfaction and reduced production costs.

BMW has also used AM to produce complex components, such as engine components and vehicle structures, which have improved vehicle performance and reduced weight. The company has also used AM to produce prototype components, which have reduced development time and costs.

Tesla

Tesla is an electric vehicle manufacturer that is using AM to produce components for electric vehicles. The company has used AM to produce battery housings, electric motors, and other components that are critical to the performance and efficiency of electric vehicles.

Tesla has also used AM to produce customized vehicle components, such as dashboards and trim, which have improved ergonomics and reduced material waste. The company has also used AM to produce prototype components, which have reduced development time and costs.

Other Companies and Organizations

In addition to these companies, many other organizations are using AM in automotive. For example, the US Department of Energy's Oak Ridge National Laboratory is using AM to produce advanced vehicle components, such as lightweight engine components and aerodynamic components.

The laboratory is also working with industry partners to develop new AM materials and processes that can be used in automotive applications. This includes developing new metal alloys and polymers that can be used in AM production, as well as developing new software and simulation tools that can be used to design and optimize AM components.

Benefits of AM in Automotive

The use of AM in automotive has several benefits, including:

- **Improved Performance**: AM components can be designed to have improved performance, such as increased strength, stiffness, and fatigue resistance.

- **Reduced Weight**: AM components can be designed to be lighter, which can improve vehicle performance and reduce fuel consumption.

- **Increased Customization**: AM allows for the production of customized components that are tailored to individual customers' preferences.

- **Reduced Production Costs**: AM can reduce production costs by reducing material waste and improving manufacturing efficiency.

- **Improved Sustainability**: AM can improve sustainability by reducing material waste and energy consumption.

Challenges and Limitations

Despite the benefits of AM in automotive, there are several challenges and limitations that must be addressed. These include:

- **Material Properties**: The properties of AM materials may not meet the required specifications for automotive applications.

- **Scalability**: AM machines may not be able to produce large quantities of components, limiting the scalability of AM in automotive.

- **Cost**: AM can be a costly process, particularly for large-scale production.

- **Regulatory Frameworks**: There is a need for standardized regulatory frameworks for AM in automotive, ensuring the safety and efficacy of AM-produced components.

By addressing these challenges and limitations, companies and organizations can fully realize the potential of AM in automotive and create innovative, sustainable, and efficient vehicles that meet the needs of the 21st century. As AM evolves from prototyping to production, it plays a crucial role in shaping the future of automotive manufacturing, especially as the industry shifts toward electric and smart mobility. Here are some notable examples and trends:

Lightweighting and Material Substitution:

The automotive sector aims to reduce fuel consumption and greenhouse gas emissions. Light-weighting, achieved through material substitution and re-designing components, is a critical solution.

Additive manufacturing allows for complex, lightweight designs that optimize material usage and reduce resource consumption during both production and use phases.

For instance, a bracket originally cast in iron was re-designed using AM (specifically, powder bed-based AM in AlSi10Mg). The result was a 69% weight reduction, leading to energy and carbon footprint savings during the use phase.

Complex Geometries and Topological Optimization:

AM enables complexity-for-free designs. Topological optimization techniques can create intricate structures that were previously impossible or cost-prohibitive.

These designs enhance performance, reduce weight, and improve overall efficiency.

Integration into Sustainable Product Development:

While light weighting positively impacts energy and carbon footprint during the use phase, it's essential to consider the entire life cycle.

AM should be an integral part of sustainable product development, balancing environmental trade-offs and long-term benefits.

In summary, additive manufacturing is revolutionizing the automotive industry by enabling innovative designs, reducing weight, and contributing to a more sustainable future. Figure 92 displayed future skills challenges faced in transition automotive industry

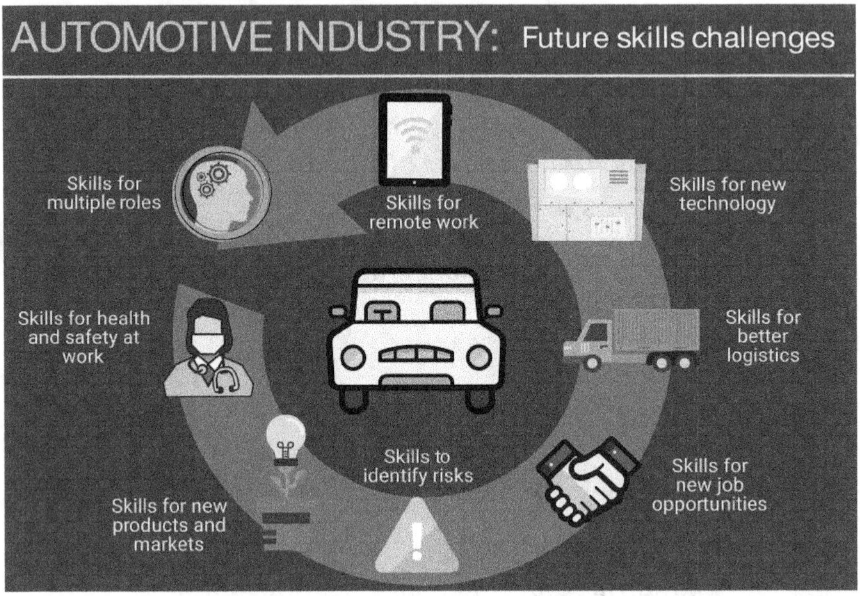

Figure 3: Future skills challenges faced in transition automotive industry [Source: https://www.cedefop.europa.eu/en/news/sectors-transition-automotive-industry]

21.5 Questions

21.5.1 Long Response Questions

1. What are the key benefits of using AM in automotive and transportation?
2. How do the properties of AM materials impact their use in automotive applications?

3. What are the challenges and limitations of scaling up AM production in automotive?
4. How can regulatory frameworks for AM in automotive be improved?
5. What are some examples of successful AM applications in automotive and transportation?

21.5.2 Multiple Choice Questions

1) **What is the primary advantage of using additive manufacturing in the automotive industry?**
   a) Increased production time
   b) Higher cost of raw materials
   c) Customization and rapid prototyping
   d) Decreased design flexibility
   Solution: - c) Customization and rapid prototyping
2) **Which of the following materials is commonly used in 3D printing for automotive parts?**
   a) Aluminum
   b) Stainless Steel
   c) Thermoplastics
   d) Glass
   Solution: - c) Thermoplastics
3) **Which technology is primarily used in additive manufacturing for creating automotive prototypes?**
   a) Injection Molding
   b) Laser Cutting
   c) 3D Printing
   d) Casting
   Solution: - c) 3D Printing
4) **What is the main purpose of 3D printing in automotive R&D?**
   a) Final production of vehicles
   b) Creating concept models and prototypes
   c) Mass production of components
   d) Vehicle assembly
   Solution: - b) Creating concept models and prototypes

5) **Which of the following is a common additive manufacturing method used in automotive applications?**
   a) Stereolithography (SLA)
   b) Injection Molding
   c) Die Casting
   d) Forging
   Solution: - a) Stereolithography (SLA)
6) **Which automotive component is commonly prototyped using 3D printing?**
   a) Engine Cylinder Head
   b) Suspension Arms
   c) Interior Trim Parts
   d) Transmission Gears
   Solution: - c) Interior Trim Parts
7) **Which additive manufacturing method is best for producing parts with high detail and resolution?**
   a) Fused Deposition Modeling (FDM)
   b) Stereolithography (SLA)
   c) Selective Laser Sintering (SLS)
   d) Direct Metal Laser Sintering (DMLS)
   Solution: - b) Stereolithography (SLA)
8) **What advantage does 3D printing offer in the design and prototyping of automotive parts?**
   a) Higher cost of production
   b) Increased time for design iterations
   c) Faster design iterations and customizability
   d) Limited material options
   Solution: - c) Faster design iterations and customizability
9) **What is a common post-processing step for improving the quality of 3D printed automotive parts?**
   a) Heat Treatment
   b) Injection Molding
   c) Metal Forging
   d) Surface Finishing
   Solution: - d) Surface Finishing
10) **Which of the following is NOT typically a material used in automotive 3D printing?**
   a) PLA (Polylactic Acid)

b) ABS (Acrylonitrile Butadiene Styrene)
c) Stainless Steel
d) Polypropylene
Solution: - d) Polypropylene

11) Which automotive component is often prototyped using 3D printing for functional testing?
a) Exhaust Manifold
b) Engine Block
c) Dashboard Trim
d) Transmission Case
Solution: - c) Dashboard Trim

12) Which material is commonly used in 3D printing for creating automotive prototypes?
a) Polyethylene
b) Carbon Fiber
c) PLA (Polylactic Acid)
d) Steel
Solution: - c) PLA (Polylactic Acid)

13) What is a common use of 3D printing in automotive manufacturing?
a) Production of final vehicle assemblies
b) Prototyping and custom part production
c) Mass production of tires
d) Casting of engine components
Solution: - b) Prototyping and custom part production

14) Which additive manufacturing technology is known for creating high-resolution parts?
a) FDM
b) SLA
c) SLS
d) DMLS
Solution: - b) SLA

15) What is one of the primary benefits of 3D printing in automotive design?
a) Increased lead times
b) Reduced design flexibility
c) Faster design iteration and prototyping
d) Higher production costs
Solution: - c) Faster design iteration and prototyping

16) Which additive manufacturing technique uses a laser to sinter powdered material?

a) Fused Deposition Modeling (FDM)
b) Selective Laser Sintering (SLS)
c) Stereolithography (SLA)
d) Direct Ink Writing (DIW)

Solution: - b) Selective Laser Sintering (SLS)

17) What is a significant benefit of using 3D printed jigs and fixtures in automotive assembly?

a) Increased cost
b) Longer assembly times
c) Improved accuracy and efficiency
d) Limited design flexibility

Solution: - c) Improved accuracy and efficiency

18) Which of the following materials is typically used in Fused Deposition Modeling (FDM) for automotive parts?

a) Nylon
b) Polycarbonate
c) Titanium
d) Graphene

Solution: - b) Polycarbonate

19) What challenge is often associated with using metal additive manufacturing in the automotive industry?

a) High material cost
b) Limited design options
c) Low accuracy
d) Slow production speed

Solution: - a) High material cost

20) How does 3D printing contribute to reducing waste in automotive manufacturing?

a) By using excess material
b) By building parts layer-by-layer, using only necessary material
c) By increasing scrap rates
d) By creating parts from solid blocks

Solution: - b) By building parts layer-by-layer, using only necessary material

21) How does additive manufacturing benefit the production of automotive prototypes?

a) By increasing production cost
b) By reducing design flexibility
c) By enabling rapid and cost-effective iterations
d) By complicating the design process

Solution: - c) By enabling rapid and cost-effective iterations

22) What is the purpose of using simulation tools in automotive additive manufacturing?

a) To reduce material costs
b) To optimize part design and performance
c) To speed up production
d) To simplify the manufacturing process

Solution: - b) To optimize part design and performance

23) Which additive manufacturing technique is ideal for creating lightweight automotive components with complex internal structures?

a) Fused Deposition Modeling (FDM)
b) Stereolithography (SLA)
c) Selective Laser Sintering (SLS)
d) Direct Metal Laser Sintering (DMLS)

Solution: - c) Selective Laser Sintering (SLS)

24) Which challenge is commonly faced when using additive manufacturing for large-scale automotive parts?

a) Limited design complexity
b) High material and equipment costs
c) Fast production speed
d) Low resolution of printed parts

Solution: - b) High material and equipment costs

25) What role does topology optimization play in automotive additive manufacturing?

a) To increase part weight
b) To improve material efficiency and strength
c) To decrease part strength
d) To simplify the design

Solution: - b) To improve material efficiency and strength

26) Which 3D printing method is ideal for creating parts with integrated electronics?

a) SLA
b) FDM
c) Inkjet Printing
d) SLS
Solution: - c) Inkjet Printing

27) What is a significant challenge associated with using 3D printing for large automotive parts?
a) High material cost
b) Limited design complexity
c) High production speed
d) Low resolution
Solution: - a) High material cost

28) How does additive manufacturing contribute to reducing waste in automotive production?
a) By creating parts from solid blocks
b) By using excess material
c) By building parts layer-by-layer, minimizing material waste
d) By increasing scrap rates
Solution: - c) By building parts layer-by-layer, minimizing material waste

29) What role does simulation play in automotive additive manufacturing?
a) To reduce material costs
b) To optimize design and performance
c) To speed up production
d) To simplify manufacturing
Solution: - b) To optimize design and performance

30) Which technique involves 3D printing metal layers with high precision for automotive components?
a) FDM
b) SLA
c) SLS
d) DMLS
Solution: - d) DMLS

31) Which additive manufacturing technique is known for producing high-strength automotive parts with complex geometries?
a) Stereolithography (SLA)

b) Direct Metal Laser Sintering (DMLS)
c) Fused Deposition Modeling (FDM)
d) Inkjet Printing
Solution: - b) Direct Metal Laser Sintering (DMLS)

32) What is the role of topology optimization in automotive additive manufacturing?
a) To increase material usage
b) To enhance part strength and reduce weight
c) To decrease design complexity
d) To simplify the manufacturing process
Solution: - b) To enhance part strength and reduce weight

33) What is a primary advantage of using additive manufacturing for producing low-volume automotive parts?
a) High setup costs
b) Reduced setup and tooling costs
c) Increased material waste
d) Longer production times
Solution: - b) Reduced setup and tooling costs

34) Which of the following is an advanced application of 3D printing in automotive engineering?
a) Producing standard engine blocks
b) Creating custom, lightweight structural components
c) Mass-producing tires
d) Assembling vehicle parts
Solution: - b) Creating custom, lightweight structural components

35) In additive manufacturing, what is a lattice structure used for in automotive parts?
a) To increase part weight
b) To reduce material usage while maintaining strength
c) To create solid, heavy parts
d) To simplify design
Solution: - b) To reduce material usage while maintaining strength

36) Which of the following techniques is used for producing high-strength polymer parts in automotive applications?
a) Fused Deposition Modeling (FDM)

b) Stereolithography (SLA)
c) Direct Metal Laser Sintering (DMLS)
d) Selective Laser Sintering (SLS)

Solution: - d) Selective Laser Sintering (SLS)

37) What is one of the main advantages of using 3D printing for custom automotive components?

a) Limited design options
b) High production costs
c) Tailored parts for specific applications
d) Decreased material properties

Solution: - c) Tailored parts for specific applications

38) Which additive manufacturing technique involves using a laser to selectively melt metal powder layers?

a) Fused Deposition Modeling (FDM)
b) Direct Metal Laser Sintering (DMLS)
c) Stereolithography (SLA)
d) Inkjet Printing

Solution: - b) Direct Metal Laser Sintering (DMLS)

39) What is a key benefit of using additive manufacturing for small-batch production of automotive parts?

a) Higher setup costs
b) Reduced design flexibility
c) Lower tooling costs
d) Increased lead times

Solution: - c) Lower tooling costs

40) Which of the following is a potential drawback of using additive manufacturing for automotive parts?

a) Customization possibilities
b) Increased material waste
c) Reduced production time
d) High cost for high-volume production

Solution: - d) High cost for high-volume production

Chapter 22

Additive Manufacturing in Consumer Products and Retail

22.1 Introduction to AM in Consumer Products and Retail

Additive manufacturing (AM) has revolutionized the consumer products and retail industries by enabling the rapid production of customized, complex, and innovative products. AM offers a range of benefits, including improved product performance, reduced lead times, and increased design flexibility. In this chapter, we will explore the applications, benefits, challenges, and limitations of AM in consumer products and retail.

The Rise of Mass Customization

One of the key drivers of AM adoption in consumer products and retail is the growing demand for mass customization. Consumers are increasingly seeking unique and personalized products that reflect their individuality and style. AM enables companies to produce customized products with complex geometries and features that would be impossible or impractical to produce using traditional manufacturing methods.

For example, companies like Nike and Adidas are using AM to produce customized athletic shoes with unique designs, colors, and features. This allows consumers to design their own shoes online and receive a customized product that meets their specific needs and preferences.

Improved Product Performance

AM also enables companies to produce products with improved performance characteristics. For example, AM can be used to produce products with complex internal structures that improve their strength, durability, and functionality. This is particularly useful in industries such as aerospace, automotive, and healthcare, where product performance is critical.

In addition, AM can be used to produce products with unique thermal, electrical, or optical properties. For example, companies like GE Appliances are using AM to produce customized refrigerator components with improved thermal insulation properties. This enables them to reduce energy consumption and improve product performance.

Reduced Lead Times

AM also enables companies to reduce lead times and improve supply chain efficiency. Traditional manufacturing methods often require long lead times for tooling, molding, and production. AM, on the other hand, enables companies to produce products on demand, reducing lead times and improving inventory management.

For example, companies like UPS and FedEx are using AM to produce customized packaging materials on demand. This enables them to reduce lead times, improve supply chain efficiency, and reduce waste.

Increased Design Flexibility

AM also enables companies to increase design flexibility and creativity. Traditional manufacturing methods often require designers to work within the constraints of traditional manufacturing processes. AM, on the other hand, enables designers to create complex geometries and features that would be impossible or impractical to produce using traditional methods.

For example, companies like Apple and Samsung are using AM to produce customized phone cases and accessories with unique designs and features. This enables them to differentiate their products and improve customer satisfaction.

Applications of AM in Consumer Products and Retail

AM has a wide range of applications in consumer products and retail, including:

- **Fashion and Apparel**: AM is being used to produce customized clothing, shoes, and accessories with unique designs, colors, and features.

- **Electronics**: AM is being used to produce customized phone cases, laptop sleeves, and other electronic accessories with unique designs and features.

- **Home Goods**: AM is being used to produce customized home decor, furniture, and appliances with unique designs and features.

- **Toys and Games**: AM is being used to produce customized toys and games with unique designs, colors, and features.

Benefits of AM in Consumer Products and Retail

The benefits of AM in consumer products and retail include:

- **Improved Product Performance**: AM enables companies to produce products with improved performance characteristics, such as strength, durability, and functionality.

- **Reduced Lead Times**: AM enables companies to reduce lead times and improve supply chain efficiency.

- **Increased Design Flexibility**: AM enables companies to increase design flexibility and creativity, enabling them to produce unique and innovative products.

- **Mass Customization**: AM enables companies to produce customized products with complex geometries and features that meet individual customers' needs and preferences.

Challenges and Limitations of AM in Consumer Products and Retail

Despite the benefits of AM in consumer products and retail, there are several challenges and limitations that must be addressed. These include:

- **Material Properties**: The properties of AM materials may not meet the required specifications for consumer products and retail applications.

- **Scalability**: AM machines may not be able to produce large quantities of products, limiting the scalability of AM in consumer products and retail.

- **Cost**: AM can be a costly process, particularly for large-scale production.

- **Regulatory Frameworks**: There is a need for standardized regulatory frameworks for AM in consumer products and retail, ensuring the safety and efficacy of AM-produced products.

By addressing these challenges and limitations, companies and organizations can fully realize the potential of AM in consumer products and retail and create innovative, customized, and high-performance products that meet the needs of the 21st century. et's explore the essential sales channels for any consumer product company. These channels play a crucial role in reaching customers and driving revenue. In Figure 93, there are the five key sales channels:

i. **Marketplace**:

Marketplaces like Amazon, eBay, and Etsy provide a platform for businesses to showcase their products to a wide audience.

Leveraging established marketplaces allows companies to tap into existing customer bases and benefit from their marketing efforts.

ii. **Offline (Physical Stores):**

Traditional brick-and-mortar stores remain essential for consumer product companies.

Physical retail locations provide face-to-face interactions, brand visibility, and immediate product availability.

iii. **Wholesale:**

Wholesale channels involve selling products in bulk to retailers or distributors.

Companies can reach a broader market by partnering with wholesalers who distribute their products to various retail outlets.

iv. **Affiliate Sales:**

Affiliate marketing involves collaborating with influencers, bloggers, or other websites.

Affiliates promote products through their channels, earning a commission for each sale generated through their referral links.

v. **E-commerce:**

E-commerce platforms (such as company websites or specialized online stores) allow direct sales to consumers.

Companies can create personalized shopping experiences, track customer data, and optimize their online presence.

Figure 1: The 5 Essential Sales Channels for Any Consumer Product Company
[Source: https://metacake.com/the-5-essential-sales-channels-for-any-consumer-product-company/]

22.2 Applications and Benefits of AM in Consumer Products

Additive manufacturing (AM) has numerous applications in consumer products, including:

Customized Products

AM enables the production of customized products, such as personalized jewelry, eyewear, and footwear. This is particularly useful in industries where customers have unique preferences or requirements. For example, companies like Warby Parker and Zenni Optical are using AM to produce customized eyewear with unique designs, colors, and features.

In addition, AM can be used to produce customized products with complex geometries and internal structures. For example, companies like Nike and Adidas are using AM to produce customized athletic shoes with unique designs, colors, and features. This enables customers to design their own shoes online and receive a customized product that meets their specific needs and preferences.

Complex Geometries

AM enables the production of complex geometries, such as intricate designs, shapes, and structures. This is particularly useful in industries where traditional manufacturing methods are limited by tooling and molding constraints. For example, companies like Apple and Samsung are using AM to produce customized phone cases and accessories with unique designs and features.

In addition, AM can be used to produce complex geometries with internal structures, such as lattice structures and honeycomb patterns. This enables companies to produce products with improved mechanical properties, such as strength, stiffness, and durability.

Functional Products

AM enables the production of functional products, such as tooling, fixtures, and machinery components. This is particularly useful in industries where traditional manufacturing methods are limited by lead times and production costs. For example, companies like GE Appliances and Whirlpool are using AM to produce customized tooling and fixtures for their manufacturing lines.

In addition, AM can be used to produce functional products with complex geometries and internal structures. For example, companies like Boeing and Airbus are using AM to produce customized aircraft components, such as engine parts and satellite components.

Sustainable Products

AM enables the production of sustainable products, such as biodegradable packaging and products made from recycled materials. This is particularly useful in industries where sustainability is a key concern. For example, companies like Procter & Gamble and Unilever are using AM to produce biodegradable packaging materials that reduce waste and improve environmental sustainability.

In addition, AM can be used to produce products with reduced material waste and energy consumption. For example, companies like IKEA and H&M are using AM to produce customized furniture and textiles with reduced material waste and energy consumption.

Electronics and Electrical Components

AM enables the production of customized electronics and electrical components, such as circuit boards, antennas, and sensors. This is particularly useful in industries where traditional manufacturing methods are limited by lead times and production costs. For example, companies like Intel and Samsung are using AM to produce customized electronics components, such as microprocessors and memory chips.

In addition, AM can be used to produce electronics and electrical components with complex geometries and internal structures. For example, companies like Tesla and SpaceX are using AM to produce customized electrical components, such as battery packs and electrical motors.

Benefits of AM in Consumer Products

The benefits of AM in consumer products include:

- **Improved Product Performance:** AM enables the production of products with improved mechanical properties, such as strength, stiffness, and durability. This is particularly useful in industries where product performance is critical. For example, companies like Nike and Adidas are using AM to produce customized athletic shoes with improved mechanical properties, such as strength and durability.

- **Reduced Lead Times:** AM enables rapid production, reducing lead times and enabling just-in-time production. This is particularly useful in industries where lead times are critical. For example, companies like Amazon and Walmart are using AM to produce customized products with reduced lead times, enabling them to respond quickly to changing customer demand.

- **Increased Design Flexibility:** AM allows for the production of complex geometries and internal structures, enabling the creation of innovative designs. This is particularly useful in industries where design flexibility is critical. For example, companies like Apple and Samsung are using AM to produce customized phone cases and accessories with unique designs and features.

- **Cost Savings:** AM can reduce material waste, energy consumption, and labor costs, resulting in cost savings. This is particularly useful in industries where cost savings are critical. For example, companies like Procter & Gamble and Unilever are using AM to produce biodegradable packaging materials that reduce waste and improve environmental sustainability.

By leveraging these benefits, companies and organizations can use AM to create innovative, customized, and sustainable products that meet the needs of the 21st century. Let's explore the link between consumer behavior and innovations in food product development, specifically focusing on ready-to-eat food products. Understanding consumer behavior is crucial for designing and marketing food products effectively. Here are some key insights related to consumer behavior in this context:

1. **Preference and Choice:**

Consumers have distinct preferences for ready-to-eat food products based on taste, convenience, and nutritional value.

Their choices are influenced by factors such as flavor variety, packaging, and ease of preparation.

2. **Purchase Decisions:**

Consumers evaluate ready-to-eat products before making a purchase. Factors like price, brand reputation, and product claims impact their decisions.

Innovations in packaging (such as resealable pouches or microwave-safe containers) can enhance the perceived value.

3. **Consumption Patterns:**

How consumers consume ready-to-eat food varies. Some prefer quick snacks, while others use them as complete meals.

Innovations in portion sizes, nutritional content, and flavor profiles cater to diverse consumption patterns.

4. **Desire and Satisfaction:**

Consumers desire tasty, healthy, and convenient options. Innovations that meet these desires lead to higher satisfaction.

Products that align with lifestyle choices (e.g., vegetarian, gluten-free) resonate well.

5. **Willingness to Pay:**

Innovations in ready-to-eat food, such as organic ingredients or sustainable packaging, influence consumers' willingness to pay.

Health-conscious consumers may prioritize premium products with added functional benefits.

In summary, understanding consumer behavior and aligning innovations with their preferences are essential for successful product development in the ready-to-eat food industry. Figure 94 shows the Consumer Behaviour Towards Ready to Eat Food Products.

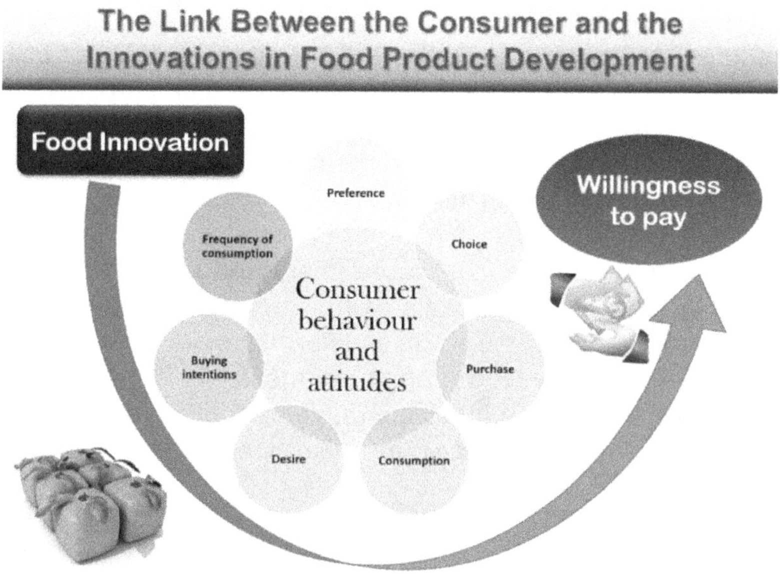

Figure 2: Consumer Behaviour Towards Ready to Eat Food Products [Source: https://mohammad-has-mcintosh.blogspot.com/2022/09/consumer-behaviour-towards-ready-to-eat.html]

22.3 Challenges and Limitations of AM in Consumer Products

Despite the benefits of additive manufacturing (AM) in consumer products, there are several challenges and limitations that must be addressed. These include:

Material Properties

One of the significant challenges of AM in consumer products is the properties of AM materials. The properties of AM materials may not meet the required specifications for consumer products, which can affect their performance, durability, and safety. For example, AM materials may not have the same strength, stiffness, or durability as traditional materials, which can limit their use in certain applications.

In addition, AM materials may have different thermal, electrical, or optical properties than traditional materials, which can affect their performance in certain applications. For example, AM materials may not be able to withstand high temperatures or electrical currents, which can limit their use in certain applications.

To address these challenges, researchers and manufacturers are developing new AM materials with improved properties. For example, companies like Carbon and Formlabs are developing new AM materials with improved strength, stiffness, and durability. These materials are being used in a wide range of applications, including consumer products, aerospace, and healthcare.

Scalability

Another significant challenge of AM in consumer products is scalability. AM machines may not be able to produce large quantities of products, limiting the scalability of AM in consumer products. This can be a significant challenge for companies that need to produce large quantities of products quickly and efficiently.

To address these challenges, manufacturers are developing new AM machines that can produce large quantities of products quickly and efficiently. For example, companies like HP and Ricoh are developing new AM machines that can produce large quantities of products at high speeds. These machines are being used in a wide range of applications, including consumer products, aerospace, and healthcare.

Cost

AM can be a costly process, particularly for large-scale production. The cost of AM machines, materials, and labor can be high, which can limit the adoption of AM in consumer products. In addition, the cost of post-processing and finishing can be high, which can add to the overall cost of AM.

To address these challenges, manufacturers are developing new AM machines and materials that can reduce the cost of AM. For example, companies like Prusa and Monoprice are developing new AM machines that are affordable and easy to use. These machines are being used by hobbyists and small businesses to produce a wide range of products, including consumer products, toys, and models.

Regulatory Frameworks

There is a need for standardized regulatory frameworks for AM in consumer products, ensuring the safety and efficacy of AM-produced products. The lack of regulatory frameworks can create uncertainty and confusion among manufacturers, consumers, and regulators.

To address these challenges, governments and regulatory agencies are developing new regulatory frameworks for AM in consumer products. For example, the US Food and Drug Administration (FDA) has developed guidelines for the use of AM in medical devices, including implants, surgical instruments, and diagnostic equipment. Similarly, the European Union has developed guidelines for the use of AM in consumer products, including toys, electronics, and furniture.

Other Challenges and Limitations

In addition to material properties, scalability, cost, and regulatory frameworks, there are several other challenges and limitations of AM in consumer products. These include:

- **Design complexity**: AM enables the production of complex geometries and internal structures, but designing these products can be challenging. Designers need to have specialized skills and software to design products for AM.

- **Material variability**: AM materials can vary in terms of their properties, which can affect the performance and quality of AM-produced products.
- **Post-processing and finishing**: AM products often require post-processing and finishing, which can be time-consuming and costly.
- **Supply chain management**: AM can disrupt traditional supply chains, requiring new logistics and distribution systems.

By addressing these challenges and limitations, companies and organizations can fully realize the potential of AM in consumer products and create innovative, customized, and sustainable products that meet the needs of the 21st century. While AM offers significant advantages, it also faces certain drawbacks which are displayed in Figure 95:

Advantages:

1. **Customization and Personalization:**
    - AM allows tailored designs for individual consumers, enhancing product uniqueness.
    - Personalized products can lead to higher customer satisfaction.

2. **Rapid Prototyping and Iteration:**
    - AM accelerates product development cycles by enabling quick prototyping and design adjustments.
    - Iterative testing and refinement are more efficient.

3. **Reduced Waste and Inventory Costs:**
    - AM produces items on demand, minimizing excess inventory and waste.
    - Just-in-time production reduces storage costs.

Limitations:

1. **Material Limitations:**

- Available materials for AM may not match the properties of traditional manufacturing materials.
- Material strength, durability, and compatibility need careful consideration.

2. **Surface Finish and Post-Processing:**
    - AM parts often requiring additional finishing steps (e.g., sanding, polishing) to achieve desired aesthetics.
    - Surface quality affects consumer perception.

3. **Production Speed and Scalability:**
    - AM processes can be slow, especially for complex or large-scale production.
    - Scaling up AM production remains a challenge.

4. **Cost and Equipment Investment:**
    - Initial setup costs for AM equipment can be high.
    - Consumer products must balance cost-effectiveness with customization.

5. **Intellectual Property and Counterfeiting:**
    - Digital files used in AM can be easily copied, raising concerns about intellectual property protection.
    - Counterfeit products could proliferate.

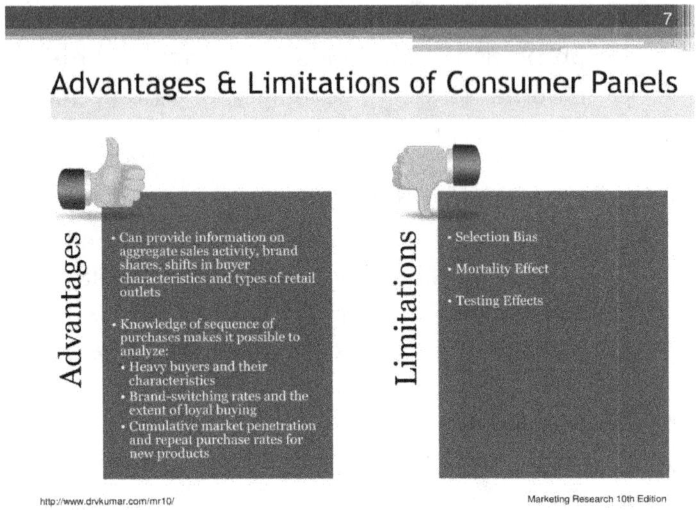

Figure 3: Advantages and limitations of AM in Consumer Products and services
[Source: https://www.slideserve.com/xavier-nielsen/marketing-research]

22.4 Case Studies and Examples

Several companies and organizations are already leveraging additive manufacturing (AM) in consumer products, including:

Nike: Customized Footwear and Athletic Equipment

Nike is using AM to produce customized footwear and athletic equipment. The company has developed a range of AM-produced products, including customized shoes, cleats, and athletic gear. Nike's AM products are designed to provide improved performance, comfort, and style for athletes and consumers.

For example, Nike's Vapor Laser Talon football cleat is produced using AM. The cleat features a customized plate that is designed to provide improved traction and speed on the field. The plate is produced using a selective laser sintering (SLS) process, which allows for the creation of complex geometries and internal structures.

Apple: Customized Electronics and Electrical Components

Apple is using AM to produce customized electronics and electrical components. The company has developed a range of AM-produced products, including customized phone cases, laptop components, and wearable devices. Apple's AM products are designed to provide improved performance, functionality, and design for consumers.

For example, Apple's iPhone cases are produced using AM. The cases feature customized designs, colors, and textures that are created using a range of AM processes, including stereolithography (SLA) and selective laser sintering (SLS). The cases are designed to provide improved protection and style for iPhone users.

Procter & Gamble: Customized Packaging and Consumer Goods

Procter & Gamble is using AM to produce customized packaging and consumer goods. The company has developed a range of AM-produced products, including customized packaging, beauty products, and household goods. Procter & Gamble's AM products are designed to provide improved performance, functionality, and sustainability for consumers.

For example, Procter & Gamble's Gillette razor handles are produced using AM. The handles feature customized designs, colors, and textures that are created using a range of AM processes, including fused deposition modeling (FDM) and SLA. The handles are designed to provide improved grip and comfort for consumers.

IKEA: Customized Furniture and Home Decor

IKEA is using AM to produce customized furniture and home decor. The company has developed a range of AM-produced products, including customized chairs, tables, and decorative items. IKEA's AM products are designed to provide improved functionality, sustainability, and style for consumers.

For example, IKEA's customized chair is produced using AM. The chair features a customized design, color, and texture that are created using a range of AM processes, including FDM and SLA. The chair is designed to provide improved comfort and support for consumers.

Other Companies and Organizations

In addition to Nike, Apple, Procter & Gamble, and IKEA, several other companies and organizations are leveraging AM in consumer products. These include:

- **Adidas**: Adidas is using AM to produce customized athletic shoes and apparel.
- **L'Oréal**: L'Oréal is using AM to produce customized beauty products and packaging.
- **Coca-Cola**: Coca-Cola is using AM to produce customized packaging and promotional items.
- **NASA**: NASA is using AM to produce customized spacecraft components and tools.

These companies and organizations are demonstrating the potential of AM in consumer products, including improved performance, functionality, and sustainability. By leveraging AM, companies can create innovative, customized, and sustainable products that meet the needs of the 21st century.

Benefits of AM in Consumer Products

The benefits of AM in consumer products are numerous. These include:

- **Improved performance**: AM enables the production of products with improved mechanical properties, such as strength, stiffness, and durability.

- **Customization**: AM enables the production of customized products that meet the specific needs and preferences of consumers.

- **Sustainability**: AM enables the production of sustainable products that reduce waste, energy consumption, and environmental impact.

- **Innovation**: AM enables the production of innovative products that were previously impossible to produce using traditional manufacturing methods.

By leveraging these benefits, companies and organizations can create innovative, customized, and sustainable products that meet the needs of the 21st century.

22.5 Questions

22.5.1 Long Response Questions

1. What are the key benefits of using AM in consumer products and retail?
2. How do the properties of AM materials impact their use in consumer products?
3. What are the challenges and limitations of scaling up AM production in consumer products?
4. How can regulatory frameworks for AM in consumer products be improved?
5. What are some examples of successful AM applications in consumer products and retail?

22.5.2 Multiple Choice Questions

1) Which of the following is a common application of 3D printing in consumer products?

a) Aerospace components
b) Medical implants
c) Custom jewelry
d) Automotive engines

Solution: - c) Custom jewelry

2) **What is a primary benefit of using 3D printing in the retail industry?**
   a) Reducing raw material waste
   b) Increasing production time
   c) Decreasing product variety
   d) Enhancing traditional manufacturing processes
   Solution: - a) Reducing raw material waste
3) **Which type of 3D printing technology is often used for creating detailed consumer product prototypes?**
   a) Fused Deposition Modeling (FDM)
   b) Selective Laser Sintering (SLS)
   c) Stereolithography (SLA)
   d) Electron Beam Melting (EBM)
   Solution: - c) Stereolithography (SLA)
4) **In which area is 3D printing least likely to be used in consumer products?**
   a) Footwear design
   b) Personalized gifts
   c) Heavy machinery parts
   d) Home décor
   Solution: - c) Heavy machinery parts
5) **What is the term for creating custom products tailored to individual preferences using 3D printing?**
   a) Mass production
   b) Customization
   c) Standardization
   d) Assembly line
   Solution: - b) Customization
6) **Which material is commonly used in 3D printing for creating consumer products like phone cases and toys?**
   a) Polyethylene
   b) ABS plastic
   c) Titanium
   d) Carbon fiber
   Solution: - b) ABS plastic
7) **What is a key advantage of using 3D printing for creating prototypes of consumer products?**
   a) Lower cost of materials

b) Faster production cycles
c) Increased complexity of designs
d) Reduced design flexibility
Solution: - b) Faster production cycles

8) **Which 3D printing technology is best suited for producing detailed and intricate consumer product designs?**
a) Inkjet Printing
b) Direct Metal Laser Sintering (DMLS)
c) Fused Deposition Modeling (FDM)
d) Stereolithography (SLA)
Solution: - d) Stereolithography (SLA)

9) **Which of the following is NOT a common use of 3D printing in retail?**
a) Customized fashion accessories
b) On-demand replacement parts
c) Large-scale industrial machinery
d) Personalized home décor
Solution: - c) Large-scale industrial machinery

10) **What is one challenge associated with using 3D printing in the consumer products industry?**
a) High speed of production
b) Limited material options
c) Low cost of manufacturing
d) High complexity of designs
Solution: - b) Limited material options

11) **How does 3D printing contribute to sustainability in consumer product manufacturing?**
a) By using more energy-intensive materials
b) By increasing the volume of waste
c) By allowing for on-demand production and reducing overproduction
d) By using more harmful chemicals
Solution: - c) By allowing for on-demand production and reducing overproduction

12) **Which 3D printing process involves melting and extruding filament to build up layers?**
a) Selective Laser Sintering (SLS)
b) Fused Deposition Modeling (FDM)
c) Stereolithography (SLA)

d) Digital Light Processing (DLP)

Solution: - b) Fused Deposition Modeling (FDM)

13) What role does post-processing play in 3D printed consumer products?

a) Reduces the strength of the product

b) Enhances the aesthetic and functional properties of the product

c) Decreases the durability of the product

d) Increases the production time significantly

Solution: - b) Enhances the aesthetic and functional properties of the product

14) Which of the following technologies allows for the use of multiple materials in a single 3D print?

a) Fused Deposition Modeling (FDM)

b) Multi-material Jetting

c) Stereolithography (SLA)

d) Selective Laser Sintering (SLS)

Solution: - b) Multi-material Jetting

15) What is a significant factor that affects the cost of 3D printed consumer products?

a) The color of the product

b) The complexity of the design and materials used

c) The shape of the product

d) The size of the product only

Solution: - b) The complexity of the design and materials used

16) In which sector has 3D printing significantly impacted product customization?

a) Automotive

b) Construction

c) Consumer electronics

d) Aerospace

Solution: - c) Consumer electronics

17) Which 3D printing technology is known for its use in creating flexible and durable consumer products?

a) Digital Light Processing (DLP)

b) Stereolithography (SLA)

c) Fused Deposition Modeling (FDM)

d) Inkjet Printing

Solution: - c) Fused Deposition Modeling (FDM)

18) **What type of data is primarily used to create a 3D model for printing?**
   a) Audio files
   b) CAD (Computer-Aided Design) files
   c) Video files
   d) Image files
   Solution: - b) CAD (Computer-Aided Design) files

19) **Which of the following consumer products has NOT typically been made using 3D printing?**
   a) Personalized phone cases
   b) Custom eyewear
   c) High-performance automotive engines
   d) Decorative home items
   Solution: - c) High-performance automotive engines

20) **Which factor is crucial for ensuring the quality of 3D printed consumer products?**
   a) Print speed
   b) Ambient temperature
   c) Material type and printer calibration
   d) Print color
   Solution: - c) Material type and printer calibration

21) **What is a major advantage of using 3D printing for low-volume consumer product manufacturing?**
   a) High initial setup cost
   b) Reduced flexibility in design changes
   c) Cost-effectiveness due to no need for expensive molds
   d) Limited customization options
   Solution: - c) Cost-effectiveness due to no need for expensive molds

22) **Which 3D printing technology is particularly effective for creating high-resolution consumer product prototypes with complex geometries?**
   a) Fused Deposition Modeling (FDM)
   b) Selective Laser Sintering (SLS)
   c) Digital Light Processing (DLP)
   d) Stereolithography (SLA)
   Solution: - d) Stereolithography (SLA)

23) What is the primary challenge associated with using metal 3D printing for consumer products?

a) High material cost

b) Limited design freedom

c) Low resolution

d) Slow printing speeds

Solution: - a) High material cost

24) Which material property is crucial for consumer products that need to withstand mechanical stress and wear?

a) Elasticity

b) Density

c) Flexibility

d) Toughness

Solution: - d) Toughness

25) What is the significance of using parametric design in 3D printing for consumer products?

a) It simplifies the design process for mass production.

b) It allows for automatic adjustment of design parameters based on user input.

c) It limits the ability to customize products.

d) It increases the time required for design changes.

Solution: - b) It allows for automatic adjustment of design parameters based on user input.

26) How do advanced 3D printing techniques like Direct Ink Writing (DIW) benefit consumer product development?

a) By increasing the cost of production

b) By enabling the creation of complex structures with functional materials

c) By reducing the accuracy of the prints

d) By limiting material options

Solution: - b) By enabling the creation of complex structures with functional materials

27) What is the primary function of a support structure in 3D printing of complex consumer products?

a) To enhance the aesthetic quality

b) To provide temporary stability during the printing process

c) To increase material usage

d) To reduce printing speed

Solution: - b) To provide temporary stability during the printing process

28) Which 3D printing technology uses a laser to fuse powdered material into solid parts?

a) Fused Deposition Modeling (FDM)

b) Selective Laser Sintering (SLS)

c) Stereolithography (SLA)

d) Inkjet Printing

Solution: - b) Selective Laser Sintering (SLS)

29) What is one reason for the high cost of industrial-grade 3D printers used in consumer product manufacturing?

a) They use inexpensive materials.

b) They require less maintenance than consumer-grade printers.

c) They have advanced features and high precision capabilities.

d) They are simpler to operate.

Solution: - c) They have advanced features and high precision capabilities.

30) What is a key factor in selecting materials for 3D printing consumer products?

a) The color of the material

b) The ease of disposal

c) The mechanical and thermal properties required for the application

d) The brand name of the material

Solution: - c) The mechanical and thermal properties required for the application

31) Which 3D printing method is best suited for producing flexible and rubber-like consumer products?

a) Stereolithography (SLA)

b) Fused Deposition Modeling (FDM) with flexible filaments

c) Digital Light Processing (DLP)

d) Selective Laser Sintering (SLS)

Solution: - b) Fused Deposition Modeling (FDM) with flexible filaments

32) What is the impact of using 3D printing on traditional supply chains in the consumer products industry?

a) It eliminates the need for inventory and warehousing.

b) It increases the complexity of logistics.
c) It requires more labor for product handling.
d) It slows down the distribution process.
Solution: - a) It eliminates the need for inventory and warehousing.

33) What is a common post-processing technique used to improve the surface finish of 3D printed consumer products?
a) Sanding and polishing
b) Painting with high-temperature paints
c) Chemical vapor smoothing
d) Thermal annealing
Solution: - c) Chemical vapor smoothing

34) How does the use of 3D printing affect the intellectual property (IP) concerns in consumer product design?
a) It eliminates IP concerns completely.
b) It simplifies the protection of IP due to easy replication.
c) It increases the risk of IP theft and counterfeiting.
d) It makes IP protection unnecessary.
Solution: - c) It increases the risk of IP theft and counterfeiting.

35) Which factor is critical for achieving high-quality 3D prints in consumer product manufacturing?
a) The resolution of the 3D printer
b) The color of the filament
c) The speed of the printing process
d) The brand of the 3D printer
Solution: - a) The resolution of the 3D printer

36) Which 3D printing technology is most appropriate for creating large-scale consumer products?
a) Fused Deposition Modeling (FDM)
b) Selective Laser Sintering (SLS)
c) Stereolithography (SLA)
d) Direct Metal Laser Sintering (DMLS)
Solution: - a) Fused Deposition Modeling (FDM)

37) What role does parametric design play in the customization of consumer products using 3D printing?
a) It restricts design changes to predefined templates.

b) It automates design adjustments based on user specifications.
c) It simplifies the design process by reducing complexity.
d) It limits customization options to basic shapes.
Solution: - b) It automates design adjustments based on user specifications.

38) What is a significant disadvantage of using 3D printing for mass production of consumer products?
a) High setup costs and slow production speeds
b) Limited design capabilities
c) Lack of material variety
d) High environmental impact
Solution: - a) High setup costs and slow production speeds

39) What advantage does using 3D printing offer in terms of product design for consumer goods?
a) It reduces the need for prototyping.
b) It allows for rapid and cost-effective iteration of designs.
c) It increases the complexity of the design process.
d) It limits design creativity due to material constraints.
Solution: - b) It allows for rapid and cost-effective iteration of designs.

40) Which technology enables the production of consumer products with embedded electronics using 3D printing?
a) Fused Deposition Modeling (FDM)
b) Inkjet Printing
c) Multi-material Jetting
d) Selective Laser Sintering (SLS)
Solution: - c) Multi-material Jetting

Chapter 23

Additive Manufacturing in Architecture and Construction

23.1 Introduction to AM in Architecture and Construction

Additive manufacturing (AM) has revolutionized the architecture and construction industries by enabling the rapid production of complex, customized, and sustainable buildings and structures. AM offers a range of benefits, including improved design flexibility, reduced construction times, and increased sustainability. In this chapter, we will explore the applications, benefits, challenges, and limitations of AM in architecture and construction.

The Need for Innovation in Architecture and Construction

The architecture and construction industries have traditionally been slow to adopt new technologies, relying on centuries-old methods and materials. However, the need for innovation in these industries has never been more pressing. The world's population is projected to reach 9.7 billion by 2050, placing immense pressure on urban infrastructure and resources. Climate change, resource depletion, and environmental degradation are also major concerns, highlighting the need for sustainable and efficient construction practices.

The Role of AM in Architecture and Construction

AM has the potential to transform the architecture and construction industries by enabling the rapid production of complex, customized, and sustainable buildings and structures. AM allows architects and engineers to design and build structures that were previously impossible to produce using traditional manufacturing methods. This includes complex geometries, curved lines, and intricate details that can be used to create unique and innovative designs.

AM also enables the use of sustainable materials and practices, reducing waste, energy consumption, and environmental impact. For example, AM can be used to produce buildings and structures using recycled materials, reducing the need for virgin resources and minimizing waste. AM can also be used to create buildings and structures that are optimized for energy efficiency, reducing energy consumption and carbon emissions.

Applications of AM in Architecture and Construction

AM has a wide range of applications in architecture and construction, including:

- **Building components**: AM can be used to produce building components such as walls, floors, and roofs. These components can be customized to meet specific design and functional requirements and can be produced using sustainable materials and practices.

- **Structural elements**: AM can be used to produce structural elements such as beams, columns, and arches. These elements can be designed to meet specific load-bearing requirements, and can be produced using high-strength materials such as steel and concrete.

- **Facade systems**: AM can be used to produce facade systems such as windows, doors, and cladding. These systems can be customized to meet specific design and functional requirements and can be produced using sustainable materials and practices.

- **Infrastructure**: AM can be used to produce infrastructure such as bridges, roads, and utilities. These infrastructure components can be designed to meet specific functional and safety requirements and can be produced using sustainable materials and practices.

Benefits of AM in Architecture and Construction

The benefits of AM in architecture and construction are numerous. These include:

- **Improved design flexibility**: AM enables the production of complex, customized, and innovative designs that were previously impossible to produce using traditional manufacturing methods.

- **Reduced construction times**: AM enables the rapid production of building components and structures, reducing construction times and costs.

- **Increased sustainability**: AM enables the use of sustainable materials and practices, reducing waste, energy consumption, and environmental impact.

- **Improved safety**: AM enables the production of structures that are optimized for safety, reducing the risk of accidents and injuries.

Challenges and Limitations of AM in Architecture and Construction

While AM has the potential to transform the architecture and construction industries, there are several challenges and limitations that must be addressed. These include:

- **Material properties**: The properties of AM materials may not meet the required specifications for building codes and regulations.

- **Scalability**: AM machines may not be able to produce large quantities of building components and structures, limiting the scalability of AM in architecture and construction.

- **Cost**: AM can be a costly process, particularly for large-scale production.

- **Regulatory frameworks**: There is a need for standardized regulatory frameworks for AM in architecture and construction, ensuring the safety and efficacy of AM-produced buildings and structures.

By addressing these challenges and limitations, companies and organizations can fully realize the potential of AM in architecture and construction, creating innovative, sustainable, and efficient buildings and structures that meet the needs of the 21st century. Let's explore how Additive Manufacturing (AM) is making an impact in the architecture and construction industry which are described below.

Prototyping and Design Iteration:

Architects and designers use AM to create rapid prototypes of building components.

Iterative design allows them to test different concepts, refine details, and visualize structures before full-scale construction.

Customization and Complex Geometries:

AM enables customized architectural elements, such as intricate facades, decorative panels, and unique shapes.

Complex geometries that were previously challenging to manufacture can now be realized.

Sustainable Construction:

AM reduces material waste by precisely depositing material where needed.

Lightweight structures with optimized material distribution enhance sustainability.

On-Site Printing and Prefabrication:

On-site 3D printing can create temporary structures, emergency shelters, or even permanent buildings.

Prefabricated components can be printed off-site and assembled quickly.

Material Innovations:

Researchers explore new materials suitable for AM in construction, including concrete, polymers, and composites.

Material properties and durability are critical considerations.

Challenges:

Scale: AM is currently more suited for smaller components; scaling up for entire buildings remains a challenge.

Structural Integrity: Ensuring AM-built structures meet safety standards is crucial.

Cost: Initial investment in AM equipment and materials can be high.

Figure 1: AM in architecture and construction
[Source: https://www.smartguy.com/article/importance-of-architecture-and-construction-built-environment]

23.2 Applications and Benefits of AM in Architecture

Additive manufacturing (AM) has numerous applications in architecture, including:

Complex Geometries

AM enables the production of complex geometries, such as curved walls, domes, and irregular shapes. These complex geometries can be used to create unique and innovative designs that were previously impossible to produce using traditional manufacturing methods. For example, AM can be used to produce buildings with curved facades, irregularly shaped roofs, or complex interior structures.

The ability to produce complex geometries has opened up new possibilities for architects and designers. They can now create buildings that are not only aesthetically pleasing but also functional and efficient. Complex geometries can also be used to create buildings that are optimized for energy efficiency, reducing energy consumption and environmental impact.

Customized Buildings

AM enables the production of customized buildings, tailored to specific client needs and preferences. This means that architects can design buildings that meet the specific requirements of their clients, whether it's a unique shape, size, or functionality. Customized buildings can also be designed to meet specific environmental or sustainability requirements, such as energy efficiency or water conservation.

The ability to produce customized buildings has revolutionized the architecture industry. Architects can now design buildings that are tailored to specific client needs, rather than relying on standardized designs. This has led to a increase in client satisfaction and a reduction in construction costs.

Sustainable Buildings

AM enables the production of sustainable buildings, using recycled materials, reducing waste, and minimizing environmental impact. This is achieved through the use of sustainable materials, such as recycled plastics, bamboo, or low-carbon concrete. AM also enables the production of buildings with optimized energy efficiency, reducing energy consumption and environmental impact.

The ability to produce sustainable buildings has become increasingly important in recent years. With the growing concern about climate change and environmental degradation, architects and designers are looking for ways to reduce the environmental impact of their designs. AM provides a solution to this problem, enabling the production of sustainable buildings that meet the needs of the 21st century.

Rapid Prototyping

AM enables rapid prototyping, allowing architects to test and refine their designs quickly and efficiently. This means that architects can produce multiple prototypes of a design, test them, and refine them before finalizing the design. Rapid prototyping has reduced the time and cost associated with traditional prototyping methods, enabling architects to work more efficiently and effectively.

Rapid prototyping has also enabled architects to test and refine their designs in a more iterative and collaborative way. They can work with clients and stakeholders to test and refine designs, ensuring that the final product meets their needs and expectations.

Construction Components

AM enables the production of construction components, such as walls, floors, and roofs, with improved mechanical properties. This means that architects can design buildings with improved structural integrity, reducing the risk of collapse or damage. AM also enables the production of construction components with improved thermal insulation, reducing energy consumption and environmental impact.

The ability to produce construction components with improved mechanical properties has revolutionized the construction industry. Architects can now design buildings that are not only aesthetically pleasing but also safe and functional.

Benefits of AM in Architecture

The benefits of AM in architecture include:

- **Improved Design Flexibility:** AM allows for the production of complex geometries and customized designs, enabling architects to push the boundaries of traditional architecture. This has led to an increase in design innovation and creativity, as architects are no longer limited by traditional manufacturing methods.
- **Reduced Construction Times:** AM enables rapid production, reducing construction times and enabling faster project completion. This has led to a reduction in construction costs and an increase in client satisfaction.
- **Increased Sustainability:** AM enables the production of sustainable buildings, reducing environmental impact and promoting eco-friendliness. This has become increasingly important in recent years, as architects and designers look for ways to reduce the environmental impact of their designs.
- **Cost Savings:** AM can reduce material waste, energy consumption, and labor costs, resulting in cost savings. This has led to a reduction in construction costs and an increase in client satisfaction.
- **Improved Collaboration:** AM enables architects to work more collaboratively with clients and stakeholders, ensuring that the final product meets their needs and expectations. This has led to an increase in client satisfaction and a reduction in construction costs.

- **Increased Accuracy:** AM enables the production of buildings with improved accuracy, reducing the risk of errors and defects. This has led to a reduction in construction costs and an increase in client satisfaction.

By leveraging these benefits, architects and designers can create innovative, sustainable, and efficient buildings that meet the needs of the 21st century. AM has the potential to transform the architecture industry, enabling the production of buildings that are not only aesthetically pleasing but also functional, sustainable, and efficient. Let's explore the applications and benefits of Additive Manufacturing (AM) in the architecture and construction industry:

Applications of AM in Architecture and Construction

1. **Prototyping and Design Iteration**:

    - Architects and designers use AM to create rapid prototypes of building components.
    - Iterative design allows them to test different concepts, refine details, and visualize structures before full-scale construction.

2. **Customization and Complex Geometries**:

    - AM enables customized architectural elements, such as intricate facades, decorative panels, and unique shapes.
    - Complex geometries that were previously challenging to manufacture can now be realized.

3. **Sustainable Construction**:

    - AM reduces material waste by precisely depositing material where needed.
    - Lightweight structures with optimized material distribution enhance sustainability.

4. **On-Site Printing and Prefabrication**:

    - On-site 3D printing can create temporary structures, emergency shelters, or even permanent buildings.
    - Prefabricated components can be printed off-site and assembled quickly.

Benefits of AM in Architecture and Construction

1. **Precision and Accuracy**:

    - AM allows precise fabrication of intricate architectural details.
    - Customized components fit perfectly, reducing errors during assembly.

2. **Time Efficiency**:

    - Rapid prototyping and on-site printing accelerate construction timelines.
    - Faster iterations lead to quicker project completion.

3. **Cost Savings**:

    - Reduced material waste and optimized designs minimize costs.
    - Prefabrication streamlines construction processes.

4. **Innovative Designs**:

    - AM encourages creativity and experimentation.
    - Architects can explore novel shapes and aesthetics.

5. **Sustainability**:

- Lightweight structures reduce environmental impact.
- AM supports sustainable building practices.

Figure 97 shows the futuristic Applications and Benefits of AM in Architecture and construction works.

Figure 2: Applications and benefits of AM in architecture
[Source: http://archicgi.com/architecture/augmented-reality-apps-architecture/]

23.3 Challenges and Limitations of AM in Architecture

Despite the benefits of additive manufacturing (AM) in architecture, there are several challenges and limitations that must be addressed. These challenges and limitations can be categorized into four main areas: material properties, scalability, cost, and regulatory frameworks.

Material Properties

One of the significant challenges of AM in architecture is the properties of AM materials. The properties of AM materials may not meet the required specifications for architectural applications. For example, the strength, durability, and weather resistance of AM materials may not be sufficient for building structures that are exposed to harsh environmental conditions.

Moreover, the material properties of AM materials can vary depending on the type of material used, the printing process, and the post-processing techniques employed. This variability can make it challenging to predict the behavior of AM materials in different environmental conditions, which is critical in architectural applications.

To address this challenge, researchers and manufacturers are working to develop new AM materials with improved properties. For example, researchers are developing new types of concrete that can be used in AM, which have improved strength, durability, and sustainability.

Scalability

Another challenge of AM in architecture is scalability. AM machines may not be able to produce large quantities of building components, limiting the scalability of AM in architecture. Currently, most AM machines are designed for small-scale production, which makes them unsuitable for large-scale architectural projects.

To address this challenge, manufacturers are developing larger AM machines that can produce larger building components. For example, some companies are developing AM machines that can produce entire walls or floors in a single print.

Cost

AM can be a costly process, particularly for large-scale production. The cost of AM machines, materials, and labor can be prohibitively expensive for many architectural projects. Moreover, the cost of post-processing and finishing AM components can add significant expenses to the overall project cost.

To address this challenge, manufacturers are working to reduce the cost of AM machines and materials. For example, some companies are developing more affordable AM machines that can be used for small-scale production.

Regulatory Frameworks

There is a need for standardized regulatory frameworks for AM in architecture, ensuring the safety and efficacy of AM-produced buildings. Currently, there is a lack of clear regulations and standards for AM in architecture, which can make it challenging to ensure the safety and quality of AM-produced buildings.

To address this challenge, governments and industry organizations are working to develop standardized regulatory frameworks for AM in architecture. For example, some countries are developing new building codes and standards that incorporate AM technologies.

Other Challenges and Limitations

In addition to the challenges mentioned above, there are several other challenges and limitations of AM in architecture. These include:

- **Design limitations**: AM requires a high level of design complexity, which can be challenging for architects and designers who are not familiar with AM technologies.

- **Labor skills**: AM requires specialized labor skills, which can be challenging to find in the current labor market.

- **Supply chain management**: AM requires a complex supply chain management system, which can be challenging to manage and coordinate.

- **Quality control**: AM requires a high level of quality control, which can be challenging to ensure in large-scale production.

Addressing the Challenges and Limitations

To address the challenges and limitations of AM in architecture, it is essential to develop new technologies, materials, and processes that can overcome these challenges. This requires a collaborative effort from researchers, manufacturers, architects, and policymakers.

Some of the strategies that can be employed to address the challenges and limitations of AM in architecture include:

- **Research and development**: Investing in research and development to improve the properties of AM materials, scalability, and cost-effectiveness of AM technologies.

- **Standardization**: Developing standardized regulatory frameworks and industry standards for AM in architecture.

- **Education and training**: Providing education and training programs for architects, designers, and laborers to develop the necessary skills and knowledge to work with AM technologies.

- **Collaboration**: Encouraging collaboration between researchers, manufacturers, architects, and policymakers to develop new solutions and address the challenges and limitations of AM in architecture.

By addressing the challenges and limitations of AM in architecture, it is possible to unlock the full potential of AM technologies and create innovative, sustainable, and efficient buildings that meet the needs of the 21st century. Sustainable architecture (As shown in Figure 98 of futuristic image of a building), while promising, faces several challenges and limitations. Let's explore them:

Challenges of Sustainable Architecture:

1. **Cost:**

- Sustainable construction often requires expensive materials and technologies (e.g., solar panels, energy-efficient windows).

- These added costs can deter developers and homeowners, especially when the benefits may not be immediately apparent.

2. **Lack of Standardization and Certification**:
   - There is no universally agreed-upon definition of sustainable buildings.

   - Multiple certification systems exist, each with its own criteria, making it confusing for developers and consumers.

3. **Limited Availability of Sustainable Materials and Technologies**:

   - Many sustainable products are still in early development stages.

   - Their higher cost or lower durability can hinder adoption, especially on tight budgets.

4. **Culture and Aesthetics**:

   - Sustainable designs may challenge traditional notions of beauty.

   - Convincing clients to prioritize sustainability over aesthetics can be a hurdle.

Limitations of Sustainable Architecture:

1. **Limited Scope**:

   - Sustainable design doesn't address all environmental impacts.

- We need holistic solutions to create a truly sustainable built environment.

Figure 3: Challenges and limitations of AM in sustainable architecture [Source: https://www.ar43.com/en/challenges-and-limitations-of-sustainable-architecture/]

23.4 Case Studies and Examples

Several companies and organizations are already leveraging additive manufacturing (AM) in architecture, pushing the boundaries of what is possible in the industry. These pioneers are demonstrating the potential of AM to transform the way we design, build, and construct buildings.

WinSun

WinSun is a Chinese company that has been at the forefront of AM in architecture. They have developed a proprietary AM technology that enables the production of entire buildings, including houses, apartments, and commercial buildings. WinSun's technology uses a mixture of cement and other materials to create a strong and durable building structure.

In 2014, WinSun made headlines by printing 10 houses in just 24 hours. Each house was printed using a massive 3D printer that stood over 20 feet tall and 30 feet wide. The houses were printed using a combination of cement, sand, and other materials, and were designed to be affordable and sustainable.

Since then, WinSun has gone on to print numerous buildings, including a 5-story apartment building and a 1,100 square meter villa. Their technology has the potential to revolutionize the construction industry, enabling the rapid production of affordable and sustainable housing.

Apis Cor

Apis Cor is a Russian company that is using AM to produce 3D-printed houses. Their technology uses a mobile printer that can print entire buildings on-site, eliminating the need for traditional construction methods. The printer uses a mixture of cement, sand, and other materials to create a strong and durable building structure.

Apis Cor's technology has been used to print houses in a variety of locations, including Russia, Europe, and the Middle East. Their houses are designed to be affordable, sustainable, and customizable, with clients able to choose from a range of designs and materials.

One of the key benefits of Apis Cor's technology is its ability to print buildings in remote or hard-to-reach locations. This makes it ideal for areas where traditional construction methods are difficult or impossible.

WATG

WATG is a global design and architecture firm that is using AM to produce customized building components. Their technology uses a combination of 3D printing and traditional construction methods to create complex and customized building designs.

WATG has used AM to produce a range of building components, including walls, floors, and roofs. Their technology enables the creation of complex geometries and customized designs, which can be used to create unique and innovative buildings.

One of the key benefits of WATG's technology is its ability to reduce waste and increase efficiency. By using AM to produce customized building components, WATG is able to reduce the amount of waste generated during the construction process.

Skanska

Skanska is a global construction company that is using AM to produce construction components. Their technology uses a 3D printer to produce concrete elements, such as walls, floors, and roofs.

Skanska has used AM to produce a range of construction components, including a 3D-printed concrete bridge in the Netherlands. The bridge was printed using a combination of cement, sand, and other materials, and was designed to be strong, durable, and sustainable.

One of the key benefits of Skanska's technology is its ability to increase efficiency and reduce costs. By using AM to produce construction components, Skanska is able to reduce the amount of labor and materials required during the construction process.

Other Examples

In addition to these companies, there are numerous other examples of AM being used in architecture and construction. For example:

- **Dubai Municipality**: Dubai Municipality has announced plans to use AM to produce 25% of the city's buildings by 2030. The municipality is working with a range of companies to develop the technology and infrastructure required to support large-scale AM construction.

- **NASA**: NASA is using AM to produce building components for its lunar and Mars missions. The agency is working with a range of companies to develop the technology and materials required to support AM construction in space.

- **University of Nantes**: The University of Nantes has developed a 3D-printed house that can be built in just 30 hours. The house is designed to be affordable, sustainable, and customizable, with clients able to choose from a range of designs and materials.

These examples demonstrate the potential of AM to transform the architecture and construction industries. By enabling the rapid production of customized and sustainable buildings, AM has the potential to address some of the world's most pressing housing and infrastructure challenges.

23.5 Questions

23.5.1 Long Response Questions

1. What are the key benefits of using AM in architecture and construction?
2. How do the properties of AM materials impact their use in architectural applications?
3. What are the challenges and limitations of scaling up AM production in architecture?
4. How can regulatory frameworks for AM in architecture be improved?
5. What are some examples of successful AM applications in architecture and construction?

23.5.2 Multiple Choice Questions

1) What is additive manufacturing commonly known as?
a) Subtractive manufacturing
b) 3D printing
c) Molding
d) Casting

Solution: - b) 3D printing

2) **Which technology is often used for 3D printing in construction?**
   a) Fused Deposition Modeling (FDM)
   b) Stereolithography (SLA)
   c) Selective Laser Sintering (SLS)
   d) Concrete printing
   Solution: - d) Concrete printing

3) **What is one of the main materials used in 3D printing for architecture?**
   a) Plastic
   b) Metal
   c) Concrete
   d) Glass
   Solution: - c) Concrete

4) **In which field is 3D printing often utilized for creating intricate building models?**
   a) Automotive
   b) Aerospace
   c) Architecture
   d) Healthcare
   Solution: - c) Architecture

5) **Which component of a 3D printer melts and extrudes the material?**
   a) Print bed
   b) Hotend
   c) Stepper motor
   d) Cooling fan
   Solution: - b) Hotend

6) **What is the primary benefit of using 3D printing in construction?**
   a) Increased material waste
   b) Slower construction time
   c) Cost efficiency and reduced waste
   d) Limited design flexibility
   Solution: - c) Cost efficiency and reduced waste

7) **Which of the following is a challenge associated with 3D printing in architecture?**
   a) High material costs

b) Limited design freedom
c) Slow production speed
d) Difficulty in scaling up

Solution: - d) Difficulty in scaling up

8) **What technology enables the use of multiple materials in a single 3D print?**
a) Single-material FDM
b) Dual-material FDM
c) SLA
d) Concrete printing

Solution: - b) Dual-material FDM

9) **Which aspect of 3D printing allows for complex geometric designs that traditional methods cannot achieve?**
a) Speed of printing
b) Layer-by-layer construction
c) Material type
d) Printer size

Solution: - b) Layer-by-layer construction

10) **Which 3D printing method is particularly suited for large-scale construction projects?**
a) Fused Deposition Modeling (FDM)
b) Stereolithography (SLA)
c) Selective Laser Sintering (SLS)
d) Concrete printing

Solution: - d) Concrete printing

11) **What is the main challenge of using concrete as a material in 3D printing for construction?**
a) Limited structural strength
b) High cost of raw material
c) Difficulty in achieving precise control over the setting time
d) Incompatibility with digital models

Solution: - c) Difficulty in achieving precise control over the setting time

12) **Which approach combines 3D printing with robotics to improve construction efficiency?**
a) Robotic-assisted additive manufacturing
b) Robotic welding
c) Automated material handling

d) CNC milling

Solution: - a) Robotic-assisted additive manufacturing

13) How does the use of 3D printing in construction impact traditional labor practices?

a) Increases the need for manual labor

b) Reduces the demand for skilled labor

c) Has no impact on labor practices

d) Increases the need for on-site training

Solution: - b) Reduces the demand for skilled labor

14) What is a common feature of 3D printed architectural structures compared to conventionally built structures?

a) Uniformity in design

b) Increased use of manual labor

c) Limited design capabilities

d) Higher material waste

Solution: - a) Uniformity in design

15) Which type of 3D printer is capable of creating full-scale buildings and infrastructure?

a) Desktop 3D printer

b) Industrial 3D printer

c) Personal 3D printer

d) Educational 3D printer

Solution: - b) Industrial 3D printer

16) Which of the following materials is often used in the construction of 3D printed buildings due to its sustainability?

a) Aluminum

b) Plastic

c) Recycled aggregates

d) Steel

Solution: - c) Recycled aggregates

17) What is a significant advantage of 3D printing in terms of reducing construction time?

a) Reduced material costs

b) Faster production speed

c) Lower design complexity

d) Increased manual labor

Solution: - b) Faster production speed

18) **Which factor is crucial for ensuring the structural integrity of 3D printed architectural elements?**
   a) Color of the material
   b) Layer adhesion and bonding
   c) Size of the printer
   d) Speed of the printer
   Solution: - b) Layer adhesion and bonding

19) **In what way can 3D printing contribute to the customization of architectural designs?**
   a) By limiting design options
   b) By providing a single material choice
   c) By enabling bespoke and intricate designs
   d) By requiring standard sizes only
   Solution: - c) By enabling bespoke and intricate designs

20) **Which recent advancement has enhanced the durability and functionality of 3D printed construction materials?**
   a) Improved CAD software
   b) Advanced composite materials
   c) Faster printing speeds
   d) Lower material costs
   Solution: - b) Advanced composite materials

21) **What factor influences the choice of material in 3D printing for architecture?**
   a) Aesthetic appeal only
   b) Cost and structural requirements
   c) Color variety
   d) Printer brand
   Solution: - b) Cost and structural requirements

22) **How does 3D printing reduce environmental impact in construction?**
   a) By increasing material consumption
   b) By using non-recyclable materials
   c) By minimizing waste and utilizing sustainable materials
   d) By requiring more energy for production
   Solution: - c) By minimizing waste and utilizing sustainable materials

23) **What role does simulation play in 3D printing for architecture?**

a) Increases material costs
b) Helps in optimizing design and predicting performance
c) Slows down the printing process
d) Reduces design flexibility
Solution: - b) Helps in optimizing design and predicting performance

24) **Which company is known for using 3D printing to produce entire buildings in record time?**
a) Apis Cor
b) Tesla
c) IBM
d) Boeing
Solution: - a) Apis Cor

25) **What is the primary advantage of using a robotic arm in 3D printing for construction?**
a) Limited material choices
b) Enhanced precision and flexibility in construction
c) Increased production cost
d) Reduced design capabilities
Solution: - b) Enhanced precision and flexibility in construction

26) **How can 3D printing contribute to disaster relief efforts?**
a) By increasing construction costs
b) By producing slow and complex structures
c) By enabling rapid construction of emergency shelters
d) By reducing design options
Solution: - c) By enabling rapid construction of emergency shelters

27) **What is a key challenge when scaling 3D printing technology for large construction projects?**
a) Material cost
b) Printer size and capacity
c) Design complexity
d) Printing speed
Solution: - b) Printer size and capacity

28) **Which feature of 3D printing allows for the creation of complex architectural forms?**
a) Use of traditional construction materials
b) Layer-by-layer additive process

c) Limited design software

d) Manual construction techniques

Solution: - b) Layer-by-layer additive process

29) What does the term "digital fabrication" refer to in the context of 3D printing for architecture?

a) Use of analog tools

b) Traditional building methods

c) Automated production of building components using digital models

d) Manual construction techniques

Solution: - c) Automated production of building components using digital models

30) Which of the following is a significant advantage of 3D printed construction over traditional methods?

a) Increased labor requirements

b) Higher material consumption

c) Faster construction time

d) Reduced design flexibility

Solution: - c) Faster construction time

31) Which company is known for its mobile 3D printing technology for on-site construction?

a) WinSun

b) Apis Cor

c) Stratasys

d) Autodesk

Solution: - b) Apis Cor

32) How does the material extrusion method work in 3D printing?

a) By using lasers to sinter materials

b) By extruding molten material layer by layer

c) By curing liquid resin with UV light

d) By melting metal powders

Solution: - b) By extruding molten material layer by layer

33) Which software is commonly used for designing 3D printable architectural models?

a) AutoCAD

b) Photoshop

c) Excel

d) Word

Solution: - a) AutoCAD

34) What is a major benefit of integrating 3D printing with BIM (Building Information Modeling)?

a) Increased manual work

b) Improved design collaboration and accuracy

c) Decreased construction speed

d) Limited design flexibility

Solution: - b) Improved design collaboration and accuracy

35) Which type of 3D printing is best suited for creating detailed and intricate architectural features?

a) Fused Deposition Modeling (FDM)

b) Stereolithography (SLA)

c) Selective Laser Sintering (SLS)

d) Digital Light Processing (DLP)

Solution: - b) Stereolithography (SLA)

36) What aspect of 3D printing technology is crucial for ensuring the strength and durability of printed structures?

a) Print speed

b) Material selection

c) Printer brand

d) Color of the material

Solution: - b) Material selection

37) What is a significant advantage of using recycled materials in 3D printing for construction?

a) Increased cost

b) Limited availability

c) Environmental sustainability

d) Reduced material strength

Solution: - c) Environmental sustainability

38) Which construction method is often combined with 3D printing to improve the efficiency of building construction?

a) Modular construction

b) Traditional brick laying

c) Wood framing

d) Stone masonry

Solution: - a) Modular construction

39) **What is the purpose of using simulation tools in the design phase of 3D printed architecture?**
   a) To increase production costs
   b) To optimize design and predict performance
   c) To limit design options
   d) To slow down the printing process
   Solution: - b) To optimize design and predict performance

40) **Which material property is critical for ensuring that a 3D printed building can withstand environmental stresses?**
   a) Color
   b) Thermal conductivity
   c) Mechanical strength
   d) Aesthetic appeal
   Solution: - c) Mechanical strength

Part 5

Business and Economics

Chapter 24

Business and Economic Implications of Additive Manufacturing

24.1 Introduction to Business and Economic Implications of AM

Additive manufacturing (AM) has the potential to transform the way businesses operate and create value. As AM continues to evolve, it is essential to understand the business and economic implications of this technology. In this chapter, we will explore the business models and strategies, economic benefits and challenges, and supply chain and logistics implications of AM. Figure 99 provides insights into the market size by technology from 2020 to 2030 in USD Billion. Here are the key takeaways:

- **Stereolithography**: Starting at around $4.4 billion in 2020, this technology shows steady growth over the years.

- **Fused Deposition Modeling (FDM)**: Also experiencing consistent growth, contributing significantly to the market.

- **Selective Laser Sintering (SLS)**: Another robust technology, with a positive trajectory.

- **Polyjet Printing**: Demonstrates promising growth potential.

- **Direct Metal Laser Sintering (DMLS)**: A technology to watch, likely to expand further.

The note on the graph indicates a **21.6% North America Market CAGR** from 2023 to 2030. Overall, this graph highlights the exciting prospects for additive manufacturing technologies in North America.

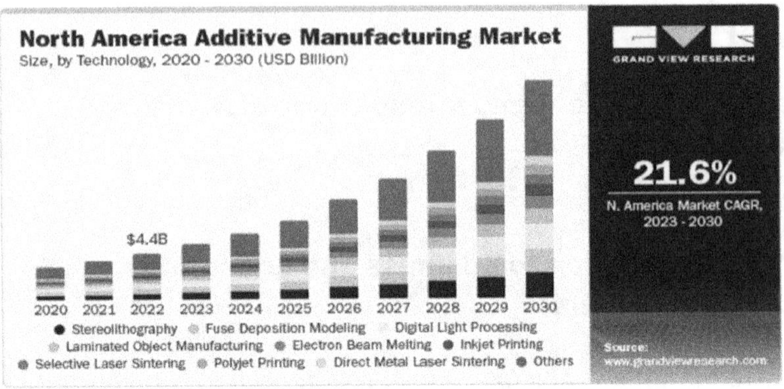

Figure 1: North America Additive Manufacturing Market
[Source: https://www.grandviewresearch.com/industry-analysis/dropshipping-market]

Business Models and Strategies

AM is enabling new business models and strategies that are disrupting traditional industries. One of the key benefits of AM is its ability to enable mass customization, which allows companies to produce customized products at a lower cost and with greater efficiency. This is enabling companies to offer personalized products and services that are tailored to individual customers' needs.

Another key benefit of AM is its ability to enable on-demand production, which allows companies to produce products only as they are needed. This is reducing inventory costs and enabling companies to respond quickly to changes in demand.

AM is also enabling new business models such as product-as-a-service, where companies offer products as a service rather than a physical product. This is enabling companies to generate recurring revenue streams and build long-term relationships with customers.

Economic Benefits

AM has the potential to generate significant economic benefits, including:

- **Job creation**: AM is creating new job opportunities in areas such as design, engineering, and production.

- **Increased productivity**: AM is enabling companies to produce products more efficiently and with greater accuracy, which is increasing productivity and reducing costs.

- **Reduced waste**: AM is reducing waste by enabling companies to produce products only as they are needed, which is reducing the environmental impact of production.

- **Increased competitiveness**: AM is enabling companies to compete more effectively in global markets by offering customized and innovative products.

Economic Challenges

While AM has the potential to generate significant economic benefits, it also poses several challenges, including:

- **Job displacement**: AM has the potential to displace jobs in traditional manufacturing industries, which could lead to unemployment and social unrest.

- **High upfront costs**: The high upfront costs of AM equipment and technology can be a barrier to adoption for many companies.

- **Regulatory uncertainty**: The regulatory environment for AM is still evolving, which can create uncertainty and risk for companies.

Supply Chain and Logistics Implications

AM is having a significant impact on supply chain and logistics operations, including:

- **Reduced inventory costs**: AM is enabling companies to produce products on-demand, which is reducing inventory costs and enabling companies to respond quickly to changes in demand.

- **Increased flexibility**: AM is enabling companies to produce products in a variety of locations, which is increasing flexibility and reducing reliance on traditional supply chain networks.

- **New logistics models**: AM is enabling new logistics models such as decentralized production and distribution, which is reducing transportation costs and increasing efficiency.

Case Studies

Several companies are already leveraging AM to transform their business models and operations, including:

- **GE Aviation**: GE Aviation is using AM to produce aircraft engine components, which is reducing production time and increasing efficiency.

- **Nike**: Nike is using AM to produce customized athletic shoes, which is enabling the company to offer personalized products to customers.

- **Siemens**: Siemens is using AM to produce industrial equipment components, which is reducing production time and increasing efficiency.

These case studies demonstrate the potential of AM to transform business models and operations, and highlight the need for companies to understand the business and economic implications of this technology.

In the next section, we will explore the business and economic implications of AM in more detail, including the impact on industry structures, the role of government policy, and the potential for AM to create new business opportunities and industries. In 2016, AM contributed significantly to various aspects of the economy and academic achievements as shown in Figure 100:

Economic Impact:

1. **State Taxes**:
    - AM activities generated approximately $241 million in state tax revenue.
    - This revenue contributes to local economies and public services.

2. **Research Projects and Funding**:
    - There were 1,700 active research projects related to AM.
    - These projects drive innovation, create jobs, and attract extramural funding.

3. **Student Involvement**:
    - A total of 7,339 students participated in AM-related research and education.
    - Student engagement fosters knowledge transfer and workforce development.

4. **Scholarly Output and Citations**:
    - Scholarly publications resulting from AM research amounted to $6.6 million.
    - These publications contribute to academic knowledge and influence future research.

5. **Patents and Licenses**:
    - Researchers were awarded 33 patents related to AM innovations.
    - Active licenses indicate commercialization potential.

6. **Doctoral Degrees Conferred**:
    - A total of 3,429 doctoral degrees were conferred in AM-related fields.

- These graduates contribute expertise to industry and academia.

Extramural Funding

- Research in AM received $274 million in extramural funding.
- This funding supports ongoing research, infrastructure, and workforce development.

Figure 2: Business and economic implications of AM
[Source: https://www.ipsinternational.org/unraveling-the-impact-capital-gains-tax-reform-act-of-1986/]

24.2 Business Models and Strategies for AM

Additive manufacturing (AM) enables new business models and strategies that are transforming the way companies operate and create value. By leveraging the unique capabilities of AM, companies can differentiate themselves, create new revenue streams, and improve their competitiveness.

Product Customization

One of the key benefits of AM is its ability to enable product customization. By producing customized products, companies can differentiate themselves from competitors and create new revenue streams. AM enables companies to produce products with complex geometries, customized shapes, and tailored properties, which is not possible with traditional manufacturing methods.

Product customization is particularly valuable in industries such as healthcare, where customized implants and prosthetics can improve patient outcomes and reduce recovery times. In the aerospace industry, customized parts can reduce weight and improve fuel efficiency, leading to cost savings and environmental benefits.

Companies such as Siemens and GE Aviation are already leveraging AM to produce customized products. Siemens is using AM to produce customized turbine blades, which are improving efficiency and reducing emissions. GE Aviation is using AM to produce customized aircraft engine components, which are reducing production time and increasing efficiency.

Service-Based Models

AM is also enabling companies to shift from product-based to service-based models. By offering customers access to products and services rather than ownership, companies can generate recurring revenue streams and build long-term relationships with customers.

Service-based models are particularly valuable in industries such as industrial equipment, where companies can offer customers access to equipment and maintenance services rather than selling them equipment outright. This approach can reduce costs, improve efficiency, and increase customer satisfaction.

Companies such as Rolls-Royce and Philips are already leveraging AM to offer service-based models. Rolls-Royce is using AM to produce customized aircraft engine components, which are offered to customers as part of a service-based model. Philips is using AM to produce customized medical devices, which are offered to customers as part of a service-based model.

Platform-Based Models

AM is also enabling companies to create platform-based models, connecting customers, designers, and manufacturers in a single ecosystem. By providing a platform for collaboration and innovation, companies can generate revenue from transaction fees, advertising, and data analytics.

Platform-based models are particularly valuable in industries such as consumer products, where companies can connect customers with designers and manufacturers to produce customized products. This approach can reduce costs, improve efficiency, and increase customer satisfaction.

Companies such as Shapeways and Sculpteo are already leveraging AM to create platform-based models. Shapeways is a 3D printing platform that connects customers with designers and manufacturers to produce customized products. Sculpteo is a 3D printing platform that connects customers with designers and manufacturers to produce customized products.

Data-Driven Models

AM is also enabling companies to leverage data and analytics to optimize production, reduce costs, and improve product performance. By collecting and analyzing data on production processes, material properties, and product performance, companies can identify areas for improvement and optimize their operations.

Data-driven models are particularly valuable in industries such as aerospace, where companies can use data analytics to optimize production processes and reduce costs. In the automotive industry, companies can use data analytics to optimize production processes and improve product performance.

Companies such as Airbus and Volkswagen are already leveraging AM to create data-driven models. Airbus is using AM to produce customized aircraft components, which are optimized using data analytics. Volkswagen is using AM to produce customized car parts, which are optimized using data analytics.

Other Business Models and Strategies

In addition to these business models and strategies, AM is also enabling other innovative approaches, including:

- **Product-as-a-Service**: Companies are offering customers access to products as a service, rather than selling them outright. This approach can reduce costs, improve efficiency, and increase customer satisfaction.

- **Design-for-Additive**: Companies are designing products specifically for AM, which is enabling new geometries, shapes, and properties. This approach can improve product performance, reduce costs, and increase efficiency.

- **Additive Manufacturing-as-a-Service**: Companies are offering AM services to customers, which is enabling companies to access AM capabilities without investing in equipment and expertise.

These business models and strategies demonstrate the potential of AM to transform industries and create new opportunities for companies. By leveraging the unique capabilities of AM, companies can differentiate themselves, create new revenue streams, and improve their competitiveness. Let's break down the key elements of Figure 101, which illustrates various business models and strategies related to additive manufacturing (AM), commonly known as 3D printing technology:

- **Bricks and Clicks**: This model combines both online and offline retail strategies. Businesses operating in this space leverage physical stores (bricks) alongside digital platforms (clicks) to reach customers effectively.

- **Value Added Reseller (VAR)**: VARs enhance existing products or services before reselling them. In the context of AM, this could involve adding features or customizations to 3D-printed items.

- **Franchise**: Franchises allow others to operate under their established brand and system. In the 3D printing industry, franchises might offer localized 3D printing services or sell 3D printers.

- **Sourcing**: Businesses focused on procuring materials or products for AM fall into this category. Sourcing is critical for maintaining a reliable supply chain.

- **Freemium**: This strategy provides basic services for free while offering premium features at a cost. In the AM realm, it could apply to software tools or design services.

- **Subscription Models**: Recurring revenue models are common in the 3D printing sector. Subscriptions might grant access to specialized software, design libraries, or ongoing support.

- **Direct Sales**: Some companies sell 3D-printed products directly to consumers without intermediaries. This approach allows for more personalized interactions.

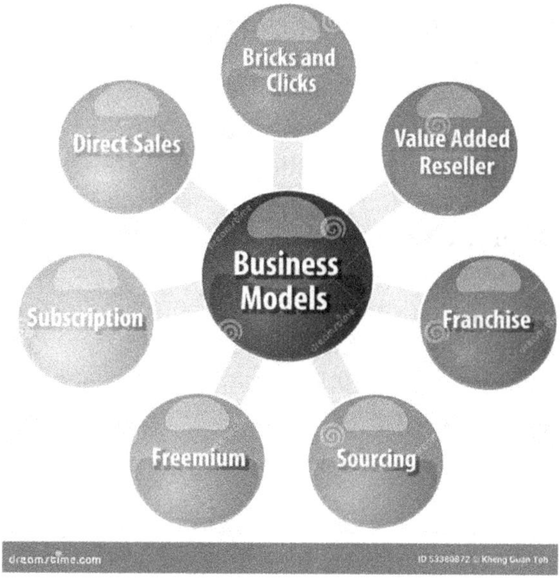

Figure 3: Business models and strategies for AM
[Source: https://www.tpsearchtool.com/images/business-model-definitions-examples-and-types]

24.3 Economic Benefits and Challenges of AM

Additive manufacturing (AM) has several economic benefits that are transforming industries and creating new opportunities for companies. However, AM also poses economic challenges that need to be addressed.

Economic Benefits

Job Creation: AM has the potential to create new jobs in design, engineering, and manufacturing. As AM continues to evolve, companies will need skilled professionals to design, engineer, and manufacture products using AM technologies. This will create new job opportunities in areas such as:

- ❖ Design for additive manufacturing
- ❖ Additive manufacturing engineering

- Materials science and engineering
- Post-processing and finishing

According to a report by Deloitte, the AM industry is expected to create over 100,000 new jobs in the United States alone by 2025.

Increased Productivity

AM can improve productivity by reducing production times and enabling just-in-time production. By producing products on-demand, companies can reduce inventory costs, minimize waste, and improve supply chain efficiency. This can lead to significant cost savings and improved competitiveness.

For example, companies such as Airbus and Boeing are using AM to produce aircraft components, which is reducing production times and improving efficiency. Similarly, companies such as Siemens and GE Aviation are using AM to produce industrial equipment components, which is reducing production times and improving efficiency.

Cost Savings

AM can reduce material waste, energy consumption, and labor costs, resulting in cost savings. By producing products with complex geometries and customized shapes, companies can reduce material waste and energy consumption. Additionally, AM can reduce labor costs by minimizing the need for manual labor and improving production efficiency.

According to a report by McKinsey, AM can reduce material costs by up to 90% and energy consumption by up to 50%. This can lead to significant cost savings and improved competitiveness.

New Revenue Streams

AM enables companies to create new revenue streams through product customization, service-based models, and platform-based models. By offering customized products and services, companies can differentiate themselves and create new revenue streams.

For example, companies such as Nike and Adidas are using AM to produce customized athletic shoes, which is creating new revenue streams and improving customer satisfaction. Similarly, companies such as Rolls-Royce and Philips are using AM to offer service-based models, which is creating new revenue streams and improving customer satisfaction.

Economic Challenges

Job Displacement

AM may displace jobs in traditional manufacturing industries. As AM continues to evolve, companies may replace traditional manufacturing methods with AM, which could lead to job displacement. This is particularly concerning for industries such as automotive and aerospace, where traditional manufacturing methods are still prevalent.

However, it's worth noting that AM is also creating new job opportunities in areas such as design, engineering, and manufacturing. Companies will need to invest in training and upskilling programs to ensure that workers have the necessary skills to work with AM technologies.

Investment Costs

AM requires significant investment in equipment, software, and training. Companies will need to invest in AM equipment, software, and training programs to ensure that they have the necessary skills and expertise to work with AM technologies.

According to a report by Grand View Research, the global AM market is expected to reach $36.6 billion by 2025, driven by increasing investment in AM equipment and software.

Regulatory Frameworks

There is a need for standardized regulatory frameworks for AM, ensuring the safety and efficacy of AM-produced products. As AM continues to evolve, regulatory frameworks will need to be developed to ensure that AM-produced products meet safety and quality standards.

This is particularly concerning for industries such as healthcare and aerospace, where safety and quality standards are critical. Regulatory frameworks will need to be developed to ensure that AM-produced products meet safety and quality standards, and that companies are held accountable for any defects or failures.

In conclusion, AM has several economic benefits, including job creation, increased productivity, cost savings, and new revenue streams. However, AM also poses economic challenges, including job displacement, investment costs, and regulatory frameworks. Companies will need to address these challenges to ensure that they can fully leverage the benefits of AM. Let's delve into the economic benefits and challenges associated with additive manufacturing (AM):

Economic Benefits of AM

- **Increased Efficiency:** AM allows for streamlined production processes, reducing waste and optimizing resource utilization.

- **Production Flexibility**: AM enables rapid prototyping and customized manufacturing, adapting to specific needs without large-scale retooling.

- **Innovation Acceleration**: AM fosters novel designs, materials, and geometries, driving innovation across industries.

- **Supply Chain Optimization**: Localized AM production reduces dependence on global supply chains, enhancing resilience.

Challenges of AM

- **Material Limitations**: Some AM materials lack mechanical properties comparable to traditional manufacturing materials.
- **Post-Processing Complexity**: Finishing, surface treatment, and quality control after printing can be intricate and time-consuming.
- **Regulatory Uncertainty**: Existing regulations may not fully address AM processes, posing challenges for certification and safety.
- **Cost and Scalability:** Initial investment in AM equipment can be high, and scaling up production remains a hurdle.

Remember, AM's transformative potential lies in balancing these benefits and addressing challenges effectively. Figure 102 shows the Economic Benefits and Challenges of AM.

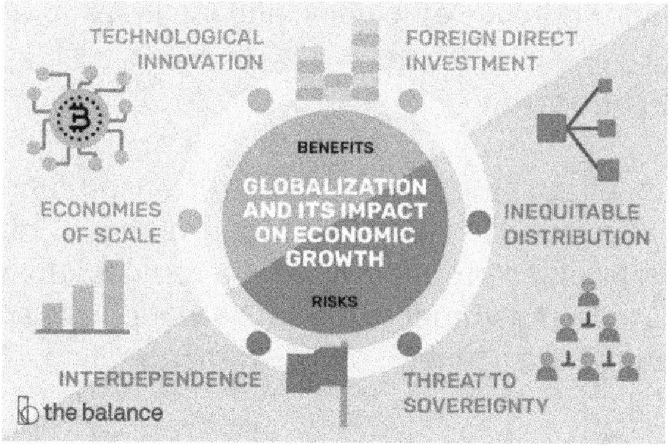

Figure 4: Economic benefits and challenges of AM
[Source: http://complianceportal.american.edu/advantages-of-economic-growth.php]

24.4 Supply Chain and Logistics Implications of AM

Additive manufacturing (AM) has significant implications for supply chain and logistics, transforming the way companies design, produce, and distribute products. By enabling just-in-time production, localized production, and new logistics models, AM is revolutionizing the supply chain and logistics landscape.

Reduced Inventor

AM enables just-in-time production, reducing the need for inventory storage and management. Traditional manufacturing methods often require companies to produce large quantities of products, which are then stored in warehouses until they are needed. This approach can lead to inventory obsolescence, waste, and unnecessary costs.

With AM, companies can produce products on-demand, reducing the need for inventory storage and management. This approach can lead to significant cost savings, improved efficiency, and reduced waste. According to a report by Deloitte, companies can reduce inventory costs by up to 50% by adopting AM.

For example, companies such as Airbus and Boeing are using AM to produce aircraft components, which are produced on-demand and delivered directly to the production line. This approach has reduced inventory costs, improved efficiency, and minimized waste.

Localized Production

AM enables localized production, reducing transportation costs and lead times. Traditional manufacturing methods often require companies to produce products in centralized locations, which are then transported to customers around the world. This approach can lead to long lead times, high transportation costs, and environmental concerns.

With AM, companies can produce products locally, reducing transportation costs and lead times. This approach can lead to improved customer satisfaction, reduced costs, and improved environmental sustainability. According to a report by McKinsey, companies can reduce transportation costs by up to 70% by adopting AM.

For example, companies such as Siemens and GE Aviation are using AM to produce industrial equipment components, which are produced locally and delivered directly to customers. This approach has reduced transportation costs, improved customer satisfaction, and minimized environmental concerns.

New Logistics Models

AM enables new logistics models, such as decentralized production and distribution networks. Traditional manufacturing methods often require companies to maintain centralized production and distribution networks, which can be inflexible and inefficient.

With AM, companies can create decentralized production and distribution networks, which are more flexible and efficient. This approach can lead to improved customer satisfaction, reduced costs, and improved environmental sustainability. According to a report by PwC, companies can reduce logistics costs by up to 30% by adopting AM.

For example, companies such as Amazon and UPS are using AM to create decentralized production and distribution networks, which are more flexible and efficient. This approach has improved customer satisfaction, reduced costs, and minimized environmental concerns.

Other Supply Chain and Logistics Implications

In addition to reduced inventory, localized production, and new logistics models, AM has several other supply chain and logistics implications, including:

- **Improved supply chain visibility**: AM enables companies to track products in real-time, improving supply chain visibility and reducing the risk of counterfeiting.

- **Increased customization**: AM enables companies to produce customized products, which can improve customer satisfaction and reduce returns.

- **Reduced packaging**: AM enables companies to produce products with minimal packaging, reducing waste and environmental concerns.

- **Improved quality control**: AM enables companies to produce products with improved quality control, reducing the risk of defects and improving customer satisfaction.

In conclusion, AM has significant implications for supply chain and logistics, transforming the way companies design, produce, and distribute products. By enabling just-in-time production, localized production, and new logistics models, AM is revolutionizing the supply chain and logistics landscape. Companies will need to adapt to these changes to remain competitive and improve customer satisfaction. Figure 103 shows the Supply Chain and Logistics Implications of AM.

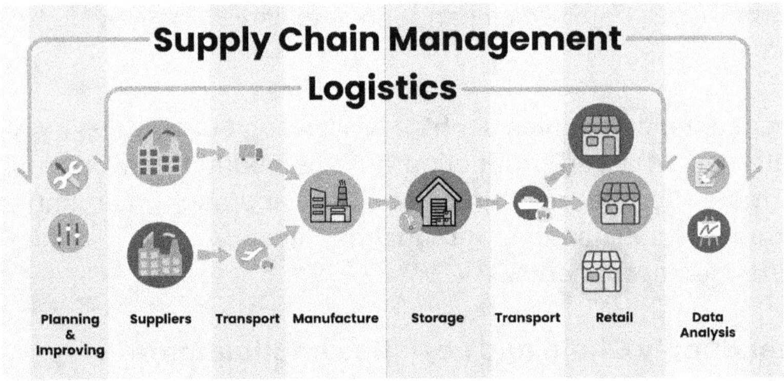

Figure 5: Supply chain and logistics implications of AM
[Source: https://www.wtagroup.com/resources-and-insights/blogs/the-difference-between-logistics-and-supply-chain-management-scm]

24.5 Questions

24.5.1 Long Response Questions

1. What are the key business models and strategies enabled by AM?
2. How can companies leverage data and analytics to optimize production and reduce costs in AM?
3. What are the economic benefits and challenges of AM, and how can companies mitigate the challenges?
4. How can companies adapt their supply chain and logistics strategies to take advantage of AM?
5. What are some examples of successful AM applications in business and economics?

24.5.2 Multiple Choice Questions

1) **What is additive manufacturing commonly known as?**
   a) Subtractive manufacturing
   b) Traditional manufacturing
   c) 3D printing
   d) Assembly line production
   Solution: - c) 3D printing
2) **Which of the following industries is NOT directly impacted by additive manufacturing?**
   a) Aerospace
   b) Healthcare
   c) Textiles
   d) Mining
   Solution: - d) Mining
3) **What is a primary advantage of additive manufacturing in prototyping?**
   a) High cost
   b) Slow production speed
   c) Design flexibility
   d) Limited material options
   Solution: - c) Design flexibility
4) **Which term describes the process of creating parts layer by layer?**

a) Subtractive manufacturing
b) Incremental production
c) Additive manufacturing
d) Bulk manufacturing
Solution: - c) Additive manufacturing

5) **What type of business model benefits greatly from on-demand production enabled by 3D printing?**
a) Mass production
b) Just-in-time manufacturing
c) Bulk inventory
d) Assembly line production
Solution: - b) Just-in-time manufacturing

6) **What is a major cost advantage of additive manufacturing in small-batch production?**
a) High labor costs
b) Low setup costs
c) Expensive raw materials
d) Large machine maintenance
Solution: - b) Low setup costs

7) **Additive manufacturing can reduce waste by using:**
a) Excessive material
b) Limited material
c) Subtractive methods
d) Layer-by-layer material deposition
Solution: - d) Layer-by-layer material deposition

8) **Which of the following is a common material used in additive manufacturing?**
a) Wood
b) Metal
c) Glass
d) Fabric
Solution: - b) Metal

9) **Additive manufacturing can be advantageous in which area of supply chain management?**
a) Global warehousing
b) On-site production
c) Long-distance shipping
d) Mass production
Solution: - b) On-site production

10) In additive manufacturing, what does 'design freedom' refer to?

a) Restriction to standard shapes
b) Ability to create complex geometries
c) Limited material choices
d) Uniform part dimensions

Solution: - b) Ability to create complex geometries

11) Which of the following is a potential economic benefit of using additive manufacturing for spare parts?

a) Increased inventory costs
b) Reduced lead times
c) High transportation costs
d) Increased production downtime

Solution: - b) Reduced lead times

12) What is a challenge associated with the widespread adoption of additive manufacturing in the automotive industry?

a) High setup costs
b) Limited material options
c) Complexity of integrating with existing processes
d) Slow prototyping speed

Solution: - c) Complexity of integrating with existing processes

13) Which economic factor is enhanced by additive manufacturing through reduced transportation needs?

a) Transportation costs
b) Raw material costs
c) Labor costs
d) Energy consumption

Solution: - a) Transportation costs

14) What does the term 'decentralized production' imply in the context of additive manufacturing?

a) Centralized factory locations
b) Production spread across multiple locations
c) Large-scale factory operations
d) Single-location production

Solution: - b) Production spread across multiple locations

15) **Which of the following is a key consideration when evaluating the economic impact of additive manufacturing?**
   a) Material cost only
   b) Labor cost only
   c) Total cost of ownership
   d) Initial setup cost only
   Solution: - c) Total cost of ownership

16) **In what way can additive manufacturing impact supply chain resilience?**
   a) By increasing supply chain complexity
   b) By reducing the need for inventory
   c) By increasing lead times
   d) By standardizing all parts
   Solution: - b) By reducing the need for inventory

17) **How does additive manufacturing influence the concept of 'mass customization'?**
   a) By providing uniform products
   b) By allowing for individualized products at scale
   c) By limiting design variations
   d) By increasing setup time
   Solution: - b) By allowing for individualized products at scale

18) **Which business model might benefit from additive manufacturing's ability to create customized products on demand?**
   a) High-volume production
   b) Custom fabrication
   c) Standardized mass production
   d) Large-scale batch production
   Solution: - b) Custom fabrication

19) **Which sector is leveraging additive manufacturing to produce highly complex and lightweight parts?**
   a) Consumer electronics
   b) Aerospace
   c) Food processing
   d) Retail
   Solution: - b) Aerospace

20) What is a potential risk associated with the adoption of additive manufacturing in terms of intellectual property?

a) Decreased innovation

b) Increased counterfeiting

c) Lower patenting costs

d) Standardization of designs

Solution: - b) Increased counterfeiting

21) How does additive manufacturing potentially disrupt traditional business models in manufacturing?

a) By increasing labor requirements

b) By enabling localized production and reducing the need for large inventories

c) By standardizing global supply chains

d) By reducing material costs only

Solution: - b) By enabling localized production and reducing the need for large inventories

22) What are the economic implications of additive manufacturing for global supply chains?

a) Increased global coordination

b) Reduced need for international shipping and warehousing

c) Higher international tariffs

d) Enhanced dependency on raw material imports

Solution: - b) Reduced need for international shipping and warehousing

23) Which factor can significantly influence the cost-effectiveness of additive manufacturing in production?

a) Cost of raw materials

b) Maintenance of traditional machinery

c) Complexity of designs and the associated processing time

d) Standardization of production techniques

Solution: - c) Complexity of designs and the associated processing time

24) What economic advantage does additive manufacturing offer in terms of product lifecycle management?

a) Increased product lifecycle duration

b) Enhanced ability to rapidly iterate and update designs

c) Decreased frequency of product updates

d) Increased manufacturing lead times

Solution: - b) Enhanced ability to rapidly iterate and update designs

25) Which of the following is a strategic consideration for businesses implementing additive manufacturing technologies?

a) Ignoring the impact on supply chain dynamics

b) Investing solely in traditional manufacturing equipment

c) Assessing the integration of additive manufacturing with existing production processes

d) Focusing only on cost reduction

Solution: - c) Assessing the integration of additive manufacturing with existing production processes

26) How can additive manufacturing contribute to environmental sustainability in production?

a) By increasing waste through excessive material use

b) By enabling the use of recycled materials and reducing waste

c) By expanding the carbon footprint of manufacturing

d) By increasing energy consumption in production

Solution: - b) By enabling the use of recycled materials and reducing waste

27) What economic factor is most directly affected by the shift from centralized to decentralized production in additive manufacturing?

a) Raw material costs

b) Shipping and logistics costs

c) Labor productivity

d) Product design costs

Solution: - b) Shipping and logistics costs

28) In the context of additive manufacturing, what does 'just-in-time' production refer to?

a) Producing in large quantities to store inventory

b) Producing items as needed, reducing inventory and storage costs

c) Increasing production lead times to anticipate demand

d) Standardizing all production schedules

Solution: - b) Producing items as needed, reducing inventory and storage costs

29) **What challenge is associated with the adoption of additive manufacturing in high-volume production scenarios?**
   a) Increased prototyping costs
   b) Limited material options
   c) Lower production speed compared to traditional methods
   d) Increased waste generation
   Solution: - c) Lower production speed compared to traditional methods

30) **How does the use of additive manufacturing affect product differentiation strategies?**
   a) By limiting customization options
   b) By allowing for highly customizable and unique products
   c) By reducing the range of available designs
   d) By focusing on mass-produced standard items
   Solution: - b) By allowing for highly customizable and unique products

31) **What role does data analytics play in enhancing the benefits of additive manufacturing?**
   a) It provides insights into reducing material costs only
   b) It helps optimize production processes and improve product quality
   c) It decreases the need for design flexibility
   d) It standardizes production techniques
   Solution: - b) It helps optimize production processes and improve product quality

32) **What is a potential drawback of additive manufacturing in terms of initial investment?**
   a) High costs of traditional manufacturing equipment
   b) Low costs of raw materials
   c) Significant capital required for advanced 3D printing technology and training
   d) Decreased need for skilled labor
   Solution: - c) Significant capital required for advanced 3D printing technology and training

33) **How does additive manufacturing influence global manufacturing strategies?**
   a) By promoting centralized production in low-cost regions

b) By supporting localized production and reducing reliance on global supply chains

c) By increasing dependency on international suppliers

d) By reducing the need for innovative manufacturing processes

Solution: - b) By supporting localized production and reducing reliance on global supply chains

34) Which economic factor can be most positively impacted by additive manufacturing in small-batch production?

a) Increased material costs

b) Reduced production flexibility

c) Decreased manufacturing costs per unit

d) Increased need for large-scale equipment

Solution: - c) Decreased manufacturing costs per unit

35) What is the primary challenge of integrating additive manufacturing into existing production processes?

a) The simplicity of the technology

b) The ability to reduce material usage

c) The complexity and cost of retrofitting current systems

d) The reduction of labor costs

Solution: - c) The complexity and cost of retrofitting current systems

36) How does additive manufacturing impact traditional supply chain management?

a) By increasing lead times and shipping costs

b) By simplifying and decentralizing supply chains

c) By maintaining large inventories

d) By standardizing production across global facilities

Solution: - b) By simplifying and decentralizing supply chains

37) In what way can additive manufacturing impact customer relationships and service models?

a) By reducing customer interaction

b) By enabling customized products and improved service offerings

c) By increasing service delivery times

d) By decreasing product personalization

Solution: - b) By enabling customized products and improved service offerings

38) **What is a potential economic disadvantage of additive manufacturing for large-scale industrial applications?**
   a) High material costs
   b) Limited design capabilities
   c) High initial investment and slower production speed
   d) Reduced flexibility in design
   Solution: - c) High initial investment and slower production speed

39) **Which of the following best describes the impact of additive manufacturing on traditional inventory management?**
   a) Increased inventory holding costs
   b) Decreased need for large inventory storage
   c) Standardized inventory levels
   d) Higher levels of product obsolescence
   Solution: - b) Decreased need for large inventory storage

40) **How does additive manufacturing contribute to sustainability in production practices?**
   a) By increasing waste and energy consumption
   b) By enabling the use of sustainable materials and reducing waste
   c) By promoting mass production methods
   d) By increasing carbon footprint
   Solution: - b) By enabling the use of sustainable materials and reducing waste

Chapter 25

Future Directions and Emerging Trends in Additive Manufacturing

25.1 Introduction to Future Directions and Emerging Trends in AM

Additive manufacturing (AM) has come a long way since its inception, and it continues to evolve and improve at a rapid pace. From its humble beginnings as a niche technology for prototyping and model making, AM has grown to become a mainstream manufacturing technology with applications in a wide range of industries. As AM technology advances, new applications, industries, and opportunities emerge, and it is essential to explore the future directions and emerging trends in AM.

Advancements in Technologies and Materials

One of the most significant areas of advancement in AM is the development of new technologies and materials. Researchers and companies are continually pushing the boundaries of what is possible with AM, developing new technologies and materials that enable faster, more accurate, and more cost-effective production.

For example, advancements in metal AM have enabled the production of complex metal parts with unprecedented accuracy and precision. New metal AM technologies such as selective laser sintering (SLS) and selective laser melting (SLM) have opened up new opportunities for the production of high-performance metal parts.

Similarly, advancements in polymer AM have enabled the production of complex polymer parts with improved mechanical properties and thermal stability. New polymer AM technologies such as fused deposition modeling (FDM) and stereolithography (SLA) have opened up new opportunities for the production of high-performance polymer parts.

In addition to advancements in metal and polymer AM, researchers are also exploring the development of new materials and technologies for AM, such as ceramics, glass, and biomaterials. These new materials and technologies have the potential to enable the production of complex parts with unique properties and functionalities.

New Applications and Industries

As AM technology advances, new applications and industries are emerging. One of the most significant areas of growth is in the production of end-use parts and products. Companies such as Airbus, Boeing, and GE Aviation are using AM to produce complex aircraft components, while companies such as Siemens and Philips are using AM to produce complex industrial equipment components.

Another area of growth is in the production of customized products, such as medical implants, prosthetics, and dental restorations. AM enables the production of customized products with complex geometries and unique properties, which is revolutionizing the healthcare industry.

In addition to these industries, AM is also being explored for its potential in other areas, such as:

- **Aerospace**: AM is being used to produce complex aircraft components, such as engine parts, satellite components, and spacecraft components.

- **Automotive**: AM is being used to produce complex automotive parts, such as engine components, transmission components, and chassis components.

- **Consumer products**: AM is being used to produce customized consumer products, such as jewelry, eyewear, and footwear.
- **Energy**: AM is being used to produce complex energy-related components, such as wind turbine blades, solar panels, and fuel cells.

Challenges and Opportunities in Education and Training

As AM technology advances, there is a growing need for education and training programs that can equip workers with the skills and knowledge needed to work with AM technologies. However, there are several challenges and opportunities in education and training that need to be addressed.

One of the biggest challenges is the lack of standardization in AM education and training. There is a need for standardized curricula and training programs that can provide workers with the skills and knowledge needed to work with AM technologies.

Another challenge is the need for continuous education and training. AM technology is constantly evolving, and workers need to be able to adapt to new technologies and materials.

Despite these challenges, there are also opportunities for education and training programs to address the growing demand for AM skills and knowledge. Online education platforms, vocational training programs, and apprenticeships can provide workers with the skills and knowledge needed to work with AM technologies.

In conclusion, the future of AM is bright, with advancements in technologies and materials, new applications and industries, and challenges and opportunities in education and training. As AM technology continues to evolve and improve, it is essential to explore the future directions and emerging trends in AM to stay ahead of the curve. Let's explore the exciting future directions and emerging trends in additive manufacturing (AM), as depicted in Figure 104:

Food Printing (FOOD)

NASA scientists are developing 3D-printed food for long space missions.

Imagine astronauts enjoying customized meals created layer by layer using AM technology.

Bio-Printing

Scientists are using 3D printing to create living tissues and organs.

Bio-printing holds immense potential for personalized medicine and regenerative therapies.

Machine and Robot Components

3D printing is revolutionizing manufacturing by producing intricate parts for machines and robots.

It enables rapid prototyping and customized designs.

Prototyping

Researchers and businesses use AM for creating 3D models. Prototyping accelerates innovation and product development cycles.

These trends showcase the versatility and transformative impact of 3D printing across diverse fields.

25.2 Advancements in AM Technologies and Materials

Several advancements in AM technologies and materials are expected to shape the future of AM, enabling the production of more complex, precise, and functional parts. These advancements will expand the range of AM applications, improve product performance, and reduce production costs.

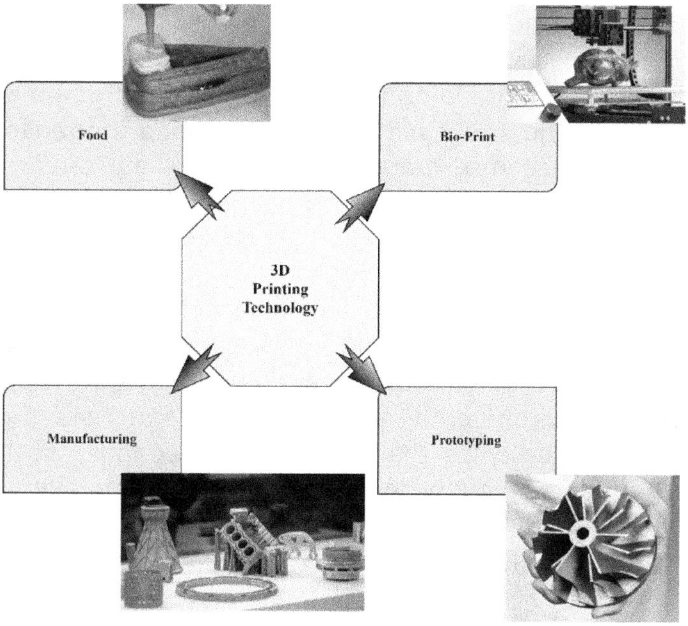

Figure 1: Future directions and emerging trends in AM
[Source: https://www.g2.com/articles/future-of-3d-printing]

Increased Resolution and Accuracy

Advancements in printer resolution and accuracy will enable the production of more complex and precise parts. High-resolution printers will be able to produce parts with intricate details, complex geometries, and precise tolerances. This will be particularly important for industries such as aerospace, automotive, and healthcare, where precision and accuracy are critical.

For example, researchers at the University of California, Los Angeles (UCLA) have developed a 3D printing technology that can produce parts with resolutions as low as 10 microns. This technology has the potential to revolutionize the production of complex parts, such as aircraft components, medical implants, and microelectronic devices.

New Materials and Properties

Research into new materials and properties, such as nanomaterials, metamaterials, and smart materials, will expand the range of AM applications. These new materials will enable the production of parts with unique properties, such as self-healing, shape-memory, and adaptive behavior.

For example, researchers at the Massachusetts Institute of Technology (MIT) have developed a new class of materials called "programmable materials" that can change shape and properties in response to environmental stimuli. These materials have the potential to revolutionize the production of adaptive structures, such as morphing aircraft wings and self-healing materials.

Multi-Material Printing

The ability to print with multiple materials will enable the creation of complex, multi-functional parts. This will be particularly important for industries such as aerospace, automotive, and consumer products, where parts often require multiple materials with different properties.

For example, researchers at the University of Michigan have developed a 3D printing technology that can print with multiple materials, including metals, polymers, and ceramics. This technology has the potential to revolutionize the production of complex parts, such as aircraft components, engine parts, and medical implants.

Bioprinting and Bioinks

Advancements in bioprinting and bioinks will enable the creation of functional, living tissues and organs. Bioprinting involves the use of living cells and biomaterials to create functional tissues and organs, which can be used for transplantation, drug testing, and tissue engineering.

For example, researchers at the Wake Forest Institute for Regenerative Medicine have developed a bioprinting technology that can create functional, living tissues, such as skin, bone, and muscle. This technology has the potential to revolutionize the field of regenerative medicine, enabling the creation of functional, living tissues and organs for transplantation and tissue engineering.

Other Advancements

In addition to these advancements, there are several other areas of research and development that are expected to shape the future of AM, including:

- **Artificial intelligence and machine learning**: The integration of artificial intelligence and machine learning into AM systems will enable the optimization of print parameters, prediction of print failures, and improvement of print quality.

- **Robotics and automation**: The integration of robotics and automation into AM systems will enable the production of complex parts with high precision and accuracy, as well as the automation of post-processing and finishing operations.

- **Sustainability and recycling**: The development of sustainable and recyclable AM materials and processes will enable the reduction of waste and environmental impact, as well as the creation of closed-loop production systems.

In conclusion, the future of AM is bright, with several advancements in technologies and materials expected to shape the industry. These advancements will enable the production of more complex, precise, and functional parts, expand the range of AM applications, and improve product performance and reduce production costs.

25.3 New Applications and Industries for AM

AM is expected to expand into new applications and industries, driving innovation and growth in various sectors. As the technology continues to evolve, it will enable the creation of complex, customized, and sustainable products that were previously impossible to produce. Here are some of the new applications and industries that AM is expected to impact:

Aerospace and Defense

AM will play a critical role in the production of lightweight, complex components for aerospace and defense applications. The ability to produce parts with complex geometries, reduced weight, and increased strength will enable the creation of more efficient and effective aircraft, spacecraft, and defense systems.

For example, NASA is already using AM to produce parts for its spacecraft, including a 3D-printed engine component that was tested in 2017. The component, which was printed using a metal alloy, was able to withstand the extreme temperatures and pressures of a rocket engine.

In addition to NASA, private companies such as SpaceX and Blue Origin are also using AM to produce parts for their spacecraft. SpaceX, for example, has used AM to produce engine components, rocket nozzles, and other parts for its Falcon 9 rocket.

Healthcare and Biomedicine

AM will enable the creation of customized, patient-specific implants, prosthetics, and medical devices. The ability to produce parts with complex geometries and customized properties will enable the creation of implants that are tailored to individual patients' needs.

For example, researchers at the University of California, Los Angeles (UCLA) have developed a 3D printing technology that can produce customized, patient-specific implants for facial reconstruction. The technology uses CT scans and 3D modeling software to create a customized implant that is tailored to the individual patient's needs.

In addition to implants, AM is also being used to produce customized prosthetics and medical devices. For example, companies such as Össur and Hocoma are using AM to produce customized prosthetic limbs and orthotics.

Energy and Environment

AM will enable the creation of sustainable, energy-efficient systems and devices. The ability to produce parts with complex geometries and customized properties will enable the creation of systems that are optimized for energy efficiency and sustainability.

For example, researchers at the Massachusetts Institute of Technology (MIT) have developed a 3D printing technology that can produce customized, energy-efficient heat exchangers. The technology uses a combination of 3D printing and machine learning to optimize the design of the heat exchanger for maximum energy efficiency.

In addition to heat exchangers, AM is also being used to produce customized, energy-efficient components for wind turbines, solar panels, and fuel cells.

Consumer Products and Retail

AM will enable the production of customized, on-demand consumer products. The ability to produce parts with complex geometries and customized properties will enable the creation of products that are tailored to individual consumers' needs.

For example, companies such as Nike and Adidas are using AM to produce customized, 3D-printed shoes. The shoes are produced using a combination of 3D printing and machine learning, which enables the creation of customized shoes that are tailored to individual consumers' foot shapes and preferences.

In addition to shoes, AM is also being used to produce customized, 3D-printed jewelry, eyewear, and other consumer products.

Other Industries

In addition to these industries, AM is also expected to impact other sectors, including:

- **Automotive**: AM will enable the production of customized, lightweight components for automotive applications, including engine parts, transmission components, and chassis components.

- **Agriculture**: AM will enable the creation of customized, sustainable agricultural systems and devices, including irrigation systems, crop monitoring systems, and farm equipment.

- **Construction**: AM will enable the creation of customized, sustainable building components, including walls, roofs, and facades.

- **Education**: AM will enable the creation of customized, interactive educational tools and models, including 3D-printed models of organs, molecules, and other complex systems.

In conclusion, AM is expected to expand into new applications and industries, driving innovation and growth in various sectors. As the technology continues to evolve, it will enable the creation of complex, customized, and sustainable products that were previously impossible to produce. The exciting future directions and emerging trends in additive manufacturing (AM), as depicted in Figure 105 with realistic figure of hand in 3D printing.

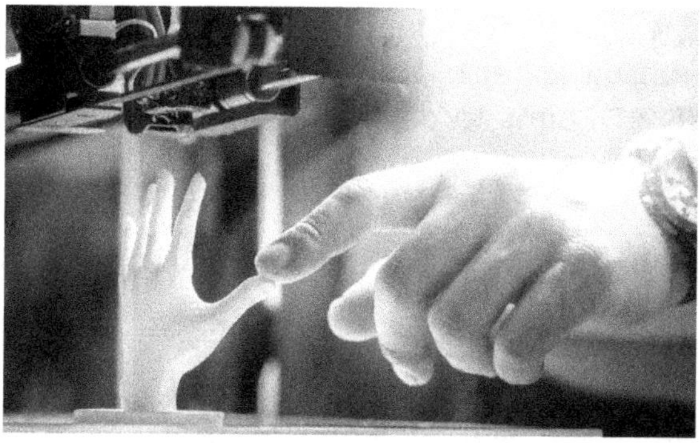

Figure 2: New applications and industries for AM
[Source: https://www.techemergence.com/artificial-intelligence-applications-additive-manufacturing-3d-printing/]

25.4 Challenges and Opportunities in AM Education and Training

As AM continues to evolve, there is a growing need for education and training programs that can prepare workers for the demands of the AM industry. The industry is facing a significant skills gap, and education and training programs must address this gap to ensure a skilled workforce. Here are some of the challenges and opportunities in AM education and training:

Developing Standardized Curricula

One of the major challenges in AM education and training is the lack of standardized curricula and training programs. Currently, there is no universally accepted curriculum for AM education, which makes it difficult for workers to transition between companies and industries.

To address this challenge, industry leaders and educators must work together to develop standardized curricula and training programs that can provide workers with the skills and knowledge required for AM. This will require collaboration between industry associations, educational institutions, and government agencies to develop a comprehensive framework for AM education and training.

For example, the Additive Manufacturing Education and Training (AMET) program, launched by the National Institute of Standards and Technology (NIST), aims to develop a standardized curriculum for AM education and training. The program brings together industry leaders, educators, and government agencies to develop a comprehensive framework for AM education and training.

Addressing the Skills Gap

The AM industry is facing a significant skills gap, with many workers lacking the skills and knowledge required for AM. This skills gap is due to a variety of factors, including the rapid evolution of AM technologies, the lack of standardized curricula, and the shortage of experienced AM professionals.

To address this skills gap, education and training programs must focus on providing workers with the skills and knowledge required for AM. This will require a combination of theoretical and practical training, as well as ongoing upskilling and reskilling opportunities.

For example, companies such as Siemens and GE Appliances are offering apprenticeships and training programs in AM to address the skills gap. These programs provide workers with hands-on experience and training in AM, as well as opportunities to work on real-world projects and collaborate with experienced AM professionals.

Providing Ongoing Training and Upskilling

The AM industry is rapidly evolving, with new technologies and techniques emerging all the time. To stay current, workers must have access to ongoing training and upskilling opportunities that can help them develop new skills and knowledge.

Education and training programs must provide workers with opportunities to upskill and reskill throughout their careers. This will require a combination of online and offline training, as well as opportunities for workers to collaborate with experienced AM professionals and work on real-world projects.

For example, online platforms such as Udemy and Coursera are offering courses and training programs in AM, which provide workers with opportunities to upskill and reskill from anywhere in the world. These platforms are particularly useful for workers who are unable to attend traditional classroom-based training programs.

Other Challenges and Opportunities

In addition to these challenges and opportunities, there are several other issues that must be addressed in AM education and training, including:

- **Diversity and inclusion**: The AM industry is facing a diversity and inclusion crisis, with many underrepresented groups lacking access to AM education and training opportunities. Education and training programs must address this issue by providing opportunities for underrepresented groups to access AM education and training.

- **Industry-academe collaboration**: The AM industry is facing a shortage of experienced AM professionals, which makes it difficult for education and training programs to provide workers with the skills and knowledge required for AM. Industry-academe collaboration is essential to address this issue, as it can provide workers with opportunities to work with experienced AM professionals and gain hands-on experience.

- **Government support**: Government support is essential for the development of AM education and training programs. Governments must provide funding and resources for AM education and training initiatives, as well as support for industry-academe collaboration and research and development in AM.

In conclusion, the AM industry is facing a significant skills gap, and education and training programs must address this gap to ensure a skilled workforce. Developing standardized curricula, addressing the skills gap, and providing ongoing training and upskilling opportunities are essential to ensure that workers have the skills and knowledge required for AM. By addressing these challenges and opportunities, we can ensure a bright future for the AM industry. Let's explore the challenges and opportunities in additive manufacturing (AM) education and training, as depicted in Figure 106:

i. **Dealing with Change**:
   - Organizational changes related to mergers, technology, and budgets pose challenges.
   - L&D professionals can craft purposeful messaging and prioritize initiatives to alleviate stress during transitions.

ii. **Developing Leaders:**
   - Leadership development is crucial at all levels.
   - Organizations must nurture leaders—from team leads to CEOs—for short- and long-term success.

iii. **Engaging Learners:**
   - Keeping learners motivated and engaged is essential.
   - Interactive, relevant content and learner-centered approaches enhance engagement.

iv. **Consistent Training Delivery:**
   - Delivering training consistently across locations and languages is a challenge.
   - Standardize content, leverage technology, and provide clear guidelines.

v. **Tracking Skills Application:**
   - Ensuring learners apply newly acquired skills is vital.
   - Regular follow-up, practical exercises, and feedback loops help reinforce learning.

vi. **Conflict-Management Skills:**
   - Instilling conflict-resolution skills is critical.
   - Training on effective communication, negotiation, and problem-solving is essential.

vii. **Demonstrating Value to Leadership:**
   - L&D professionals must showcase the impact of training.

- Metrics, data-driven insights, and success stories demonstrate value.

viii. **Quantifying Training Effectiveness:**
- Measuring training impact is challenging.
- Use pre- and post-assessments, performance metrics, and learner feedback.

ix. **Improving Learning Effectiveness:**
- Enhance learning outcomes through personalized approaches.
- Adapt content, delivery methods, and assessments to individual needs.

x. **Adapting to a Varied Audience:**
- Design modules that resonate with diverse learners.
- Consider cultural differences, learning preferences, and accessibility.

Remember, addressing these challenges proactively ensures effective AM education and empowers learners for the future

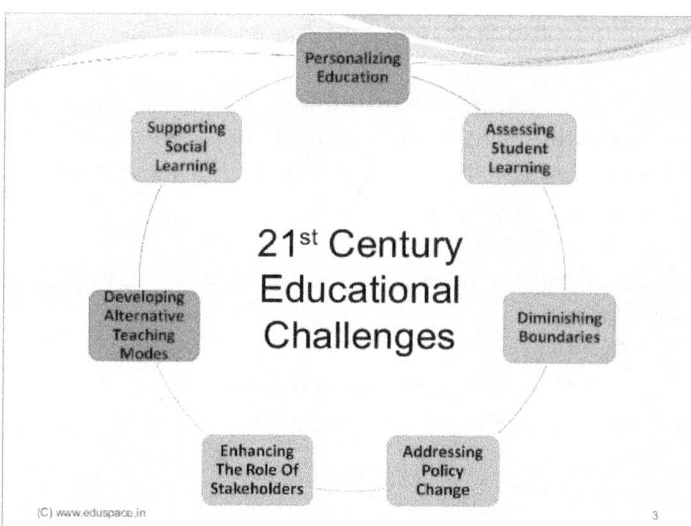

Figure 3: Challenges and opportunities in AM education and training [Source: https://www.slideshare.net/Geo_mathew/21st-century-educational-challenges]

25.5 Questions

25.5.1 Long Response Questions

1. What are the key advancements in AM technologies and materials, and how will they shape the future of AM?
2. What are the new applications and industries for AM, and how will they drive growth and innovation?
3. What are the challenges and opportunities in AM education and training, and how can they be addressed?
4. How can companies and governments invest in AM education and training to ensure a skilled workforce?
5. What are some examples of successful AM applications in emerging industries?

25.5.2 Multiple Choice Questions

1) **What is the primary benefit of 3D printing technology?**
   a) Reduced material waste
   b) Faster transportation
   c) Increased manual labor
   d) Higher energy consumption
   Solution: - a) Reduced material waste
2) **Which sector is heavily investing in 3D printing for prototyping and production?**
   a) Food Industry
   b) Entertainment Industry
   c) Aerospace Industry
   d) Textile Industry
   Solution: - c) Aerospace Industry
3) **What is one common material used in 3D printing?**
   a) Glass
   b) Metal
   c) Plastic
   d) Wood
   Solution: - c) Plastic
4) **Which technology is often combined with 3D printing for enhanced applications?**
   a) Artificial Intelligence

b) GPS
c) Cloud Computing
d) Blockchain
Solution: - a) Artificial Intelligence

5) **In which field is 3D printing used to create prosthetic limbs?**
a) Healthcare
b) Automotive
c) Agriculture
d) Architecture
Solution: - a) Healthcare

6) **What is the primary goal of using 3D printing in manufacturing?**
a) To increase production time
b) To simplify the supply chain
c) To reduce operational costs
d) To decrease product quality
Solution: - c) To reduce operational costs

7) **Which technology is NOT typically associated with 3D printing?**
a) Stereolithography (SLA)
b) Fused Deposition Modeling (FDM)
c) Ultrasonic Welding
d) Selective Laser Sintering (SLS)
Solution: - c) Ultrasonic Welding

8) **What is a major challenge in the adoption of 3D printing for large-scale production?**
a) High cost of materials
b) Limited design options
c) Slow production speed
d) Lack of skilled operators
Solution: - c) Slow production speed

9) **Which aspect of 3D printing is often improved by emerging technologies?**
a) Print resolution
b) Software complexity
c) Color variety
d) Printer weight
Solution: - a) Print resolution

10) What is "additive manufacturing"?

a) Adding additional steps in the manufacturing process
b) Increasing the number of production lines
c) Building objects layer by layer from digital models
d) Reducing the size of products

Solution: - c) Building objects layer by layer from digital models

11) Which of the following is a potential future application of 3D printing technology?

a) Space travel
b) In-flight meals
c) Mass communication
d) Agricultural harvesting

Solution: - a) Space travel

12) What does "customization" in 3D printing typically refer to?

a) Using standard designs for all products
b) Modifying products to meet specific customer requirements
c) Increasing the size of printed objects
d) Using fewer materials

Solution: - b) Modifying products to meet specific customer requirements

13) Which industry is exploring 3D printing for creating food items?

a) Automotive
b) Healthcare
c) Food Industry
d) Education

Solution: - c) Food Industry

14) Which 3D printing method is known for its use of liquid resins?

a) Fused Deposition Modeling (FDM)
b) Stereolithography (SLA)
c) Selective Laser Sintering (SLS)
d) Digital Light Processing (DLP)

Solution: - b) Stereolithography (SLA)

15) In the context of 3D printing, what is "bioprinting"?

a) Printing using biodegradable materials

b) Creating biological tissues and organs

c) Printing designs for biological research

d) Using biological agents as ink

Solution: - b) Creating biological tissues and organs

16) **What is the significance of “meta-materials” in 3D printing?**

a) They are standard materials used in printing.

b) They have unique properties derived from their structure.

c) They are materials with high thermal conductivity.

d) They are commonly used in food industry applications.

Solution: - b) They have unique properties derived from their structure.

17) **Which of the following is a key advantage of using metal 3D printing in aerospace applications?**

a) Lower material strength

b) High production cost

c) Complex geometries and reduced weight

d) Reduced precision in parts

Solution: - c) Complex geometries and reduced weight

18) **How does “distributed manufacturing” relate to 3D printing?**

a) Manufacturing is centralized in one location.

b) Manufacturing is done by multiple large factories.

c) Production is spread across multiple locations closer to consumers.

d) Manufacturing processes are simplified.

Solution: - c) Production is spread across multiple locations closer to consumers.

19) **Which factor is crucial for improving the speed of 3D printing processes?**

a) Material density

b) Layer height and print speed

c) Printer temperature

d) Print resolution

Solution: - b) Layer height and print speed

20) **What does “process intensification” mean in the context of 3D printing?**

a) Increasing the number of printers

b) Enhancing the efficiency and productivity of the printing process

c) Adding more steps to the production process

d) Using more complex designs

Solution: - b) Enhancing the efficiency and productivity of the printing process

21) What role does "AI" play in the future of 3D printing?

a) It reduces the need for automation.

b) It enhances design optimization and defect detection.

c) It decreases the overall quality of printed parts.

d) It simplifies material selection.

Solution: - b) It enhances design optimization and defect detection.

22) Which of the following is a limitation of current 3D printing technology in industrial applications?

a) Limited material options

b) Low customization potential

c) Low precision in part dimensions

d) High operational costs

Solution: - a) Limited material options

23) What is the primary use of "generative design" in additive manufacturing?

a) To create designs based on traditional methods

b) To develop complex, optimized structures through algorithmic processes

c) To simplify existing designs

d) To standardize design processes

Solution: - b) To develop complex, optimized structures through algorithmic processes

24) What is a significant challenge associated with the adoption of 3D printing in the medical field?

a) High cost of materials

b) Need for highly specialized equipment

c) Limited research on biocompatibility

d) Complexity of regulatory approval

Solution: - d) Complexity of regulatory approval

25) Which advanced material is being researched for 3D printing to enhance strength and durability?

a) Graphene

b) Polystyrene

c) Polyethylene

d) Nylon

Solution: - a) Graphene

26) How does "digital twin technology" benefit additive manufacturing?

a) It reduces the need for digital models.

b) It allows for real-time simulation and optimization of physical assets.

c) It simplifies traditional manufacturing processes.

d) It increases the physical inventory of products.

Solution: - b) It allows for real-time simulation and optimization of physical assets.

27) What does "self-replicating 3D printing" refer to?

a) Printers that can create new printers.

b) Printing multiple copies of the same object.

c) Using 3D printers to replicate themselves in a factory.

d) Creating objects that replicate biological systems.

Solution: - a) Printers that can create new printers.

28) Which of the following is an emerging trend in the field of 3D printing?

a) Manual assembly of parts

b) Integration with IoT (Internet of Things) for smart manufacturing

c) Reducing automation

d) Using only traditional materials

Solution: - b) Integration with IoT (Internet of Things) for smart manufacturing

29) What is "sustainability" in the context of future 3D printing developments?

a) Increasing waste generated by the process

b) Reducing energy consumption and environmental impact

c) Using non-recyclable materials

d) Expanding the use of non-renewable resources

Solution: - b) Reducing energy consumption and environmental impact

30) Which factor is critical for the future of bioprinting technologies?

a) Printer size
b) Material cost
c) Biocompatibility of materials
d) Print speed
Solution: - c) Biocompatibility of materials

31) In which area is "hyper-customization" of 3D printed products most beneficial?
a) Mass production of identical items
b) Creating tailored products for individual needs
c) Standardizing product designs
d) Reducing product variety
Solution: - b) Creating tailored products for individual needs

32) What role do "smart materials" play in 3D printing?
a) They are used for high-temperature applications.
b) They adapt their properties in response to environmental changes.
c) They increase the weight of printed parts.
d) They simplify the printing process.
Solution: - b) They adapt their properties in response to environmental changes.

33) How is "robotic 3D printing" advancing the field?
a) By reducing the need for complex designs
b) By enabling large-scale, precise, and efficient printing
c) By limiting the types of materials used
d) By decreasing the range of printable geometries
Solution: - b) By enabling large-scale, precise, and efficient printing

34) What is a major factor driving the use of 3D printing in construction?
a) Lower material costs
b) Ability to print large-scale structures
c) Simplicity of designs
d) Traditional building techniques
Solution: - b) Ability to print large-scale structures

35) Which area of 3D printing is expected to benefit most from advances in machine learning?
a) Basic design modeling
b) Print quality optimization and defect detection

c) Traditional manufacturing methods

d) Manual assembly

Solution: - b) Print quality optimization and defect detection

36) What is the impact of "distributed manufacturing" on supply chains?

a) It centralizes production to one location.

b) It shortens supply chains by localizing production.

c) It increases the complexity of supply chains.

d) It reduces the need for local manufacturing.

Solution: - b) It shortens supply chains by localizing production.

37) What is the significance of "open-source" in the context of 3D printing?

a) It restricts access to 3D printing designs.

b) It allows sharing and collaboration on designs and processes.

c) It reduces the quality of printed parts.

d) It increases manufacturing costs.

Solution: - b) It allows sharing and collaboration on designs and processes.

38) How does "predictive maintenance" relate to additive manufacturing?

a) It involves predicting when parts will fail and scheduling maintenance.

b) It focuses on extending the lifespan of 3D printers.

c) It reduces the need for preventative maintenance.

d) It increases the frequency of machine breakdowns.

Solution: - a) It involves predicting when parts will fail and scheduling maintenance.

39) What is the role of "digital twins" in 3D printing?

a) They are physical copies of digital designs.

b) They provide virtual models for real-time simulation and analysis.

c) They replace traditional manufacturing techniques.

d) They limit the use of digital models.

Solution: - b) They provide virtual models for real-time simulation and analysis.

40) What is a potential future trend for 3D printing in terms of sustainability?

a) Increased use of non-recyclable materials

b) Development of biodegradable and recyclable printing materials

c) Higher energy consumption

d) Reduced focus on environmental impact

Solution: - b) Development of biodegradable and recyclable printing materials

References

1. American Institute of Architects. (2020). Additive manufacturing in architecture.
2. Apis Cor. (2020). Additive manufacturing in architecture.
3. ASTM International. (2012). Standard terminology for additive manufacturing technologies.
4. ASTM International. (2018). Standard terminology for additive manufacturing technologies.
5. Autodesk. (2020). Autodesk Fusion 360.
6. Autodesk. (2020). ReCap: 3D scanning and reverse engineering software.
7. BMW. (2020). Additive manufacturing.
8. Boeing. (2020). Additive manufacturing.
9. Consumer Technology Association. (2020). Additive manufacturing in consumer products.
10. Deloitte. (2020). Additive manufacturing: A new era of supply chain management.
11. FDA. (2020). Additive manufacturing in healthcare.
12. Ford Motor Company. (2020). Additive manufacturing.
13. General Motors. (2020). Additive manufacturing.
14. Gibson, I., Rosen, D. W., & Stucker, B. (2015). Additive manufacturing technologies: 3D printing, rapid prototyping, and direct digital manufacturing. Springer.
15. Harvard Business Review. (2020). The business of additive manufacturing.
16. International Journal of Additive Manufacturing. (2020). Emerging trends in additive manufacturing.
17. ISO/ASTM 52910:2018. Standard terminology for additive manufacturing technologies.
18. Johnson & Johnson. (2020). Additive manufacturing.
19. Lockheed Martin. (2020). Additive manufacturing.
20. McKinsey & Company. (2020). The future of additive manufacturing.
21. Medtronic. (2020). Additive manufacturing.
22. NASA. (2020). Additive manufacturing in aerospace.
23. National Science Foundation. (2020). Future directions in additive manufacturing.
24. Nike. (2020). Additive manufacturing.

25. Procter & Gamble. (2020). Additive manufacturing.
26. SAE International. (2019). Additive manufacturing in automotive.
27. Skanska. (2020). Additive manufacturing in construction.
28. SolidWorks. (2020). SolidWorks CAD software.
29. Stryker. (2020). Additive manufacturing.
30. Tesla. (2020). Additive manufacturing.
31. Tinkercad. (2020). Tinkercad CAD software.
32. UCLA. (2020). Additive manufacturing in healthcare.
33. WATG. (2020). Additive manufacturing in architecture.
34. WinSun. (2020). Additive manufacturing in architecture.
35. Wohlers, T. (2020). Wohlers report 2020: 3D printing and additive manufacturing state of the industry. Wohlers Associates.
36. World Economic Forum. (2020). Additive manufacturing: A new era of economic growth.
37. Kumar, L. J., Pandey, P. M., & Wimpenny, D. I. (Eds.). (2019). *3D printing and additive manufacturing technologies* (Vol. 311). Singapore: Springer.
38. Hausman, K. K., & Horne, R. (2014). *3D printing for dummies*. John Wiley & Sons.
39. Zukas, V., & Zukas, J. A. (2015). *An introduction to 3D printing*. First Edition Design Pub.
40. Redwood, B., Schffer, F., & Garret, B. (2017). *The 3D printing handbook: technologies, design and applications*. 3D Hubs.
41. Horvath, J., & Cameron, R. (2014). *Mastering 3D printing* (p. 85). Berkeley, CA: Apress.
42. Tsoulfas, G., Bangeas, P. I., & Suri, J. (Eds.). (2019). *3D Printing: Application in Medical Surgery E-Book*. Elsevier Health Sciences.
43. Horne, R. (2023). *3D printing for dummies*. John Wiley & Sons.
44. Barnatt, C. (2014). *3D Printing*. Wroclaw, Poland: ExplainingTheFuture. com.
45. Singh, S., Prakash, C., & Singh, R. (Eds.). (2020). *3D printing in biomedical engineering*. Singapore: Springer.

46. Evans, B. (2012). *Practical 3D printers: The science and art of 3D printing*. Apress.
47. Lipson, H., & Kurman, M. (2013). *Fabricated: The new world of 3D printing*. John Wiley & Sons.
48. Wake, N. (Ed.). (2021). *3D printing for the radiologist, E-Book*. Elsevier Health Sciences.
49. Wimpenny, D. I., Pandey, P. M., & Kumar, L. J. (Eds.). (2017). *Advances in 3D printing & additive manufacturing technologies* (Vol. 1, pp. 1-184). Singapore: Springer.
50. France, A. K. (2013). *Make: 3D printing: the essential guide to 3D printers*. Maker Media, Inc.
51. Coward, C. (2015). *3D Printing*. Dorling Kindersley Ltd.
52. Ian Gibson, I. G. (2015). Additive manufacturing technologies 3D printing, rapid prototyping, and direct digital manufacturing.
53. Heinrich, A. (Ed.). (2021). *3D Printing of Optical Components* (Vol. 233). Cham: Springer International Publishing.
54. Horvath, J., & Cameron, R. (2015). *3D Printing with mattercontrol*. Apress.
55. Sandhu, K., Singh, S., Prakash, C., Subburaj, K., & Ramakrishna, S. (Eds.). (2022). *Sustainability for 3D printing*. Berlin: Springer.
56. Kalaskar, D. M. (Ed.). (2022). *3D printing in medicine*. Woodhead Publishing.
57. Basit, A. W., & Gaisford, S. (Eds.). (2018). *3D printing of pharmaceuticals*. Berlin, Germany: Springer International Publishing.
58. Chua, C. K., Vadakke Matham, M., & Kim, Y. J. (2017). *Lasers in 3D printing and manufacturing*.
59. Fink, J. K. (2018). *3D industrial printing with polymers*. John Wiley & Sons.
60. Bañón, C., & Raspall, F. (2021). *3D printing architecture: workflows, applications, and trends*. Berlin/Heidelberg, Germany: Springer.
61. Godoi, F. C., Bhandari, B., Prakash, S., & Zhang, M. (Eds.). (2018). *Fundamentals of 3D food printing and applications*. Academic press.

62. Ahmad, N., Packirisamy, G., & Dutta, R. (Eds.). (2019). *3D printing technology in nanomedicine*. Elsevier.
63. McMills, A. E. (2017). *3D printing basics for entertainment design*. Routledge.
64. Sharma, N. R., Subburaj, K., Sandhu, K., & Sharma, V. (Eds.). (2021). *Applications of 3D printing in Biomedical Engineering*. Springer.
65. Kelly, J. F. (2013). *3D printing: build your own 3D printer and print your own 3D objects*. Que Publishing.
66. Muralidhara, H. B., & Banerjee, S. (Eds.). (2021). *3D printing technology and its diverse applications*. CRC Press.
67. Eyers, D. (Ed.). (2020). *Managing 3D printing: Operations management for additive manufacturing*. Springer Nature.
68. Fisher, G. (2013). *Blender 3D printing essentials*. Packt Publishing Ltd.
69. Cvetković, D. (Ed.). (2018). *3D Printing*. BoD–Books on Demand.
70. Bell, C. (2015). *3D printing with delta printers*. Apress.
71. Ballardini, R. M., Norrgård, M., & Partanen, J. (Eds.). (2016). *3D Printing, Intellectual Property and Innovation*. Kluwer Law International BV.
72. Dunham, S., Mosadegh, B., Romito, E. A., & Zgaren, M. (2018). Applications of 3D printing. In *3D printing applications in cardiovascular medicine* (pp. 61-78). Academic Press.
73. Soloman, S. (2020). *3D Printing & Design*. Khanna Publishing House.
74. Singh, S. (2011). *Beginning google sketchup for 3D printing*. Apress.
75. Bernier, S. N., Luyt, B., & Reinhard, T. (2015). *Design for 3D printing: scanning, creating, editing, remixing, and making in three dimensions*. Maker Media, Inc.
76. Sheng, R. (2022). *3D Printing: A Revolutionary Process for Industry Applications*. Woodhead Publishing.
77. Kerr, T. (2023). *3D Printing*. Springer International Publishing AG.

78. Gonzalez, S. R., & Bennett, D. B. (2016). *3D printing: A practical guide for librarians* (Vol. 22). Rowman & Littlefield.
79. Yang, W., & Jian, R. (2023). *Polymer 3D Printing and 3D Copying Technology*. Springer.
80. Dipaola, M. (2018). *3D printing in orthopaedic surgery*. Elsevier Health Sciences.
81. Shishkovsky, I. (Ed.). (2016). *New trends in 3D printing*. BoD–Books on Demand.
82. Noorani, R. (2017). *3D printing: technology, applications, and selection*. CRC Press.
83. Cano, L. M. (2015). *3D printing: a powerful new curriculum tool for your school library*. Bloomsbury Publishing USA.
84. Horvath, J., & Cameron, R. (2018). *Mastering 3D printing in the classroom, library, and lab*. Apress.
85. Larson, J. (2013). *3D Printing Blueprints*. Packt Publishing Ltd.
86. Mongeon, B. (2015). *3D technology in fine art and craft: Exploring 3D printing, scanning, sculpting and milling*. Routledge.
87. Ovsianikov, A., Yoo, J., & Mironov, V. (2018). *3D printing and biofabrication*. Springer Publishing Company, Incorporated.
88. Haq, M. I. U., Raina, A., & Naveed, N. (Eds.). (2023). *3D Printing and Sustainable Product Development*. CRC Press.
89. Sandhu, K., Singh, S., Prakash, C., Sharma, N. R., & Subburaj, K. (Eds.). (2022). *Emerging Applications of 3D Printing During CoVID 19 Pandemic*. Berlin, Heidelberg: Springer.
90. Sandhu, K., Singh, S., Prakash, C., Subburaj, K., & Ramakrishna, S. (Eds.). (2022). *3D Printing in Podiatric Medicine*. Academic Press.
91. Van Wijk, A. J. M., & van Wijk, I. (2015). *3D printing with biomaterials: Towards a sustainable and circular economy*. IOS press.

92. Kumar, L. J., Pandey, P. M., & Wimpenny, D. I. (Eds.). (2019). *3D printing and additive manufacturing technologies* (Vol. 311). Singapore: Springer.
93. Srivastava, M., Rathee, S., Maheshwari, S., & Kundra, T. K. (2019). *Additive manufacturing: fundamentals and advancements*. CRC Press.
94. Godec, D., Gonzalez-Gutierrez, J., Nordin, A., Pei, E., & Ureña Alcázar, J. (2022). *A guide to additive manufacturing* (p. 324). Springer Nature.
95. Gibson, I., Rosen, D. W., Stucker, B., Khorasani, M., Rosen, D., Stucker, B., & Khorasani, M. (2021). *Additive manufacturing technologies* (Vol. 17, pp. 160-186). Cham, Switzerland: Springer.
96. Pou, J., Riveiro, A., & Davim, J. P. (Eds.). (2021). *Additive manufacturing*. Elsevier.
97. Kumar, S. (2020). *Additive manufacturing processes* (p. 205). Cham: Springer.
98. Wimpenny, D. I., Pandey, P. M., & Kumar, L. J. (Eds.). (2017). *Advances in 3D printing & additive manufacturing technologies* (Vol. 1, pp. 1-184). Singapore: Springer.
99. Khan, M. A., & Jappes, J. W. (Eds.). (2022). *Innovations in additive manufacturing*. Berlin/Heidelberg, Germany: Springer.
100. Natarajan, J. (Ed.). (2021). *Advances in Additive Manufacturing Processes*. Bentham Science Publishers.
101. Killi, S. W. (2017). *Additive manufacturing: design, methods, and processes*. CRC Press.
102. Leary, M. (2019). *Design for additive manufacturing*. Elsevier.
103. Froes, F. H., & Boyer, R. (Eds.). (2019). *Additive manufacturing for the aerospace industry*. Elsevier.
104. Celik, E. (2020). *Additive manufacturing: science and technology*. Walter de Gruyter GmbH & Co KG.
105. Pei, E., Monzón, M., & Bernard, A. (Eds.). (2019). *Additive Manufacturing-Developments in training and education* (Vol. 235). Cham: Springer International Publishing.

106. Balasubramanian, K. R., & Senthilkumar, V. (Eds.). (2020). *Additive manufacturing applications for metals and composites*. IGI Global.
107. Shi, Y., Yan, C., Zhou, Y., Wu, J., Wang, Y., Yu, S., & Ying, C. (2021). *Materials for additive manufacturing*. Academic Press.
108. Banga, H. K., Kumar, R., Kalra, P., & Belokar, R. M. (Eds.). (2022). *Additive Manufacturing with Medical Applications*. CRC Press.
109. Zhou, K. (Ed.). (2022). *Additive Manufacturing Technology: Design, Optimization, and Modeling*. John Wiley & Sons.
110. Singh, R., & Kumar, R. (Eds.). (2022). *Additive manufacturing for plastic recycling: efforts in boosting a circular economy*. CRC Press.
111. Kirihara, S., & Nakata, K. (Eds.). (2021). *Multi-dimensional additive manufacturing*. Berlin/Heidelberg, Germany: Springer.
112. Shishkovsky, I. V. (Ed.). (2022). *Advanced Additive Manufacturing*. BoD–Books on Demand.
113. Dutta, B., Babu, S., & Jared, B. H. (2019). *Science, technology and applications of metals in additive manufacturing*. Elsevier.
114. Kamara, S., & Faggiani, K. S. (2021). *Fundamentals of Additive Manufacturing for the Practitioner*. John Wiley & Sons.
115. Mukherjee, T., & DebRoy, T. (2023). *Theory and practice of additive manufacturing*. John Wiley & Sons.
116. Youssef, H., El-Hofy, H., & Ahmed, M. (2024). *Fundamentals of Additive Manufacturing: Principles, Technologies, and Applications*. CRC Press.

www.ingramcontent.com/pod-product-compliance
Lightning Source LLC
LaVergne TN
LVHW052204200926
848481LV00001B/27
* 9 7 8 1 9 2 2 6 1 7 7 0 5 *